Computational Social Sciences

Computational Social Sciences

A series of authored and edited monographs that utilize quantitative and computational methods to model, analyze and interpret large-scale social phenomena. Titles within the series contain methods and practices that test and develop theories of complex social processes through bottom-up modeling of social interactions. Of particular interest is the study of the co-evolution of modern communication technology and social behavior and norms, in connection with emerging issues such as trust, risk, security and privacy in novel socio-technical environments.

Computational Social Sciences is explicitly transdisciplinary: quantitative methods from fields such as dynamical systems, artificial intelligence, network theory, agent-based modeling, and statistical mechanics are invoked and combined with state-of-the-art mining and analysis of large data sets to help us understand social agents, their interactions on and offline, and the effect of these interactions at the macro level. Topics include, but are not limited to social networks and media, dynamics of opinions, cultures and conflicts, socio-technical co-evolution and social psychology. Computational Social Sciences will also publish monographs and selected edited contributions from specialized conferences and workshops specifically aimed at communicating new findings to a large transdisciplinary audience. A fundamental goal of the series is to provide a single forum within which commonalities and differences in the workings of this field may be discerned, hence leading to deeper insight and understanding.

More information about this series at http://www.springer.com/series/11784

Tamás Rudas • Gábor Péli
Editors

Pathways Between Social Science and Computational Social Science

Theories, Methods, and Interpretations

Springer

Editors
Tamás Rudas
Department of Statistics
Eötvös Loránd University
Budapest, Hungary

Gábor Péli
Centre for Social Sciences
Hungarian Academy of Sciences
Centre of Excellence

Department of Sociology Gáspár Károli
University of the Reformed Church in
Hungary

ISSN 2509-9574 ISSN 2509-9582 (electronic)
Computational Social Sciences
ISBN 978-3-030-54935-0 ISBN 978-3-030-54936-7 (eBook)
https://doi.org/10.1007/978-3-030-54936-7

This Springer imprint is published by the registered company Springer Nature Switzerland AG
The registered company address is: Gewerbestrasse 11, 6330 Cham, Switzerland

Preface

The chapters in this volume represent a variety of forms and phases of transition from traditional social science to computational social science, illustrating that the latter, though characterized by distinct research methods, is an integral part of the former. The volume extends social science to new domains, raising new questions or offering new solutions for insights into old ones. What kind of new theoretical models, approaches, or even more importantly, new mindsets could or should be constructed by researchers to capture the context of these investigations? What kind of methodological advances are triggered when implementing these investigations? How do the traditional and computational aspects of social science interface with each other? The three parts of the volume visit these issues. The first part discusses theoretical models and approaches constructed to back up such investigations and also a critical view of the usefulness of these models. Kertész et al. review 10 years of experience in data-driven modeling of social networks. The next chapter by Troitzsch revisits the problem of coupling between formal model specifications and the theoretical insights these models reflect upon. The second part of the volume visits specific research domains where applying tailor-made computational social science approaches provide new insights: the extraction of sociological knowledge through automated text analytics (Németh and Koltai), the combination of scientific and non-scientific surveys to improve estimates and to reduce costs (Sakshaug et al.), social media image analysis to obtain insights into human cognition and emotions (Dudău), and the validation of model outcomes by comparison with social reality (Troitzsch). The chapters in the third part exemplify how computational social science interfaces with extant social science practices so that the state-of-the-art interpretative power of the latter complements the data-processing capabilities of the former. The topics considered are gender gap detection (Crippa et al.), market structuration and evolution (García-Díaz), political network reconstruction (Gulyás et al), participatory budgeting (Aziz), and the content analysis of social media discourse (Koltai et al.).

In an Epilogue to the volume, the editors discuss how computational social science can contribute to exploring an emergent new social reality in algorithmic societies.

When editing this volume, we wanted to emphasize a number of key points that recur in the chapters in various ways:

The emergence of computational social science is not an exogenous development imposed upon the social sciences; rather, it is an endogenous phenomenon.

Troitzsch (Formal . . .) emphasizes that the endogenous development of computational social science within traditional social science domains has already began some six decades ago. Crippa et al. point out that a massive number of measures developed to describe the multifaceted nature of gender gap can be handled, in an integrative way, with specially developed computational techniques. The agent-based simulation research line put forward by García-Díaz is endogenous on two accounts. First, that sort of research has emerged endogenously by the need to study the behavior among interacting agents in markets. A major goal was transcending traditional market models that assume optimizing behavior. As agents behave erratically, the optimization now resides at the systemic level: the members of these firm ecologies possess different adaptive capabilities and are confronted with market selection forces. A second endogeneity arises within the simulation process itself, in the form of the emergent market structure patterns. We encounter further streams of endogeneous computational social science involvement when the amalgamation methods of computational and traditional research aspects are optimized for some chosen criteria. This criterion can be sociological content validity maximization (Németh and Koltai), cost and error minimization (Sakshaug et al.), or the implementation and emergence of democratic values like in the case of the participatory budgeting of civil projects (Aziz). The cumulative series of *in silico* experiments on network tie formation (Kertész et al.) demonstrates an endogenous way to explain emergent behavioral patterns in large-scale social networks. The experiments could reproduce, in a robust and replicable manner, some well-known social phenomena like the one captured by Granovetter's *Strength of weak ties* thesis.[1]

There are different – though interrelated – pathways between traditional social science and computational social science, including theoretical, methodological and topical approaches.

In her chapter, Dudău departs from the technical issues of image analysis in order to arrive at semantic issues of interpreting social media content. Doing so, she exemplifies the linking between cutting-edge methods and state-of-the-art socio-psychological interpretations. García-Díaz demonstrates how a methodological choice—endowing model agents with the ability of inductive reasoning—can lead to emergent market structure patterns, which we can identify with real-world

[1]Granovetter, Mark S. (1973) "The Strength of Weak Ties." *American Journal of Sociology*, 78(6): 1360–1380.

market patterns. Troitzsch (Validating . . .) discusses the relationship between real societal phenomena and the stylized facts, which agent-based simulation models can reproduce, also warning that the obtained similarities between the two do not necessarily imply that the models correctly describe social reality.

There are concrete and practical examples as to how social scientists with a traditional training may approach computational social science through these pathways.

Németh and Koltai introduce their methods by heavily building on the concepts and terminology of classical quantitative research methodologies, so inviting sociologists with a classical training to join the computational social science pool. Crippa et al. provide an accessible computational way of handling a vast array of gender disparity measures for scholars with traditional quantitative training. By feeding longitudinal study data into simulation models and by comparing their outcomes with real data, Troitzsch (Formal . . .) provides a practical example of how traditional social science research obtains insights from computational approaches. The same chapter also provides a description of how empirical research can be interfaced with a theoretical machinery operating with the symbol system of computer science, thus making the empirical inputs tractable for a broad range of computational techniques.

The exploration of these pathways improves the interpretative power of traditional social science by incorporating results from computational social science.

Crippa et al. demonstrate how the interpretative power of traditional gender statistics improves by incorporating results from the computational methods they propose. Analyzing a corpus of 2 million Instagram posts, Koltai et al. put two methods forward to detect discourse types concerning suicide and mental health. On the basis of this typology, their research also reveals how interpretations of these two concepts are constructed in the social media. The respective simulation results of García-Díaz and Troitzsch (Formal . . .) help better understand the difficult-to-observe micro-level processes that underlie visible and robust macro-level outcomes. Gulyás et al. take a highly important political science topic that notoriously defies traditional survey-and interview-based exploration: the interpersonal dynamics in top political executive committees. Based on the careful coding of the written meeting reports, they reconstructed the personal networks that had been influencing the historically known, observable political decisions of these executive bodies. Finally, the series of network simulations reported by Kertész et al. boost sociological research by reproducing a variety of well-known social phenomena (typically given in a qualitative way, in the form of stylized facts), associating them with numerically tractable representations.

We owe thanks to many people for their contributions to this book. In addition to being indebted to our devoted authors and to the professional and helpful staff at Springer, we are extremely grateful to James Kitts for inviting this volume for consideration for the Computational Social Science Series of Springer. We are deeply indebted to Vinicius Gorczeski for handling the several complex technical and organizational aspects of the editorial work, including communication with the authors and managing several rounds of internal reviews.

Tamás Rudas

Budapest, Hungary
January 2020

Gábor Péli

Contents

Part I Theory: Dilemmas of Model Building and Interpretation

Modeling the Complex Network of Social Interactions 3
János Kertész, János Török, Yohsuke Murase, Hang-Hyun Jo, and Kimmo Kaski

Formal Design Methods and the Relation Between Simulation Models and Theory: A Philosophy of Science Point of View 21
Klaus G. Troitzsch

Part II Methodological Toolsets

The Potential of Automated Text Analytics in Social Knowledge Building 49
Renáta Németh and Júlia Koltai

Combining Scientific and Non-scientific Surveys to Improve Estimation and Reduce Costs 71
Joseph W. Sakshaug, Arkadiusz Wiśniowski, Diego Andres Perez Ruiz, and Annelies G. Blom

Harnessing the Power of Data Science to Grasp Insights About Human Behaviour, Thinking, and Feeling from Social Media Images 95
Diana Paula Dudău

Validating Simulation Models: The Case of Opinion Dynamics 123
Klaus G. Troitzsch

Part III New Look on Old Issues: Research Domains Revisited by Computational Social Science

A Spatio-Temporal Approach to Latent Variables: Modelling Gender (im)balance in the Big Data Era 159
Franca Crippa, Gaia Bertarelli, and Fulvia Mecatti

Agent-Based Organizational Ecologies: A Generative Approach to Market Evolution 179
César García-Díaz

Networks of the Political Elite and Political Agenda Topics: Creation and Analysis of Historical Corpora Using NLP and SNA Methods 197
Attila Gulyás, Martina Katalin Szabó, Orsolya Ring, László Kiss, and István Boros

Participatory Budgeting: Models and Approaches 215
Haris Aziz and Nisarg Shah

From Durkheim to Machine Learning: Finding the Relevant Sociological Content in Depression and Suicide-Related Social Media Discourses 237
Júlia Koltai, Zoltán Kmetty, and Károly Bozsonyi

Epilogue 259
Tamás Rudas and Gábor Péli

Index 273

Author Biographies

Haris Aziz is a Scientia Fellow and Senior Lecturer at UNSW Sydney, Team Leader at Algorithmic Decision Theory group, and Director of the Sydney EconCS network. His research interests lie at the intersection of artificial intelligence, theoretical computer science, and mathematical social sciences, especially computational social choice and algorithmic game theory. He is a recipient of the Scientia Fellowship (2018–), CORE Chris Wallace Research Excellence Award (2017), and the Julius Career Award (2016–2018). In 2015, he was selected by the Institute of Electrical and Electronics Engineers (IEEE) for the AI 10 to Watch List. He is on the Board of Directors of IFAAMAS, is an Associate Editor of JAIR, and is on the International Advisory Board of Advanced Intelligent Systems. He completed his Ph.D. at the University of Warwick in 2009, M.Sc. at Oxford University, and B.Sc. (Honors) at Lahore University of Management Sciences. He undertook postdoctoral research at the Ludwig Maximilian University of Munich and Technical University of Munich in Germany. He has held visiting scientist/academic roles at Oxford University, Harvard University, and University Paris Dauphine.

Gaia Bertarelli is a Ph.D. in Methodological Statistics. She is a Research Fellow in Statistics in the Department of Economics and Management at the University of Pisa. She is involved in several European research projects and collaborates with the Italian National Statistics Institute. Her research focuses mainly on sampling statistics, latent variables models, data integration, new indicators, small area estimation of poverty and inequality, and living conditions measures.

Annelies G. Blom is Full Research Professor of Data Science at the School of Social Sciences, University of Mannheim, Germany. She is Head of the German Internet Panel (GIP) at the Collaborative Research Center 884 "Political Economy of Reforms" and Project Leader of several methodological research projects funded by the German Research Foundation (DFG). Previously, she was an Assistant Professor at the University of Mannheim; set up her own consulting business (Survex – Survey Methods Consulting); was the Head of Unit Survey Methods at the Survey of Health, Ageing, and Retirement in Europe (SHARE); a Doctoral Researcher at the European Social Survey (ESS); and a Researcher at the National

Centre for Social Research (NatCen), London. She studied in degree programs at University College Utrecht (B.A.), the Conservatory Utrecht, the University of Oxford (M.Phil.), the University of Essex (Ph.D.), and the University of Leuven. Her research concentrates on survey data collection processes, associated errors, and error correction as well as various applications of artificial intelligence to the collection of social scientific data, such as voice data collection, data fusion, and predictive analytics for attrition processes.

István Boros is a Research Assistant in the CSS-RECENS group at the Hungarian Academy of Sciences. He is a Market Researcher at the State Lottery Corporation and a member of World Lottery Association. He holds a position of Assistant lecturer at the Budapest University of Technology and Economics. He has a B.A. degree in Sociology and an M.A. degree in Big Data and Network Analysis.

Károly Bozsonyi is an Associate Professor, sociologist, and expert in quantitative methodology of social sciences and mathematical modeling of social processes. He is the Director of the Institute of Social Sciences at Károli Gáspár University, Budapest, Hungary.

Franca Crippa is an Associate Professor of Social Statistics, currently in the Department of Psychology, University of Milano-Bicocca, Italy, where she teaches Statistics for Social Research. Her research interests embrace gender studies, higher education, and random effects models in the psychological field. Her research activities involve participation in research groups and scientific committees, both at the national level and with colleagues in the European Union. She has organized and chaired several sessions in international conferences.

Diana Paula Dudău is currently a Ph.D. candidate in Psychology at the West University of Timisoara under the supervision of Prof. Florin Alin Sava. Her thesis is situated at the interplay between clinical psychology and data science, with a focus on the linguistic and non-linguistic digital traces of depression and anxiety. She is working as a cognitive-behavioral psychotherapist and is a former research assistant in the Department of Psychology, West University of Timisoara. Diana is a University of Bucharest alumna (B.Sc. in Psychology and M.A. in Clinical Psychology) and has an M.Sc. in Biostatistics from the Polytechnic University of Timisoara. Also, she is a self-taught data scientist. In January 2018, she attended, as a student, the 4th International Winter School on Big Data (BigDat2018). She wrote and published several scientific papers independently and in co-authorship. Her current main research interests revolve around the goal of harnessing cutting-edge technology and data science to advance research in psychotherapy, clinical, health, and social psychology.

César García-Díaz is an Associate Professor in the Department of Business Administration of Pontificia Universidad Javeriana (Colombia). He holds a Ph.D. in Economics and Business from the University of Groningen (the Netherlands). Dr. García-Díaz studies the evolution of organizational and economic systems and is interested in the link between micro-level rules, structural interdependence, and macro-level outcomes in a variety of settings (e.g., organizational dynamics, industry evolution, competitive spatial location in markets). He is also interested in the use of computational models for better policy design.

Attila Gulyás is a Research Fellow in the CSS-RECENS group at the Hungarian Academy of Sciences, since 2016. He is engaged in multiple projects dealing with network analysis his main project being the NLP-based research of political data. Besides these activities, he provides technical assistance for many research projects mainly using his computational and analytical skills. He graduated as an electrical engineer in 2004 and obtained a Ph.D. degree in that field in 2011 alongside with a second Ph.D. in Sociology, which he complemented with an M.Sc. in Artificial Intelligence in 2013. His main research interests within social sciences are in fairness theories, network analysis, and behavioral economics.

Hang-Hyun Jo is currently a Junior Research Group Leader/Assistant Professor at Asia Pacific Center for Theoretical Physics in the Republic of Korea. He received his M.Sc. (2001) and Ph.D. (2006) in Statistical Physics from Korea Advanced Institute of Science and Technology. He has previously been a Research Fellow at Korea institute for Advanced Study, a Postdoctoral Researcher at Aalto University in Finland, and a Research Assistant Professor at Pohang University of Science and Technology in the Republic of Korea. He was awarded Yong-Bong Prize by The Korean Physical Society in 2017 for his achievements in statistical physics. His research interests cover various topics in statistical physics, complex systems, and computational social science, including complex networks and bursty time series analysis.

Kimmo Kaski has pioneered computational complex systems research as Academy Professor (1996–2006) and as Director of two CoEs (2000–2011). He is a Supernumerary Professorial Fellow of Wolfson College, Oxford, Fellow of APS, IOP, the Finnish Academy of Sciences and Letters, and Follow of the Academia Europeae. His current research interests are in the complexity of physical, economic, social, and information systems, earning him a pioneering role in complex social network research. He has supervised a large number of Ph.D. theses in Finland, the UK, and the USA. Kimmo has recently acted as the coordinator of EU's FP7 FET Open ICT Collective project He is also a partner in EU's Horizon 2020 FET Open IBSEN project as well as in EU's Horizon 2020 SoBigData project.

János Kertész as a Postdoctoral Researcher at the University of Cologne and TU Munich and Researcher in different positions at the Institute of Technical Physics in Budapest. Since 1991, he has been Professor at the Budapest University of Technology and Economics, where he was Director of the Institute of Physics from 2001 to 2012. Since 2012, he has been Professor at the Central European University (CEU) where he is presently Head of the Department of Network and Data Science and Director of the Ph.D. Program in Network Science, the first of this kind in Europe. His main interest is in interdisciplinary applications of statistical physics. He has contributed to the fields of percolation theory, fractals, granular media, econophysics, network science, and computational social science. He is

Elected Member of the Hungarian Academy of Sciences. He has been awarded with several recognitions, including "Finland Distinguished Professorship" of the Finnish Academy/TEKES, the "Széchenyi Prize" of the Hungarian State, and "Benjamin Lee Professorship" of the Asia Pacific Center for Theoretical Physics.

László Kiss is a Research Fellow in the CSS-RECENS group at the Hungarian Academy of Sciences. He is a former Assistant Lecturer at the Eötvös Loránd University at the University of Pécs, a college Senior Lecturer at the Budapest College of Communication and Business, and a Researcher at Educatio Llc. He graduated as a sociologist from Eötvös Loránd University in 2001, and obtained a Ph.D. degree in Sociology in 2016. His main research interests are Hungarian social history in the socialist era, historical elite research, and history of the higher education system.

Zoltán Kmetty is a Postdoctoral Researcher at the Centre for Social Sciences, Hungarian Academy of Sciences. He is also a Senior Lecturer at the Faculty of Social Sciences at the Eötvös Loránd University. He has diverse research interests including political sociology, network studies, and suicide studies. He is an expert in methodology, survey design, and quantitative analysis.

Júlia Koltai is a premium Postdoctoral Researcher at the Eötvös Loránd Research Network, Centre for Social Sciences, and an Assistant Professor at Eötvös Loránd University. She obtained her Ph.D. in Sociology in 2013. Her main research focus is on quantitative methodology and statistics. In the past few years, her interest turned to large-scale (big) data and network analysis and natural language processing. She works with large amounts of text data from sources like Twitter, Instagram, and Facebook. Júlia is author of more than 30 book chapters and papers published in journals including *Social Networks* and *International Journal of Sociology*.

Fulvia Mecatti is a Professor of Statistics in the Department of Sociology and Social Research, University of Milano-Bicocca, and a former Director of the Ph.D. program in Statistics. She is a former President of the Survey Sampling Group of the Italian Statistical Society and an elected member of the Equal Opportunities Committee at the University of Milano-Bicocca. Her research focuses on methodological developments, with primary interests in Sampling Statistics, Computer Intensive Inference, and Gender Statistics.

Yohsuke Muraze is a Research Scientist at RIKEN Center for Computational Science (R-CCS) in Japan. He received his M.Sc. (2007) and Ph.D. (2010) in Applied Physics from the University of Tokyo. He was awarded the Docomo Mobile Science Outstanding Performance Prize in Social Science by Mobile Communication Fund in 2019 for his achievements in computational social science. His research interests include complex systems, network science, computational social science, and agent-based simulations.

Renáta Németh is Head of Statistics Department at Eötvös Loránd University, Faculty of Social Sciences. She holds an M.Sc. in Applied Mathematics and an M.A. and Ph.D. in Sociology. She is also the Director of the Survey Statistics and Data Analytics M.Sc. Program. Her research interests are primarily in the field of graphical modeling, marginal modeling, and survey statistics, with emphasis on the problem of causation. More recently, her interest has turned to automated text analytics. She is particularly inspired by discovering different application areas of statistics, understanding their epistemological differences, and adapting methodological knowledge from the different fields. She is the author of more than 30 book chapters and papers published in journals including *Sociological Methodology*, *Biometrika*, *Field Methods*, *Journal of Epidemiology and Community Health*, and American Journal of Public Health.

Gábor Péli graduated in mathematics, physics, and sociology at the ELTE University in Budapest. Between 1991 and 2016, he worked in several research and teaching positions in the Netherlands, at the Universities of Amsterdam, Groningen, and Utrecht. Currently, he is a Senior Researcher at the Centre for Social Sciences, Hungarian Academy of Sciences Centre of Excellence in Budapest and Professor of Sociology at the Gáspár Károli University of the Reformed Church in Hungary. His research interests include logical modeling in organization science, social network analysis, and the dynamics of socio-economic transition processes. He published several papers on organizational topics in journals including *Organization Science*, *Social Networks*, *American Journal of Sociology*, *American Sociological Review*, *Academy of Management Journal*, *Sociological Methodology*, *Journal of Mathematical Sociology*, and *Sociological Theory*.

Orsolya Ring is a Research Fellow in the CSS-RECENS group at the Hungarian Academy of Sciences since 2018. She is an Archivist-Historian (Ph.D., 2011, ELTE University of Budapest). She was working at the National Archives of Hungary from 2000 to 2018 where was the Head of the Department of Government Organs and Workers' Parties' Documents after 1945 (from 2016 to 2018). She is a Visiting Lecturer at the University of Theatre and Film Arts. Orsolya is an Editor of *Korall Journal for Social History*. Her research interests includes society and cultural history of the twentieth century, historical elite research, and theater history.

Tamás Rudas is Professor in the Department of Statistics, Faculty of Social Sciences, at the Eötvös Loránd University in Budapest. Formerly, he was Founding Dean of the Faculty of Social Sciences of ELTE and Director General of the Centre for Social Sciences at the Hungarian Academy of Sciences. Tamás Rudas is also an Affiliate Professor in the Department of Statistics, University of Washington, Seattle. He is an Elected Fellow of the European Academy of Sociology and Past President of the European Association of Methodology. He has held several visiting positions in universities in Europe and the United States. His research concentrates on statistical methods for the analysis of categorical data and, in general, on the methodology of the social sciences. He has published in *The Annals of Statistics*, *Journal of the Royal Statistical Society*, *Biometrika*, *Sociological Methodology*, and *Quality and Quantity*. He authored/ edited volumes published by Springer and Sage.

Diego Perez Ruiz is a Research Associate in Statistics at the University of Manchester. He completed his B.Sc. in Applied Mathematics and Computer Science at the National Autonomous University of Mexico (UNAM) in Mexico City before completing his M.Sc. in Probability and Statistics at the Center for Mathematical Research (CIMAT) in Guanajuato and his Ph.D. in Mathematical Science at the University of Manchester. At UNAM, he awarded the Gabino Barreda medal for academic excellence, for being the best student of his generation in the Applied Mathematics degree program. In 2019, he was awarded with the prestigious Cochran-Hansen Prize for his paper, Supplementing Small Probability Samples with Nonprobability Samples: A Bayesian Approach, awarded by the International Association of Survey Statistician (IASS). He is also a member of various statistical societies, including the Royal Statistical Society, the American Statistical Society, the International Society for Bayesian Analysis, and the International Statistical Society. He has experience in statistical computing, data visualization, nonparametric statistics, machine learning, time series analysis, and forecasting. In the recent years, he worked on statistical methods for modeling social processes, on case studies of the Bangladesh and UK labor markets and on other topics in survey methodology.

Joseph W. Sakshaug is Acting Head of the Statistical Methods Research Department and Head of the Data Collection and Data Integration Unit at the Institute for Employment Research (IAB) in Nuremberg, Germany. He is also University Professor of Statistics in the Department of Statistics at the Ludwig Maximilian University of Munich, and Honorary Full Professor in the School of Social Sciences at the University of Mannheim. Previously, he was Associate Professor in Social Statistics at the University of Manchester (UK), and Assistant Professor of Statistics and Social Science Methodology at the University of Mannheim. He received his M.S. and Ph.D. in Survey Methodology from the University of Michigan-Ann Arbor and B.A. in Mathematics from the University of Washington-Seattle. From 2011 to 2013, he was an Alexander von Humboldt Postdoctoral Research Fellow at the IAB and at the Ludwig Maximilian University of Munich. He is a faculty member in the International Program in Survey and Data Science and an Adjunct Research Assistant Professor at the Institute for Social Research at the University of Michigan. His research focuses on data quality issues in complex surveys, the integration of multiple data sources, and empirical research methods.

Nisarg Shah is an Assistant Professor in the Department of Computer Science at the University of Toronto. He is broadly interested in theory and applications of algorithmic economics. His research spans topics such as computational social choice, multi-agent systems, game theory, and incentives in machine learning. He particularly explores theoretical definitions of fairness in algorithmic decision-making. He is the Co-founder of two not-for-profit websites, Spliddit.org and RoboVote.org, which provide free access to provably optimal solutions to everyday fair allocation and voting problems, and have been used by more than 200,000 people to date. He is the winner of the IFAAMAS Victor Lesser Distinguished Dissertation Award (2016), and the Facebook Fellowship (2014–2015).

Martina Katalin Szabó is an Assistant Research Fellow in the CSS-RECENS group at the Centre for Social Sciences, Hungarian Academy of Sciences since

2016. She is an Assistant Lecturer in the Department of Russian Philology, Faculty of Humanities, University of Szeged since 2015. She graduated from Russian and Hungarian linguistics and literature (2010) and from Hungarian as a foreign language (2012) at the University of Szeged. She defended her Ph.D. thesis in (2018) sentiment analysis of Hungarian texts (a field of computational linguistics). Her research interests include computational linguistics, corpus linguistics, content analysis with computational linguistic methods, and lexical pragmatics.

János Török is an Associate Professor in the Department of Theoretical Physics at Budapest University of Technology and Economics. He completed his Ph.D. in Physics at Université Paris Sud and the Budapest University of Technology and Economics. His main areas of research are granular materials, opinion and social network modelling, with focus on non-equilibrium phase transitions.

Klaus G. Troitzsch was a Professor of Computer Applications in the Social Sciences at the University of Koblenz-Landau and since 1986 until he officially retired in 2012 (but continues his academic activities). He earned his first degree in political science. After 8 years in active politics in Hamburg and after having completed his Ph.D., he returned to academia, first as a Senior Researcher in an election research project at the University of Koblenz-Landau, and from 1986, as a Full Professor of Computer Applications in the Social Sciences. His main interests in teaching and research are Social Science methodology, modeling and simulation. He worked on several EU-funded projects devoted to social simulation and policy modeling. He authored, co-authored, and co-edited books and articles on social simulation, and organized several national and international conferences in this field.

Arkadiusz Wisniowski is a Senior Lecturer in Social Statistics in the Social Statistics Department, School of Social Sciences, University of Manchester. He is also a Lead of the Statistical Modelling Group at the Cathie Marsh Institute (CMI); associate of the ESRC Research Centre for Population Change (CPC) at the University of Southampton; Visiting Fellow to the Research School of Social Sciences, Australian National University; and Adjunct Associate Professor at the Asian Demographic Research Institute, University of Shanghai. From 2009 to 2015, he worked as a Research Fellow at the ESRC Centre for Population Change and Southampton Statistical Sciences Research Institute, University of Southampton. From 2007 to 2009, he was a Researcher at the Central European Forum for Migration and Population Research in Warsaw, and Teacher and Research Assistant in the Department of Applied Econometrics, Warsaw School of Economics. He received his Ph.D. in Economics from the Warsaw School of Economics. His research concentrates on developing statistical methods for modeling and forecasting complex social processes, with a particular focus on migration and mobility, and combining various sources of data. He also has a general interest in data science, time series analysis, Bayesian computational methods, micro-econometrics, opinion polls, and ageing.

Part I
Theory: Dilemmas of Model Building and Interpretation

Modeling the Complex Network of Social Interactions

János Kertész, János Török, Yohsuke Murase, Hang-Hyun Jo, and Kimmo Kaski

1 Introduction

One of the great scientific missions of mankind is to understand itself. A number of disciplines cope with this challenge, from biology and medicine to psychology and sociology. Natural sciences have developed their methodology following the example of physics, relying on the interaction between empirical observations (and, if possible, experimental studies) and theoretical approaches, where the former is always decisive. Modeling is a crucial step in this process. Models are based on the

J. Kertész (✉)
Department of Network and Data Science, CEU, Budapest, Hungary
e-mail: KerteszJ@ceu.edu

János Török
Department of Network and Data Science, CEU, Budapest, Hungary

MTA-BME Morphodynamics Research Group, and Department of Theoretical Physics, Budapest University of Technology and Economics, Budapest, Hungary

Y. Murase
RIKEN Center for Computational Science, Kobe, Hyogo, Japan

H.-H. Jo
Asia Pacific Center for Theoretical Physics, Pohang, Republic of Korea

Department of Physics, Pohang University of Science and Technology, Pohang, Republic of Korea

Department of Computer Science, Aalto University, Espoo, Finland

K. Kaski
Department of Computer Science, Aalto University, Espoo, Finland

The Alan Turing Institute, British Library, London, UK

T. Rudas, G. Péli (eds.), *Pathways Between Social Science and Computational Social Science*, Computational Social Sciences,
https://doi.org/10.1007/978-3-030-54936-7_1

distinction between more and less important factors with focus on such mechanisms that could explain or even predict observations.

Social sciences have been in a much more difficult situation for many reasons. Just to name a few: the observer and the observed are of the same kind; the system is adaptive, i.e., the results are influenced by the observation process; the quantification of relevant features is highly nontrivial; the availability of empirical data is restricted. However, much progress has been achieved in strengthening the quantitative and computational aspects of social science research. The recent development due to the digital revolution and the related data deluge have enormously accelerated this process. "Being able to automatically and remotely obtain massive amounts of continuous data opens up unprecedented opportunities for social scientists to study organizations and entire communities or populations" (Butler 2007). We are now in the position to find answers to questions, which could earlier be asked at most at a speculative level, while today empirical studies are possible.

One of such questions is about the structure of the network of human interactions at the societal scale. While this has been at the focus of interest of social scientists at least from the 1970s (Granovetter 1973), large-scale empirical investigations seemed to be hopeless for a long period of time. This situation has changed as electronic communication records such as call detail records (CDRs) of mobile phones (Onnela et al. 2007), online social network (OSN) data (Ugander et al. 2011), and tracking people by electronic devices (Cattuto et al. 2010; González et al. 2008) provide reliable information about social interactions between people. The networks mapped out using these sources and methods are considered as good proxies for the social network at large scale.[1] Having such data opens up the opportunity to test earlier hypotheses and formulate a theory. An important step toward this goal is the construction of models, which are able to reflect the empirical observations. The aim of this paper is to summarize our activities in this direction.

2 Characterization of Large-Scale Social Networks

Digital footprints tell detailed information about the structural and temporal aspects of human interactions. Emails (Eckmann et al. 2004; Kossinets and Watts 2006), mobile phone calls (Blondel et al. 2015; Onnela et al. 2007), datasets from OSNs like Facebook (Ugander et al. 2011) and Twitter (Kwak et al. 2010), and even data of face-to-face proximity (Isella et al. 2011; Zhao et al. 2011) provide ample empirical data to analyze the social network and deduce its main features. We certainly do not possess a complete description, but we are in the position of listing a number of important features, so called stylized facts[2] (Jo et al. 2018). We summarize some of them below.

[1]The use of networks based on communication data as a proxy for the real social network has its limitations which we will discuss later.

[2]We call *stylized facts* the set of established properties, which have been found repeatedly as characteristic for social networks.

2.1 Degree Distribution

The degree of a node in a network is the number of its neighbors and its distribution over the nodes is one of the basic characteristics of networks in general. Human society is extremely inhomogeneous, and this is also reflected in the broadness of the degree distribution of the social network. All big data-based empirical studies point in this direction, irrespective of what communication channel was used. The specific form of the distribution may be a result of the details of the service, e.g., on Twitter the incoming and outgoing degree distributions differ strongly, while this is not so if mutuality is requested. In this sense perhaps the best data can be obtained from CDRs (Blondel et al. 2015; Onnela et al. 2007). While some results show a "scale-free," power-law tail[3] in the distribution (Barabási and Bonabeau 2003), e.g., in Twitter (Aparicio et al. 2015), deviations are also well known, for Facebook see Ugander et al. (2011). According to Dunbar's social brain hypothesis the number of *real social* interactions is limited by finite resources like brain capacity and available time. Therefore, the degree distribution should have a characteristic value, namely, the Dunbar number (known to be about 150) (Dunbar 2011), indicating the average size of egocentric social relationships. Such a characteristic value or typical scale contradicts the power-law tailed, "scale-free" degree distribution.

An important common feature is that the degree distributions empirically observed on communication and OSN data are monotonically decreasing. It has been argued (Török et al. 2016) that this must be an artifact due to the sampling, because it is unrealistic that the maximum probability is that people have one social contact, as even patients with strong autistic disorder do have usually more. In fact, big communication data comes from single channel observations due to technical and privacy reasons. Using a model of channel selection, it could be demonstrated that indeed, even starting from peaked degree distribution the single channel information leads to monotonically decreasing degree distributions (Murase et al. 2019; Török et al. 2016). In summary, social networks are peaked and right-skewed, broad degree distribution with a characteristic cutoff value around the Dunbar number.

2.2 Assortative Mixing by Degree

In social networks, homophily constitutes one of the basic link formation mechanisms (McPherson et al. 2001). Accordingly, high-degree nodes (i.e., popular people) have the tendency to be linked to other high-degree nodes, which is a kind

[3]Power-law tail of a distribution of a variable x means that for large values the distribution behaves like $x^{-\gamma}$. As a power-law function is scale invariant, no characteristic scale occurs in it—hence the "scale free" terminology.

of correlation called assortative mixing by degree (Fisher et al. 2017). This can be measured, e.g., by the dependence of average degree of the neighbors $\langle k_{nn}\rangle(k)$ of a node of degree k: In case of assortative mixing, the function $\langle k_{nn}\rangle(k)$ depends monotonically on the degree k.

2.3 Clustering

Another type of correlations manifests itself as the abundance of triangles in social networks, which is a consequence of the effect that friends of friends easily become friends. Globally, this feature can be quantified by the so-called global clustering coefficient, the ratio of the total number of triangles, to the third of the number of connected triples of nodes. Locally, the clustering coefficient C_i of a node i with degree k_i is the ratio of the triangles having i as a vertex and the maximum possible number of such triangles, $k_i(k_i - 1)/2$ (Newman 2010). A general observation is that the local clustering coefficient $C(k)$ averaged over the nodes of degree k is a monotonically decreasing function with a dependence of $\sim 1/k$ (Szabó et al. 2004). The decay of the clustering coefficient with the degree is natural; as with increasing degree, the number of relationships closing the triangles cannot increase quadratically; therefore the density of triangles around a high-degree node will be lower than around a small degree node. Nevertheless, the global or average clustering coefficient remains large even in a highly connected network.

2.4 Granovetterian Community Structure

In many networks, including social ones, there appears yet another type of correlation, due to the formation of mesoscopic structures or communities. These are densely wired parts of the network that are loosely connected to the rest of the network, and they are present in all complex empirical networks, not only in social ones (Newman 2010). It is assumed that they have particular roles in the functioning of the network as has been discussed for social networks since the seminal work by Mark Granovetter (1973). He first defined the strength of a tie or link, indicating the intimacy of the relationship and then put forward the hypothesis that the human society consists of communities, which are strongly wired with high weight links and these communities are then connected by weak ties. This hypothesis was proved at the societal level by using mobile phone CDRs (Onnela et al. 2007) and other communication data (Weng et al. 2018). This community structure severely influences information transfer, which was Granovetter's main focus. Since within the communities connections are tight and interactions are frequent, information is practically instantaneously shared. Therefore, the chance to acquire new information, e.g., for a job offer, is much higher through a weak link, bridging two different communities. But the Granovetterian local structure has important consequences also on the global scale. The title of Granovetter's original paper ("The strength of weak ties") refers to the fact that weak ties hold the network together. The latter can be best demonstrated by a percolation test (Onnela et al. 2007): Removing links one by one in ascending or descending order of the link weights up to the fraction f of links, the system undergoes percolation (fragmentation) transitions, which is characterized by well-defined transition points or percolation thresholds, f_{desc} for the descending order and f_{asc} for the ascending order. For the case of the ascending order, the weak ties are first removed; therefore the fragmentation occurs earlier than for the descending order as the links between the communities are cut first. For the descending order, fragmentation occurs much later as the strongly wired communities are first diluted, but the network at that stage still holds together. The positive difference $\Delta f = f_{desc} - f_{asc}$ in the fragmentation thresholds indicates the existence of the Granovetterian structure.

2.5 Tie Formation and Fading

What makes a social tie? This has been a central issue in social sciences from the very beginning. There is consensus that homophily is a major tie forming factor (McPherson et al. 2001) as similarity in a broad variety of social traits such as gender, age, status, taste, etc. may bring people closer to each other. For example, a study of CDRs has shown how high intensity relationships depend on gender and age (Palchykov et al. 2012) and confirmed that—besides kinship—homophily plays an important role in tie formation in communication networks.

It is of not less interest to investigate the reason, why relationships break or fade. There are trivial causes, like one of the partners passing away or moving far (geographic proximity is still an important factor in maintaining social ties (Lengyel et al. 2015)). In other cases the social dynamics results in the decay or end of a relationship. In summary, friendships can be broken up or fade away with time leading to a dynamically changing social network.

2.6 Multiplex Structure and Overlapping Communities

Social relationships can be of various kinds, depending on the context, like friendship, co-working, doing sport together, etc. Therefore, the social network is a multiplex (Bianconi 2018; Kivelä et al. 2014), i.e., a multilayer network, where links in different layers represent connections of the same person in the different contexts (see Fig. 1 as an illustration). Not only the context can serve as the basis for considering the social network as a multiplex: Division of the interactions into categories and corresponding layers according to communication channels (Török et al. 2016) or interaction strengths (Unicomb et al. 2019) are further possibilities.

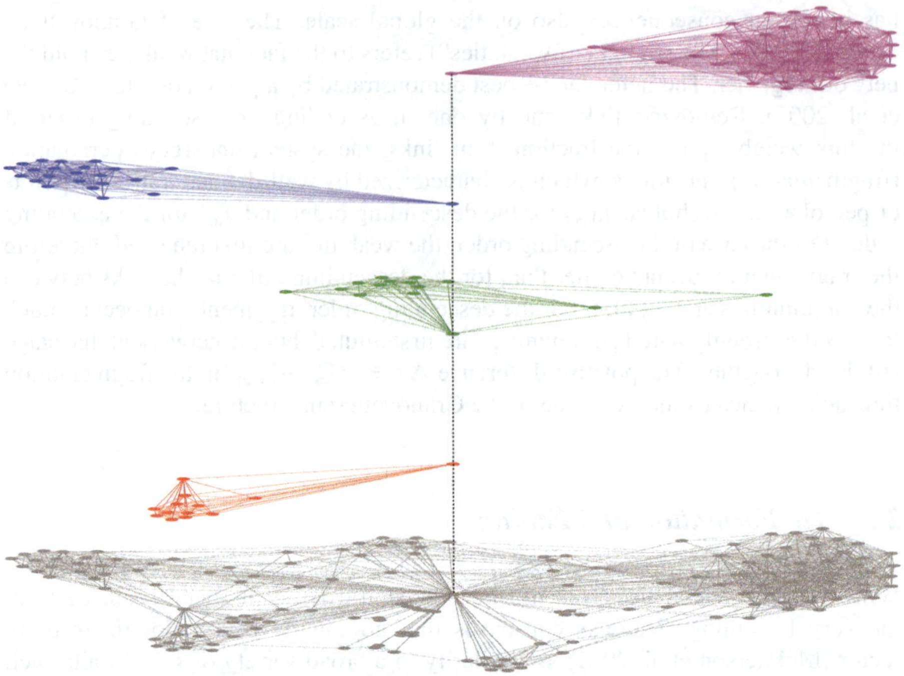

Fig. 1 Schematic picture of the multiplex structure of an egocentric network. The ego (central node) is in different types of relationships (kinship, schoolmates, sport, working, etc.) with the alters, which are represented at different layers. The whole social network—in this case that of the ego—is the aggregate of the layers indicated at the bottom. (Data taken from an OSN (Kertész et al. 2018))

The fact that individuals are involved in different contexts has the consequence that the community structure is overlapping. Therefore most of the community detection methods (Fortunato 2010) based on partitioning the network are inadequate, and special ones targeted at overlapping communities have to be applied (Ahn 2010; Palla et al. 2005).

3 Modeling Granovetterian Community Structure

The study of community structure has largely focused on detecting it (Fortunato 2010), and it is a highly non-trivial problem. Much less attention has been paid to the relevant question of the formation of the communities, in particular to the question: What is the mechanism leading to the peculiar community structure characteristic for the human society? Clearly, this is closely related to the way ties are formed and broken. In a simple model by Kumpula et al. (2007) puts forward a mechanism based on basic link formation mechanisms and an oversimplified process for eliminating interactions to reproduce the Granovetterian community structure, expressing link weight vs. topology correlation.

3.1 Weighted Social Network

The Weighted Social Network (WSN) model (Kumpula et al. 2007) is formulated as follows. There are N nodes (individuals), who may be connected to other nodes by weighted links. A node i has strength $s_i = \sum_j w_{ij}$, with w_{ij} being the weight of a link (i, j). The two main link formation mechanisms are considered to be (i) random linkage or global attachment (GA) and (ii) triadic closure or local attachment (LA). The GA is or represents a crude approximation to the "focal closure," while LA takes the dominant part of the "cyclic closure" (Kossinets and Watts 2006).

Nodes of the network are visited at random, and the time is measured in terms of sweeps through all the nodes. If a node i is picked, first the random linkage is realized: If the node i has neighbors, then with probability p_r, a link with weight w_0 to a random node is created. Otherwise, the link is created with probability 1. As a next step, a triadic closure follows for node i in which a triangle is either formed or strengthened: a neighbor j of the node i is selected with probability w_{ij}/s_i and the j's neighbor k (if there exist any) with probability $w_{jk}/(s_j - w_{ij})$. If the link (k, i) does not exist, it is created with the probability p_Δ and with weight w_0. The weights of all the old links involved in this process are incremented by a reinforcement term δ. This step reflects that high weight links are used more often and that using a relationship strengthens it. In addition, triangles are formed with preference, as it is characteristic for social networks ("friends of friends get easily friends"). In order to reach a stationary state, nodes are deleted, and new ones with no links are introduced with probability p_d, i.e., all the links of a randomly selected node are deleted. For more realistic link deletion processes, see Murase et al. (2015) and later part of this paper.

Among the parameters of p_r, p_Δ, w_0, p_d, and δ, the reinforcement term δ is the most relevant to obtain the Granovetterian community structure. By tuning δ between 0 and 1, the resulting networks change from one without any community structure to a very inhomogeneous network with communities and clear Granovetterian correlation between the topology and the link weights (see Fig. 2). This model succeeded in reproducing various stylized facts including the community structure, Granovetterian weight-topology relation, assortative mixing, decreasing local clustering spectrum, and the relationship between the node strength and degree. However, the WSN model has only a single layer; thus important aspects of the multilayer structure of social networks are missing. As mentioned above, the indicator of the Granovetterian community structure is Δf, the difference between the fragmentation thresholds when removing the links in the order of descending and ascending weights, which has the rather high value $\Delta f \approx 0.35$ for $\delta = 1$.

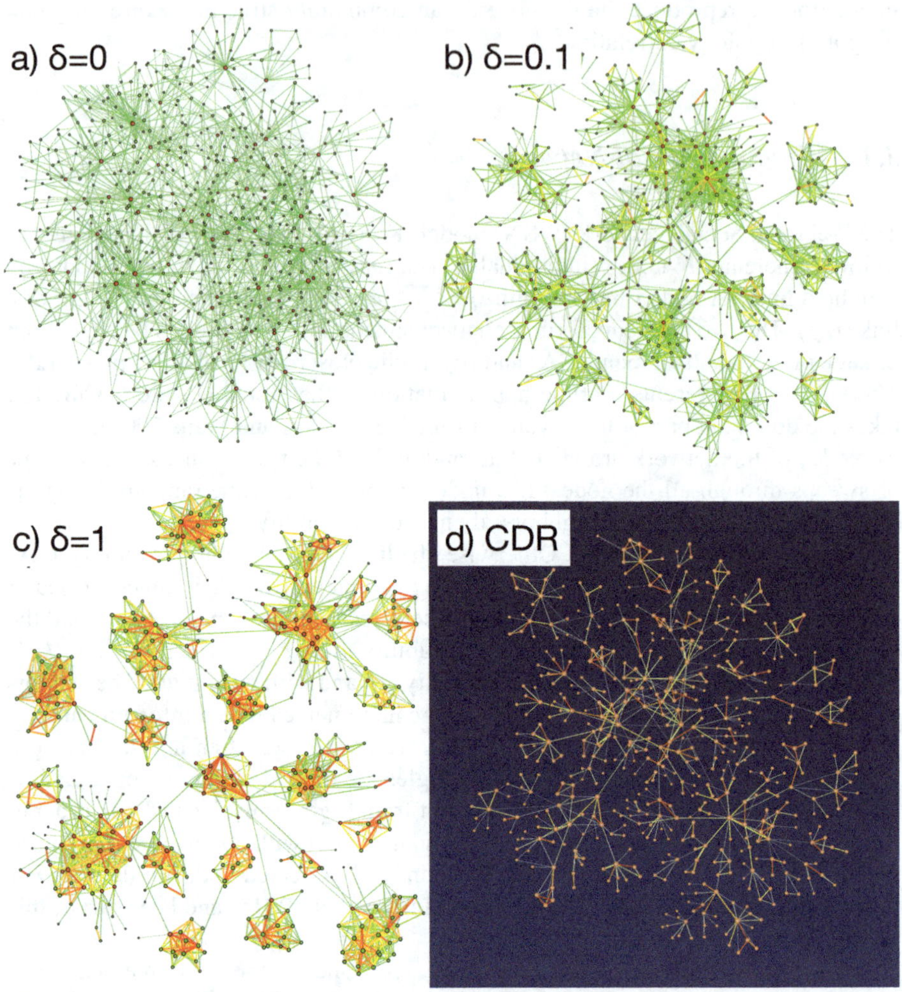

Fig. 2 (**a**)–(**c**): Results from simulations of the WSN model. The parameters are $p_r = 10^{-4}$, $p_d = 10^{-3}, and w_0 = 1$, and p_Δ was set such that the average degree is kept constant (~10). The reinforcement term δ is indicated in the figures. For $\delta = 0$, no community structure is seen. It emerges as δ is increased, and for $\delta = 1$, strongly wired communities connected by weak links are visible as required by the Granovetter picture (Kumpula et al. 2007). (**d**) Part of the network as obtained from CDR data (Onnela et al. 2007). Color (gray-scale) coding illustrates the strength of the ties. (After Onnela et al. 2007 and Kumpula et al. 2007)

3.2 Link Expiration

As mentioned above, in the simplest version of the model, the node deletion was used to drive the system into a stationary state. This corresponds to, e.g., the death of a person or the person's departure from an online service, when all the connections are broken at once. However, the termination of a relationship may occur in various ways: A relationship may end abruptly, for example, when an intimate relationship suddenly breaks up but a friendship can also fade gradually if not enough effort is put into it, e.g., because of changing location, interest, or simply due to the emergence of new friendships (Saramäki et al. 2014). These three categories of ending relationships can be easily implemented in the WSN model (Murase et al. 2015). Besides node deletion, one can define link deletion with a corresponding rate p_{dl} and an aging mechanism, which is essentially the opposite of the reinforcement mentioned earlier: If a link is not used, its weight is multiplied by an aging factor a, and it is removed if the link weight gets smaller than a threshold value (taken as 0.01). This decay of the link weight reflects the common experience that relationships decline unless they are fostered.

3.3 Multiplex Structure

It is interesting to investigate the impact of the different types of link deletion mechanisms on the topology of the resulting network (Murase et al. 2015). The following qualitative observations can be made on simulations of the model. While in all cases the Granovetterian structure is present ($\Delta f > 0$), it is most pronounced for the aging model ($a = 0.9$, $\Delta f \sim 0.8$), and it is weakest for the link deletion case ($p_{dl} = 0.0035$, $\Delta f \sim 0.1$) with other model parameters chosen such that the average degrees of the generated networks are all $\langle k \rangle \sim 11$. Other network characteristics are also influenced. The degree distribution for the networks with node deletion mechanism is broad and peaked, while it is much narrower with considerably smaller variance for link deletion and aging. Node deletion and link deletion lead to weak assortativity ($\langle k_{nn} \rangle (k)$ is a slightly increasing function), which is much more pronounced for the aging model, where the slope of the $\langle k_{nn} \rangle (k)$ is large.

The interesting observation from this study is that (a) Granovetterian structure is robust and can be considered as a consequence of both the focal and cyclic closure mechanisms (here approximated by GA and LA, respectively); (b) aging leads to more separated communities with rather homogeneous degrees; (c) node deletion seems responsible for the broadness of the degree distribution. In reality, all these mechanisms coexist, and their contributions manifest themselves in some of the stylized facts. It would be an interesting research goal to investigate empirically with traditional methods and using Big Data the ratio of the above mechanisms (and possibly others) in link termination.

While in reality we all belong to several communities, the WSN model in its original form is not suitable to reflect this aspect. The origin of this overlapping structure of the social network is mainly such that the individuals are involved in different types of relationships each possibly forming communities. In a proper representation, these different relationships form separate layers in a multiplex (see Fig. 1).

An important aspect is missing from the above model, namely, the multiplex character of the network, and, consequently the overlap of the communities is small. This can be measured by applying an appropriate community detection method (Ahn 2010) and calculating the average number of communities a node belongs to.

It is a real challenge to construct a multiplex model, where Granovetterian structure is present both in the layers and in the aggregate. The plain WSN generates single-layer samples and puts such layers simply on the top of each other destroys the Granovetterian structure. This indicates that some correlation between the layers is needed. As geographic proximity is still an important principle in the tie formation (Lengyel et al. 2015), we introduced the following model (Murase et al. 2014).

The model is embedded into a two-dimensional space on which the N nodes are distributed at random. Periodic boundary conditions are used in both directions, meaning that the system is put onto a torus so that it does not have boundary effects.[4] A node has connections within the different m layers. We apply the WSN model for $N = 50{,}000$, $p_r = 0.0005$, $p_\Delta = 0.05$, $p_d = 0.001$, $\delta = 1$, and $w_0 = 1$. We use the following constraint (Murase et al. 2014): Focal attachment is not made entirely randomly, but we assume that the probability for making a new connection is higher if the two nodes are geographically closer. The probability that a node i makes a new connection to a node j during such a step is proportional to

$$p_{ij} \sim r_{ij}^{-\alpha}, \tag{1}$$

where r_{ij} is a distance between nodes i and j and α is the parameter controlling the dependence on the geographic distance. There are no correlations between the layers if $\alpha = 0$ and as its value increases the correlations become larger. Figure 3 illustrates how the topology depends on α. It is clear from Fig. 3 that for small α no Granovetterian structure can be observed: Δf is very small. It may be surprising that the average number of communities a node belongs to is rather high in this limit. This is an artifact due to the community detection: By its nature the method leads to communities even in a totally random network like the Erdős–Rényi one. These do not have physical meaning, and so their overlap does not have either. The overlap gets filled with meaning in parallel with the emergence of real communities, which happen to show the Granovetterian weight-topology correlations. Therefore,

[4] *Periodic boundary conditions: nodes on the left (top) side of the space are also close to the ones on the right (bottom) edge.*

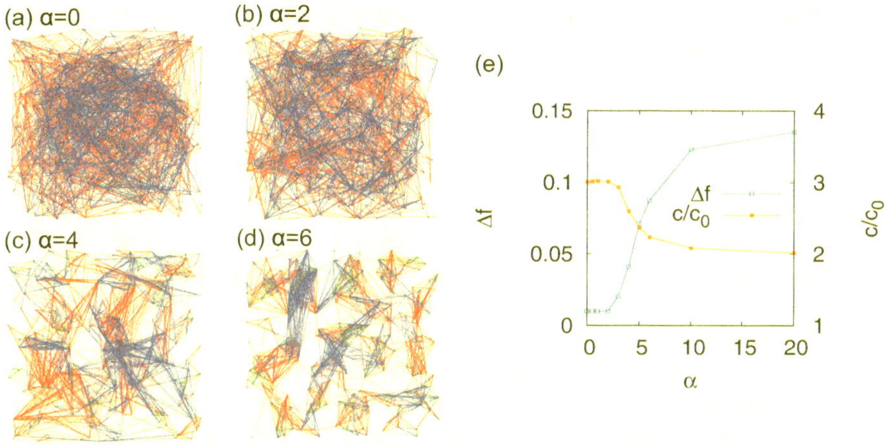

Fig. 3 (**a**)–(**d**): Snapshots of network configurations on a two-layer WSN with different values of α (the parameter controlling the dependence of the global attachment on the spatial separation of the nodes, see Eq. (1)). In the panels (**a**) and (**b**), the correlation is zero or very small; thus the two layers destroy the communities, and no Granovetterian structure is visible, while for larger correlations, as showin in (**c**) and (**d**), the typical strongly correlated communities connected by weak ties emerge. (**e**) Variation of the threshold difference Δf and the average number of communities c a node belongs to (normalized by the single-layer value $c_0 \approx 2.9$). (After Murase et al. 2014)

at large enough α, we have very pronounced Granovetterian communities with still considerable average overlap.

4 Homophily and Structural Changes

In real social tie formation, homophily is one of the main mechanisms, but in the WSN model, it is not explicitly included. However, it is known that focal closure is governed to a large extent by homophily and the random selection we used above in the Global Attachment step is only a crude approximation to this type of link formation. The situation is somewhat better in the multiplex model with geographic correlations, as proximity is one possible basis for homophilic attraction. However, there are many other features like gender, age, race, political attitude, religion, economic and social status, interest, etc. that can act as bases for homophilic attraction. When studying the cultural drift, the tendency of culture traits to move throughout an area, Axelrod introduced a model, which was able to treat a diversity of cultural features (Axelrod 1997) characterizing a single person. We are going to use the same representation, but from a different point of view. While the main question in the studies by Axelrod and Maxi San Miguel's group (e.g., see Toral et al. 2003) has been the vulnerability of the cultural diversity with respect to globalization and they considered the underlying network as given, we are interested in how the network emerges as a consequence of homophilic interactions (Murase et al. 2019).

Our model is yet another generalization of the WSN model in which individuals are characterized by a vector containing F features. For simplicity, we assume that F is the same for all members of the population and each feature can take one of the same number of q values. Thus the feature vector of node i is a set of F integers: $(\sigma_i^1, \sigma_i^2, \ldots, \sigma_i^F)$ with $\sigma_i^f \in \{1, \ldots, q\}$ for each $f \in \{1, \ldots, F\}$. The size F of the vector characterizes the social complexity of the population, and the number q possible values per feature represents its heterogeneity. In the homophilic WSN model, the network is updated such that the features of nodes are taken into account, when ties are formed or strengthened. In each step, a feature f of node i is randomly chosen from its F features, and it can make a link with or chooses a neighbor from nodes sharing the same value of the feature f, i.e., $\sigma_i^f = \sigma_j^f$, where j is the index of the node selected for tie formation (in case of Global Attachment) or of the neighbor (in case of Local Attachment).

We have simulated this model with the following WSN parameters $N = 50{,}000$, $p_\Delta = 0.02$, $p_r = 0.001$, $p_d = 0.005$, and $\delta = 1$ (Murase et al. 2019). Homophily, as it is considered in this generalized WSN model, changes the topology substantially, showing an interesting dependence on the parameters F and q. For example, keeping $q > 1$ fixed, the average degree or the clustering coefficient shows a non-monotonic dependence on F, with a minimum for the average degree or a maximum for the clustering coefficient at $F_c(q)$. Similar behavior can be observed for fixed F as a function of q defining $q_c(F)$. In order to understand the nature of this non-monotonic behavior, it is convenient to define the feature overlap between the nodes i and j:

$$o_F(i, j) = \frac{1}{F} \sum_{f=1}^{F} \delta(\sigma_i^f, \sigma_j^f), \tag{2}$$

where $\delta(x, y)$ is the Kronecker delta. It turns out that $\langle o_F \rangle$, the average of o_F over connected neighbors, is close to 1 for $F < F_c$, at which point it starts decreasing monotonically. In parallel to this, the average number of communities a node belongs to is 1 for $F < F_c$, beyond which it increases monotonically. These properties of the model strongly resemble the phase transition in statistical physics (Reichl 2016). Here the transition can be understood in the following qualitative terms. If F is small, it is easy to find partners with large overlap. As F increases, the possibilities shrink, and the set of linked nodes becomes more closed (increasing clustering). Beyond the transition point, it is not possible any more to select the partners with large overlap, making the system more inhomogeneous and the clustering decreases. This is illustrated in Fig. 4.

To understand what is behind this transition, we start by estimating the average number $\tilde{N}$ of nodes sharing a specific set of traits: $\tilde{N} \approx N/q^F$. The transition is estimated to occur when $\tilde{N}$ becomes less than the typical degree for the

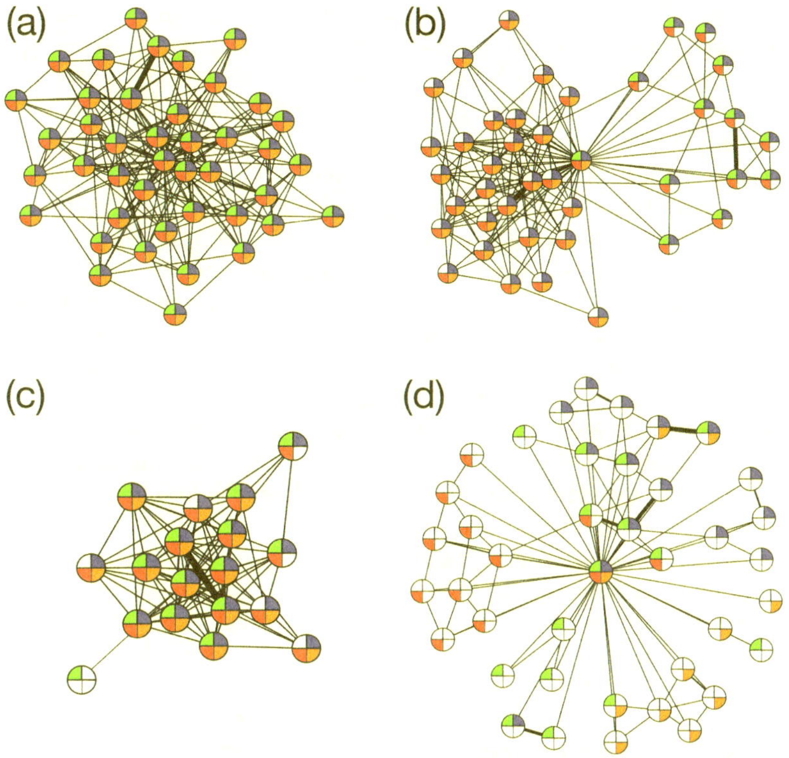

Fig. 4 Typical egocentric networks (ego: central node) for fixed $F = 4$ as a function of q. In panels (**a**, **b**) $q = 4$ (segregated phase), (**c**) $q = 7$ (near the transition point), and (**d**) $q = 20$ (diverse phase). When a feature value of a neighbor is identical to that of the ego, it is colored the same; otherwise it is left white. In the segregated phase, a node is typically connected to its matching nodes (**a**); only a small fraction of nodes have links to a few other types of nodes, bridging communities (**b**). (**c**): A node has a smaller degree, and some of its neighbors are not matching nodes. (**d**): Feature overlap between neighboring nodes is even less, and the ego belongs to more diverse communities. (Figure in Murase et al. (2019) by Yohsuke Murase et al. is licensed under CC BY 4.0)

original WSN model, because beyond that a node is not able to find appropriate "partners" any more. According to the simulations, this is a correct order of magnitude estimate, and it also describes correctly the scaling of the average link density (Murase et al. 2019). This means that plotting the average degree as a function of N/q^F, the curves fall on top of each other for values above $F_c(q)$.

The following picture emerges: Let us call two nodes matching, if they have the same values for all the features. Below F_c (or q_c), the average feature overlap is $\langle o_F \rangle \approx 1$ meaning that nodes form homogeneous communities of size $\sim N/q^F$ from mostly matching nodes, indicating segregation due to homophily. We can only occasionally find nodes, which do not match perfectly with their neighbors, and these serve as bridges between rather homogeneous communities. At F_c a transition from the large homogeneous, segregated communities to smaller, more heterogeneous ones occurs. The origin of this transition is that as the space of opportunities increases, either by increasing F or q or both, the nodes will have more and more chance to get connected to other nodes that overlap with them only in the feature, which was just asked for.

One may wonder about the relevance of this finding, as all individuals have a large number of features and the possible choices within one feature may also be large or even continuous. Our interpretation is that the features are not equally important in the formation of ties. Especially, there are constellations, when some features become extremely relevant, like in times of sharpening political situations, social turmoils, or wars. Another related example is the community formation in online social networks, where people make contacts by considering few topical issues. According to our model, this leads already to segregation, which is further amplified by algorithmic bias (Sîrbu et al. 2019) leading to the well-known effect of "echo chambers." The few considered features result already in rather segregated communities, and the mechanism, by which the flow of information is directed to the users, enhances this effect.

The phase with more heterogeneous communities is then a model of society, where diversity is more common and individuals have close relationships of different gender, age, location, religion, political view, etc. Each of these traits may give rise to its own community in which there is no perfect feature overlap.

The relationship between homophily and segregation is not new to social scientists. Already Schelling (1969) showed that segregation can occur without aversion against people with different features—homophily alone can lead to it. Here our aim was to investigate how homophily influences the social network if elementary tie formation mechanisms are taken into account represented here by GA and LA mechanisms. Our results demonstrate that in a society, where only a few features matter or where the possible value of features is strongly reduced, segregation is dramatically enhanced.

5 Summary and Outlook

Modeling the society at large scale is a real challenge, and we are only at the very beginning of this endeavor. The large and increasing amount of data, which are becoming available for this type of research, gives a solid basis for research and provides possibilities to quantitatively test various theoretical approaches. Models can be of different kind, depending on their motivation, tools, and goals. Here we have shown two kinds of them: (i) Models using simple, established social mechanisms to reproduce and interpret empirical observations (e.g., WSN); (ii) models highlighting plausible mechanisms using simplified settings (e.g., homophilic WSN). Both approaches constitute substantial part of computational social science and have their own role: While the former should provide the basis for detailed quantitative understanding with possible predictive power, and the latter gives insight into unexplored mechanisms.

While enjoying the broad possibilities opened by data-based modeling, we have to be aware of the limitations. For example and as mentioned above, the reduction of the data to one channel of communications leads to artifacts, when the networks created by such one channel data are used as a proxy for the whole social network. Although data are abundant, in contrast to targeted data collection for scientific purposes, they underlie legal restrictions mostly due to ethical, privacy concerns, and the patterns are often hidden. An additional ethical aspect is related to modeling itself. Since the system under consideration consists of humans, the question rises, how granular such an approach should be and what should be the relationship between the target of the model and the model itself. A whole area of research is devoted to the interesting ethical issues of computational social science, which have been fully ignored here.

There are many very interesting related scientific questions for further research, and we mention here some. From empirical point of view, the collection of stylized facts should be continued. One of the burning problems is the selection of communication channels. Once we understood this problem better, we could resolve the abovementioned limitations more easily. Furthermore, bringing the simplified models closer to reality is a promising and important avenue for applications. Having well-functioning social models could be used as a test bed for policy makers with reduced risk. These models could be interactive in the sense that a part of the population could participate in the validation and testing and give feedback to the decision makers. This could provide a more solid ground for new measures than just using surveys.

Acknowledgments Research reported here was supported by European grants FP7 SIMPLEX (JK), H2020 MULTIPLEX (JK), Hungarian grant OTKA K129124 "Uncovering patterns of social inequalities and imbalances in large-scale networks" (JK, JT), Aalto AScI (JT) the Academy of Finland (KK), The Alan Turing Institute (KK). YM acknowledges support from MEXT as "Exploratory Challenges on Post-K computer (Studies of multi-level spatiotemporal simulation of socioeconomic phenomena)" and from Japan Society for the Promotion of Science (JSPS) (JSPS KAKENHI; Grant No. 18H03621). H-HJ acknowledges financial support by Basic Science Research Program through the National Research Foundation of Korea (NRF) grant funded by the Ministry of Education (NRF-2018R1D1A1A09081919). The systematic simulations in this study used OACIS (Murase 2014).

References

Y.-Y. Ahn, J.P. Bagrow, S. Lehmann, Link communities reveal multiscale complexity in networks. Nature **466**(7307), 761–764 (2010)

S. Aparicio, J. Villazón-Terrazas, G. Álvarez, A model for scale-free networks: application to twitter. Entropy **17**, 5848–5867 (2015)

R. Axelrod, The dissemination of culture: a model with local convergence and global polarization. J. Confl. Resolut. **41**(2), 203–226 (1997)

A.-L. Barabási, E. Bonabeau, Scale-free networks. Sci. Am. **288**(5), 60–69 (2003)
G. Bianconi, *Multilayer Networks* (Oxford University Press, Oxford, 2018)
V.D. Blondel, A. Decuyper, G. Krings, A survey of results on mobile phone datasets analysis. EPJ Data Sci. **4**(10) (2015)
D. Butler, Data sharing threatens privacy. Nature **449**, 644–645 (2007)
C. Cattuto, W. Van den Broeck, A. Barrat, V. Colizza, J.-F. Pinton, A. Vespignani, Dynamics of person-to-person interactions from distributed RFID sensor networks. PLOS ONE **5**, e11596 (2010)
R.I.M. Dunbar, Constraints on the evolution of social institutions and their implications for information flow. J. Inst. Econ. **7**(Special Issue 03), 345–371 (2011)
J.-P. Eckmann, E. Moses, D. Sergi, Entropy of dialogues creates coherent structures in e-mail traffic. Proc. Natl. Acad. Sci. U. S. A. **101**(40), 14333–14337 (2004)
D.N. Fisher, M.J. Silk, D.W. Franks, The perceived assortativity of social networks: methodological problems and solutions, in *Trends in Social Network Analysis: Information Propagation, User Behavior Modeling, Forecasting, and Vulnerability Assessment*, ed. by R. Missaoui et al. Lecture Notes in Social Networks (Springer, Berlin, 2017), pp. 1–19
S. Fortunato Community detection in graphs. Phys. Rep. **486**, 75–174 (2010)
M.C. González, C.A. Hidalgo, A.-L. Barabási. Understanding individual human mobility patterns. Nature **453**(7196), 779–782 (2008)
M.S. Granovetter, The strength of weak ties. Am. J. Sociol. **78**(6), 1360–1380 (1973)
L. Isella, J. Stehlé, A. Barrat, C. Cattuto, J.-F. Pinton, W. Van den Broeck, What's in a crowd? Analysis of face-to-face behavioral networks. J. Theor. Biol. **271**(1), 166–180 (2011)
H.-H. Jo, Y. Murase, J. Török, J. Kertész, K. Kaski, Stylized facts in social networks: Community-based static modeling. Physica A **500**, 23–39 (2018)
J. Kertész, J. Török, Y. Muraze, H.-H. Jo, K. Kaski, Multiplex modeling of society, in *Multiplex and Multilevel Networks*, ed. by S. Battiston et al. (Oxford University Press, Oxford, 2018)
M. Kivelä, A. Arenas, M. Barthelemy, J.P. Gleeson, Y. Moreno, M.A. Porter, Multilayer networks. J. Complex Netw. **2**(3), 203–271 (2014)
G. Kossinets, D.J. Watts, Empirical analysis of an evolving social network. Science **311**(5757), 88–90 (2006)
J.M. Kumpula, J.P. Onnela, J. Saramäki, K. Kaski, J. Kertész, Emergence of communities in weighted networks. Phys. Rev. Lett. **99**(22), 228701 (2007)
H. Kwak, C. Lee, H. Park, S. Moon. What is twitter, a social network or a news media? in *Proceedings of the 19th International Conference on World Wide Web*, WWW'10, New York (ACM, 2010), pp. 591–600
B. Lengyel, A. Varga, B. Ságvári, Á. Jakobi, J. Kertész, Geographies of an online social network. PLoS ONE **10**(9), e0137248 (2015)
M. McPherson, L. Smith-Lovin, J.M. Cook, Birds of a feather: homophily in social networks. Annu. Rev. Sociol. **27**, 415–444 (2001)
Y. Murase, T. Uchitane, N. Ito, A tool for parameter-space explorations. Phys. Proc. **57**, 73–76 (2014)
Y. Murase, J. Török, H.-H. Jo, K. Kaski, J. Kertész, Multilayer weighted social network model. Phys. Rev. E **90**(5), 052810 (2014)
Y. Murase, H.-H. Jo, J. Török, J. Kertész, K. Kaski, Modeling the role of relationship fading and breakup in social network formation. PLoS ONE **10**(7), e0133005 (2015)
Y. Murase, H.-H. Jo, J. Török, J. Kertész, K. Kaski, Sampling networks by nodal attributes. Phys. Rev. E **99**, 052304 (2019)
Y. Murase, H.-H. Jo, J. Török, J. Kertész, K. Kaski, Structural transition in social networks: the role of homophily. Sci. Rep. **9**, 4310 (2019)
M.E.J. Newman, *Networks: An Introduction* (Oxford University Press, Oxford, 2010)
J.P. Onnela, J. Saramäki, J. Hyvönen, G. Szabó, D. Lazer, K. Kaski, J. Kertész, A.L. Barabási, Structure and tie strengths in mobile communication networks. Proc. Natl. Acad. Sci. **104**(18), 7332–7336 (2007)

J.-P. Onnela, J. Saramäki, J. Hyvönen, G. zabó, M.A. de Menezes, K. Kaski, A.-L. Barabási, J. Kertész, Analysis of a large-scale weighted network of one-to-one human communication. New J. Phys. **9**(6), 179 (2007)

V. Palchykov, K. Kaski, J. Kertész, A.-L. Barabási, R.I.M. Dunbar, Sex differences in intimate relationships. Sci. Rep. **2**, 370 (2012)

G. Palla, I. Derényi, I. Farkas, T. Vicsek, Uncovering the overlapping community structure of complex networks in nature and society. Nature **435**, 814–818 (2005)

L.E. Reichl, *A Modern Course in Statistical Physics* (Wiley, Weinheim, 2016)

J. Saramäki, E.A. Leicht, E. López, S.G.B. Roberts, F. Reed-Tsochas, R.I.M. Dunbar, Persistence of social signatures in human communication. Proc. Natl. Acad. Sci. **111**(3), 942–947 (2014)

T.E. Schelling, Models of segregation. Am. Econ. Rev. **59**, 488–493 (1969)

A. Sîrbu, D. Pedreschi, F. Giannotti, J. Kertész, Algorithmic bias amplifies opinion fragmentation and polarization: a bounded confidence model. PLoS ONE **14**, e0213246 (2019)

G. Szabó, M. Alava, J. Kertész, Clustering in complex networks, in *Complex Netowrks*, ed. by E. Ben-Naim et al. Lecture Notes in Physics, vol. 650 (Springer, Berlin, 2004), pp. 139–162

R. Toral, M. San Miguel, K. Klemm, V.M. Eguíluz, Global culture: a noise-induced transition in finite systems. Phys. Rev. E **67**, 045101 (2003)

J. Török, Y. Murase, H.-H. Jo, J. Kertész, K. Kaski, What big data tells: sampling the social network by communication channels. Phys. Rev. E **94**, 052319 (2016)

J. Ugander, B. Karrer, L. Backstrom, C. Marlow, The anatomy of the facebook social graph, Nov 2011

S. Unicomb, G. Iñiguez, J. Kertész, M. Karsai, Reentrant phase transitions in threshold driven contagion on multiplex networks, ArXiv, 1902.04707 (2019)

L. Weng, M. Karsai, N. Perra, F. Menczer, A. Flammini, Attention on weak ties in social and communication networks, in *Computational Social Sciences* (Springer, Berlin, 2018), pp. 213–228

K. Zhao, J. Stehlé, G. Bianconi, A. Barrat, Social network dynamics of face-to-face interactions. Phys. Rev. E **83**(5), 056109 (2011)

Formal Design Methods and the Relation Between Simulation Models and Theory: A Philosophy of Science Point of View

Klaus G. Troitzsch

1 Introduction

Computational models of human societies are highly formalised models written in some programming language, i.e. in "the third symbol system, computer simulation" (Ostrom 1988), and hence are most likely to express exactly what modellers had in mind when they formulated their models. One could therefore assume that their mental models and the software written by them and executed by their computers, delivering simulation results, coincide to the extreme grade—if one could be sure that the simulation model does not contain any involuntary bugs and that the modellers succeeded in expressing their mental model without any aberration. Even if this could be guaranteed (and it is difficult to guarantee as modellers might usually find it necessary to make compromises with respect to the peculiarities of the chosen simulation toolbox which might make it difficult to express exactly what one wants to programme), readers of a simulation model might still be at a loss to reconstruct what exactly the authors of the model code had in mind when they wrote it. This is typically the case when the simulation model was written in a programming language unfamiliar to the reader or when only incomprehensible or even corrupted code is available yet (as, e.g. the FORTRAN IV code of the garbage can model in Cohen et al. (1972), see Inamizu (2015)[1]).

[1] Beside the "mysteries in the original simulation model" identified by Inamizu (2015), the German translation of the original paper (Cohen et al. 1990) contained a number of typos in the programme code which made it uncompilable.

K. G. Troitzsch (✉)
Institut für Wirtschafts- und Verwaltungsinformatik (retired), Universität Koblenz-Landau, Mainz, Germany
e-mail: klaus.g.troitzsch@bluewin.ch

T. Rudas, G. Péli (eds.), *Pathways Between Social Science and Computational Social Science*, Computational Social Sciences,
https://doi.org/10.1007/978-3-030-54936-7_2

This makes it necessary to find ways to describe a computer simulation model in a way that makes understanding and—at least as important—replication easier. In the case of Cohen et al. (1972), replications have been done several times—for a structuralist reconstruction along the lines of Sect. 3, see Troitzsch (2008, 2013)—and this is also true for some even earlier models such as Schelling's segregation model (Schelling 1971) and its predecessors (for a discussion of the early history of segregation models, see Hegselmann (2017), and for a structuralist reconstruction, see Troitzsch 2017). In both of these reconstructions, the aim was to make the description of the theory behind the simulation models more explicit than a natural language description could have been and, at the same time, to avoid the technicalities which a procedural simulation language usually needs to tell the computer how exactly it has to achieve its results. As Ostrom (1988, p. 384) stated, "programs also contain some bookkeeping information that is irrelevant to the process being modelled, but these can be easily be separated from the key theoretical propositions"—but this was certainly not true for "BASIC, FORTRAN or LISP" which he uses as examples, and it is not even true for more modern languages like JAVA which nowadays is used (and even NetLogo programmes still contain some bookkeeping). Some declarative languages (like the obsolete MIMOSE (Möhring 1996; Troitzsch 1992) or Prolog (Balzer et al. 2014)) allow to reduce a simulation programme to the pure theory. Hence it is appropriate to use the unified language of logic and mathematics—despite the "animadversions of many people about the inadequacy of mathematical work in the social sciences to express the full complexity of human behavior" (Suppes 1968, p. 284).

There are a lot of models deserving attention, replication and extension which are still difficult to understand (particularly because the original code is lost), such that it is desirable to have some standard for describing design and implementation of and experimentation with simulation models. One of these is "A Computer Simulation Model of Community Referendum Controversies" (Abelson and Bernstein 1963), slightly more complex than the garbage can model or the segregation model and less often replicated than these, but a decade older and republished as one of "the key articles in the emerging field of computational social science" in the section on "Precursors and Early Work" in a recent collection (Gilbert 2010), and as this paper and a companion paper (Abelson 1964) is often cited (most thoroughly by Alker Jr. 1974) in the context of opinion dynamic models, it lends itself as an example for this chapter. It can be seen as a middle range theory of opinion formation in a rather large group of individuals. This theory was expressed in natural language (as the first symbol system in the terms of Ostrom 1988) and in a computer code (as Ostrom's third symbol system) which is unfortunately lost without traces. This middle range theory was applied to the process in which citizens of a county in the USA formed their opinions with respect to a local political issue—the fluoridation of their drinking water—but could also have been, or be, applied to any other (preferably local) issue in a community referendum.

This chapter describes a number of approaches to establish such standards (next section) and then offers another such approach which is highly formalised and even discusses the connection between the simulated (target) system and the simulating model (Sect. 3) which lends itself (but has rarely been used) to describe the core of a simulation model, the theory behind it and the intended applications of its theory. Section 4 introduces a replication of Abelson's and Bernstein's model both using the ODD description and the structuralist reconstruction of the theories behind it. The final section discusses the results.

2 Formal Design Methods for Agent-Based Simulation

In the past decade, a number of strategies or protocols to support formal descriptions of simulation models have been developed, of which the ODD protocol (Grimm et al. 2006, 2010; Müller et al. 2013) seems to be most widely accepted although there is still a minority of papers which use it. Partly this is due to the fact that an ODD description will usually cover a dozen pages, so conference papers which are normally restricted to 12 pages or even less cannot contain an ODD description; journal articles are somewhat less restricted, but even then ODD description are still rare; hence ODD is more often found in PhD theses or in the supplementary material of journal articles.

2.1 An ODD Protocol for Abelson's and Bernstein's Early Work

The following list is a complete ODD+D description reconstructed from Abelson and Bernstein (1963) which should serve to allow readers to understand their model without reading the original paper.[2]

I. **Overview**

I.i. ***Purpose***

I.i.a *What is the purpose of the study?* "to describe the specific features of this particular simulation model, bringing several levels of theory and both experimental and field phenomena to bear upon the total conception; to illustrate the properties of the model by giving some results of a preliminary trial upon artificial, albeit realistic, data; to discuss some of the broad problems that are likely to be encountered in this type of approach; and finally, thus, to elucidate the general character of simulation technique, which seems to offer eventual promise of uniting theories of individual behavior with theories of group behavior" [p. 93].

I.i.b *For whom is the model designed?* Scientists, students/teachers

[2]Page references in this subsection refer to Abelson and Bernstein (1963).

I.ii. ***Entities, state variables and scales***

I.ii.a *What kinds of entities are in the model?* Citizen agents, news channels, sources, places (where citizens meet and exchange information) are the active entities, i.e. agents, of the model. Beside these, there are assertions (as passive objects, for short called memos in Table 1 and in the NetLogo model) [pp. 94–95]. These are implemented as a list of the following structure: `[from` *S* `via` *X* `at` *t* `opinion` *o* `aspect` *a* `state` *s* `forgettability` *f*`]` where the *S* denotes the source or citizen which generated the memo, *X* is the channel or place between sender and receiver, *t* is the time of generation, *o* denotes whether the memo is pro or con, *a* is the aspect of the issue which the memo refers to, whereas *s* shows whether the memo was accepted or rejected (or not yet decided upon) and *f* is an auxiliary item which carries on whether the memo can be forgotten later within the current period. Hence a complete assertion or memo might represent a sentence spoken by a natural person S_1 at place *X* and understood by another person S_2 with the following content: "S_1 told me (S_2) that her current opinion at $t =$ five minutes ago was $o =$ pro with respect to aspect $a =$ harmfulness, and $s = 1$, i.e. I agree with her and I am unlikely ($f = 0.2$) to forget about her opinion". More details can be found in Figs. 1 and 2.

I.ii.b *By what attributes (i.e. state variables and parameters are these entities characterised?* Of citizens: "demographic characteristics; predisposing experiences and attitudes toward the referendum campaign arguments; frequency of exposure to the several news channels; attitudes toward well-known persons and institutions in the community (who might subsequently prove pivotal in the campaign); knowledge, if any, and acceptance of various standard assertions, pro or con, on the referendum issue; frequency of conversation about local politics, and the demographic characteristics of conversational partners, if any; initial interest in the referendum issue, initial position on the issue, and voting history in local elections" [p. 95].

Of sources: these have "attitude positions" (cf. rule A18) which control the assertions which they distribute over channels.

Of channels: these have a "bias" (cf. rule B27): "Some of the channels, usually the specialized ones, will be clearly biased toward one or the other side of the issue (This information will be in the input to the computer.)" [p. 112].

More details can be found in Fig. 1 (see also Alker Jr. 1974, p. 144, Figure 3) which gives all information about the entities, their instance variables and their methods, in terms, however, of the NetLogo replication, as the original code is lost (and would obviously not have lent itself to an UML description, as it was very machine-near code).

I.ii.c *What are the exogenous drivers of the model?* "The standard local communication channels are represented in the computer, and can be loaded each simulated week with appropriate assertions from sources" [p. 95].

I.ii.d *If applicable, how is space included in the model?* "Each individual specifies on the initial survey the places where he is likely to hold conversations on community issues" [p. 113].

I.ii.e *What are the temporal and spatial resolutions and extents of the model?* "Following the conversational cycle, a new 'week' is in effect, and the individuals are exposed anew to the channels … " [p. 112]. In the first "half week" citizens are only exposed to sources via channels, in the second "half week" they communicate with each other; the rules controlling the 2 "half weeks" are fairly similar.

Fig. 1 Class diagram for base sets, their attributes and their functions in the Abelson-Bernstein model and its NetLogo replication

I.iii. *Process overview and scheduling*

I.iii.a *What entity does what, and in what order?* "... the individuals are exposed anew to the channels ... [p. 112]; "Each individual in turn is confronted with each of his potential conversational partners in turn" [p. 108]. Nothing is said about the order in which citizens are exposed to source and channels in the first "half-week" and to each other in the second "half week".

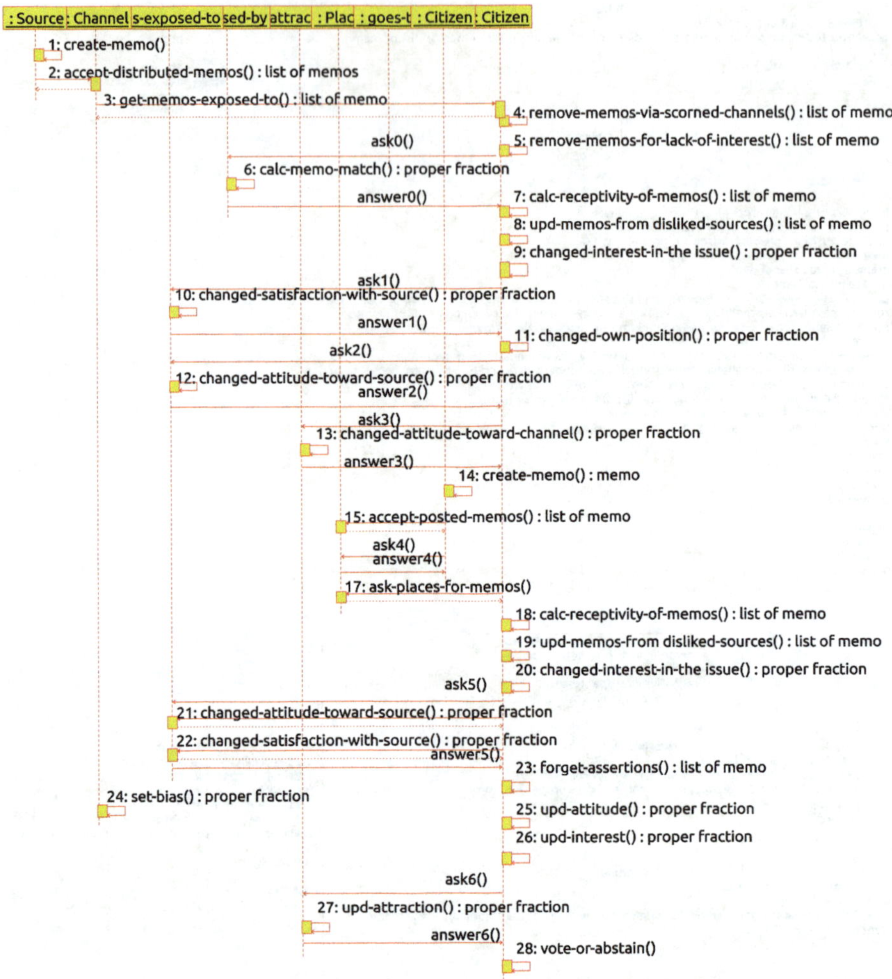

Fig. 2 Sequence diagram for the Abelson Bernstein model and its NetLogo replication

II. Design Concepts

II.i. *Theoretical and Empirical Background*

II.i.a *Which general concepts, theories or hypotheses are underlying the model's design at the system level or at the level(s) of the submodel(s) ...* Abelson and Bernstein specify several theoretical approaches underlying their model, for instance

- Thurstone's theory according to which "attitudes are measurable" (Thurstone 1928) mentioned in Abelson (1964, p. 142)

- theoretical and empirical findings reported in Tannenbaum (1956) about "initial attitude toward source and concept as factors in attitude change through communication" which are used to support the "assumptions ... made about receptivity" [p. 99]—see the method `calc-receptivity-of memos` in the class diagram Fig. 1 (see also Alker Jr. 1974, p. 144, Figure 3) and the sequence diagram Fig. 2 (see also Alker Jr. 1974, p. 145, Figure 4) as well as the function c_r in the functions list on Table 1

but most of the sources on which Abelson and Bernstein rely are not listed in this ODD description, but nearly all of their assumptions are supported by the cited literature.

II.i.b *On what assumptions are the agents' decision models based* See II.i.a.

II.i.c *Why are certain decision models chosen?* See II.i.a.

II.i.d *If the model ... is based on empirical data, where does the data come from?* "On the basis of actual survey data gathered in the community by a probability sample ten or more weeks prior to the scheduled referendum, a large number of actual people (e.g. 500) are anonymously represented in the computer" [p. 94].

II.i.e *At which level of aggregation were the data available?* At the individual level.

II.ii. ***Individual Decision-Making***

II.ii.a *What are the subjects and objects of decision-making? On which level of aggregation is decision-making modelled?* Subjects: citizen agents; objects: assertions to be endorsed or not, votes cast at time step ten of the simulation.

II.ii.b *What is the basic rationality behind agents' decision-making in the model? Do agents pursue an explicit objective or have other success criteria?* Agents follow a detailed set of rules, but do not have objectives or other success criteria.

II.ii.c *How do agents make their decisions?* They follow the detailed set of rules laid out on pp. 98–113 of the article.

II.ii.d *Do the agents adapt their behaviour to changing endogenous and exogenous state variables?* No, the set of rules remains constant, but the agents' attitudes change according to the rules and to the endorsed assertions which they receive from different sources. "Changes can be effected in two general ways: (1) by exposure to public 'assertions' from 'sources', these appearing in communication 'channels' (broadly defined); (2) via conversations with others who have some stand on the issue and who may also make assertions" [p. 95].

II.ii.e *Do social norms or cultural values play a role in the decision-making process?* No.

II.ii.f *Do spatial aspects play a role in the decision process?* Only in so far as agents meet each other in fixed places, which reduce the scope of agents with which they can communicate.

II.ii.g *Do temporal aspects play a role in the decision process?* "The first three [rules, B22–B24] deal with the probability that *i* will 'forget' assertions previously accepted, that is, act in future interactions as though he had never encountered them" [p. 110].

II.ii.h *To which extent and how is uncertainty included in the agents' decision rules?* Several rules define probabilities of endorsing and forgetting (A1, A2, B1, B22, C2).

II.iii. ***Learning***

II.iii.a *Is individual learning included in the decision process?* No.

II.iii.b *Is collective learning implemented in the model?* No.

Table 1 Synopsis between $\mathbf{M}_p(\mathbf{CR})$ and the rules listed in Abelson's and Bernstein's article

$\mathbf{M}_p(\mathbf{CR})$			
Base sets	Explanation		
$\mathcal{C}$	A non-empty finite set [of citizens]		
$\mathcal{S}$	A non-empty finite set [of sources like TV or radio stations or newspaper journalists producing information]		
$\mathcal{X}$	A non-empty finite set [of channels which transmit information from sources to citizens]		
$\mathcal{P}$	A non-empty finite set [of places where citizens typically meet and exchange information]		
$\mathcal{M}$	Memos, short for Abelson's and Bernstein's assertions (see Sect. 2.1, I.ii.a), [as instance variables of citizens, sources, channels and places]		
$\mathbf{M}_p(\mathbf{CR})$			
	Functions		Explanation
Name	Domain	Range	
i_i	$\mathcal{C} \times \mathcal{T}$	(0, 1)	$i_i(\gamma, t)$ yields the current interest in the issue
p_d	$\mathcal{C}$	(0, 1)	$p_d(\gamma)$ yields the (constant) predisposition on the issue
a_p	$\mathcal{C} \times \mathcal{T}$	(0, 1)	$a_p(\gamma, t)$ yields the current attitude position on in the issue
m_m	$\mathcal{C} \times \mathcal{T}$	$\mathbb{P}(\mathcal{M})$	$m_m(\gamma, t)$ yields the subset of assertions (memos) currently in γ's memory
m_p	$\mathcal{C} \times \mathcal{T}$	$\mathbb{P}(\mathcal{M})$	$m_m(\gamma, t)$ yields the subset of assertions (memos) having been posted by γ
a_s	$\mathcal{S}$	(0, 1)	$a_s(\sigma)$ yields the (constant) attitude position of source σ (which, according to rules A18–A20, "is estimated [empirically] and input to the computer")
m_c	$\mathcal{X} \times \mathcal{T}$	$\mathbb{P}(\mathcal{M})$	$m_m(\chi, t)$ yields the subset of assertions (memos) currently available from channel χ
m_p	$\mathcal{P} \times \mathcal{T}$	$\mathbb{P}(\mathcal{M})$	$m_m(\pi, t)$ yields the subset of assertions (memos) having been posted at place π
μ	$\mathcal{C} \times \mathcal{S} \times \mathcal{T}$	(0, 1)	$\mu(\gamma, \sigma, t)$ yields the current value of the assertion match between citizen γ and source σ (see equation 1)
a_{tt}	$\mathcal{C} \times \mathcal{S} \times \mathcal{T}$	(0, 1)	$a_{tt}(\gamma, \sigma, t)$ yields the current value of the attitude of citizen γ towards source σ
r	$\mathcal{C} \times \mathcal{S} \times \mathcal{T}$	(0, 1)	$r(\gamma, \sigma, t)$ yields the current value of the receptivity of citizen γ to source σ
$s_a t$	$\mathcal{C} \times \mathcal{S} \times \mathcal{T}$	(0, 1)	$s_{at}(\gamma, \sigma, t)$ yields the current value of the satisfaction of citizen γ with source σ
$a_t r$	$\mathcal{C} \times \mathcal{X} \times \mathcal{T}$	(0, 1)	$a_{tr}(\gamma, \chi, t)$ yields the current value of the attraction channel χ has for citizen γ
μ	$\mathcal{C} \times \mathcal{C} \times \mathcal{T}$	(0, 1)	$\mu(\gamma_i, \gamma_j, t)$ yields the current value of the assertion match between citizen γ_i and citizen γ_j (see equation 1)
a_{tt}	$\mathcal{C} \times \mathcal{C} \times \mathcal{T}$	(0, 1)	$a_{tt}(\gamma_i, \gamma_j, t)$ yields the current value of the attitude of citizen γ_i towards citizen γ_j

(continued)

Table 1 (continued)

Synopsis continued

Functions	Explanation

M_p(CR)

Functions			Rules	Explanation
Name	Domain	Range		
g_m	$C \cup \mathcal{S}$	$\mathcal{M}$	(A1, A2)[a]	Generates a new memo for a citizen to post or for a source to distribute it over a channel
a_{ms}	$\mathcal{X} \times \mathcal{S}$	$\mathcal{M}$	(A1, A2)	Moves a memo from a source to a channel
g_m	$C \times \mathcal{X}$	$\mathcal{M}$	(A1, A2, B1)	Moves a memo from a channel to a citizen
r_u	$\mathcal{M}^n$	$\mathcal{M}^m$	A1	Removes memos that were transferred via unattractive channels
r_ℓ	$\mathcal{M}^n$	$\mathcal{M}^m$	A2, B1	Removes memos for lack of interest
c_m	$C \times \mathcal{S}$	(0, 1)	(A5)	Calculates the match memo opinions between a source and a citizen (for details see equations 1–7)
c_r	C	(0, 1)	A3–A10, B2–B5	Calculates the receptivity of a source for a citizen
c_d	C	$\mathcal{M}$	A11	Changes the state field for memos which came from disliked sources
χ_{i1}	C	(0, 1)	A12–A14	Changes a citizen's interest in the issue (for details see equations 1–7)
χ_{s1}	$C \times \mathcal{S}$	(0, 1)	A15–A17	Changes a citizen's satisfaction with a source
χ_p	C	(0, 1)	A18–A20	Changes a citizen's own position
χ_{as}	C	(0, 1)	A21	Changes a citizen's attitude towards a source
$\chi_{a\chi 1}$	C	(0, 1)	A22	Changes the attractivity of a channel for a citizen
a_{mc}	$\mathcal{P} \times C$	$\mathcal{M}$	(B1)	Moves a citizen-posted memo from a place blackboard to a citizen
f	$\mathcal{M}^n$	$\mathcal{M}^m$[b]	B22–B24	Removes forgettable memos from a citizen's memory
χ_{s2}	$C \times \mathcal{S}$	(0, 1)	B25	Changes a citizen's attitude towards a source
χ_{i2}	C	(0, 1)	B26	Changes a citizen's interest in the issue
$\chi_{a\chi 2}$	$C \times \mathcal{X}$	(0, 1)	B27	Changes the attractivity of a channel for a citizen
v		$\{-1, 0, 1\}$	C1–C2	Determines whether and how a citizen will vote

[a]Rules in parentheses only mention such a function
[b]In all $\mathcal{M}^n \rightarrow \mathcal{M}^m$ cases $m \leq n$

II.iv. ***Individual Sensing***

II.iv.a *What endogenous and exogenous state variables are individuals assumed to sense and consider in their decisions? Is the sensing process erroneous?* Citizen agents sense assertions from peers and from channels, with a certain probability which depends on their interest, but without error.

II.iv.b *What state variables of which other individuals can an individual perceive? Is the sensing process erroneous?* None. Citizen agents communicate only via assertions, and these are either accepted as they are or rejected.

II.iv.c *What is the spatial scale of sensing?* Citizen agents perceive other citizens' assertions only when they meet at their usual meeting place; assertions from public channels have no spatial restriction.

II.iv.d *Are the mechanisms by which agents obtain information modelled explicitly, or are individuals simply assumed to know these variables?* The former.

II.iv.e *Are costs for cognition and costs for gathering information included in the model?* Not explicitly, the role of interest could be thought to be similar to the role of costs.

II.v. ***Individual Prediction*** There are no predictions.

II.vi. ***Interaction***

II.vi.a *Are interactions among agents and entities assumed as direct or indirect?* Indirect: Citizen receive assertions from sources via channels; whether the interaction is direct between two citizens is not entirely clear, it rather seems that this interaction is bound to places where they meet.

II.vi.b *On what do the interactions depend?* Channel attraction in the case of source assertions, interest in the issue in the case of assertions from peers.

II.vi.c *If the interactions involve communication, how are such communications represented?* Explicit messages (assertions).

II.vi.d *If a coordination network exists, how does it affect the agent behaviour? Is the structure of the network imposed or emergent?* Citizens receive assertions only via channels and at places where they meet peers. The structure of the network does not change over time.

II.vii. ***Collectives*** There are no collectives or other aggregations.

II.viii. ***Heterogeneity*** Agents of each type are homogeneous as they have the same structure, processes are equal among them, but some state variables are randomly assigned during the initialisation and change over time.

II.ix. ***Stochasticity*** Assertions are randomly generated according to the attitude state variable of sources and citizens, assertions are accepted, rejected and forgotten with a probability depending mainly on interest in the issue, but also on the attractivity of a channel or on the assertion match between citizens or citizens and sources.

II.x. ***Observation***

II.x.a *What data are collected from the ABM for testing, understanding and analysing it, and how and when are they collected?* Abelson and Bernstein report only very little about their results. The NetLogo model offers a lot of test outputs and several plots and monitors showing how the model develops over time, including the attitudes of citizens and the instance variables of the relations between citizens and between citizens and channels/sources.

II.x.b *What key results, outputs or characteristics of the model are emerging from the individuals? (Emergence)* As there is no emerging structure beside the frequency distributions of attitudes and channel-citizen and source-citizen relations, nothing important can be observed in this respect.

III. **Details**

III.i. ***Implementation Details***

III.i.a *How has the model been implemented?* On an IBM 7090, using FAP, The FORTRAN Assembly Program Ferguson and Moore (1961). The replication is implemented in NetLogo.

III.i.b *Is the model accessible and if so where?* No, and if it were, it would not be useful after more than half a century. But see the reimplementation in Sect. 4.

III.ii. ***Initialisation***

III.ii.a *What is the initial state of the model world, i.e. at time $t = 0$ of a simulation run?* For the citizens' communication network, this is specified as follows [p. 113]: "Each individual specifies on the initial survey the places where he is likely to hold conversations on community issues. For each appropriate place the individual thinks of an actual conversational partner (if any), and tells the interviewer the key demographic characteristics of this actual partner. This information then serves as a template in locating pseudo-partners". For the sources and channels, their attitude positions [p. 106] and biases [p. 112] as well as the "assertion value for each source within each channel" [p. 100] are "input to the computer", i.e. derived from empirical data. For the connection of citizens to sources and channels, there is no precise specification. In the NetLogo model, citizens are distributed over the world, as are places and citizens are assigned to their two nearest places where they meet their peers; the number of peers per citizen depends on the structure of the citizens-places network, but citizens never have more than six communication partners. For the connection between sources, channels and citizens, all channels are open to all citizens, and all sources distribute over all channels.

III.ii.b *Is the initialisation always the same, or is it allowed to vary among simulations?* Presumptively the latter.

III.iii. ***Input Data***

III.iii.a *Does the model use input from external sources such as data files or other models to represent processes that change over time?* Presumptively yes, at least for the assertions from the media channels. "The two columnists make opposing assertions for all ten weeks of the campaign. The mayor intervenes in the fourth week and is mildly pro-fluoridation. The data were manufactured from a combination of demographic information, intuition, and actual survey statistics on fluoridation gathered by Arnold Simmel and others" [p. 115].

III.iv. ***Submodels***

III.iv.a *What, in detail, are the submodels that represent processes listed in "Process overview and scheduling"?* There are no submodels; the scheduling of events in Abelson's and Bernstein's model seems to be a round-robin process; the NetLogo model uses the usual `ask` strategy.

III.iv.b *What are the model parameters, their dimension and reference values?* In the original model, parameters are not mentioned. In the NetLogo replication, a number of model parameters are hidden in the code, mainly those which parameterise the "direct functions" and the "inverse functions" mentioned in Sect. 3 with respect to rules A2, A3 and B22.

III.iv.c *How were submodels designed or chosen, and how were they parameterised and then tested?* Does not apply.

2.2 Other Approaches

Beside the ODD protocol already mentioned and exemplified above for the example used in this chapter, there are a few similar approaches:

- The approach described in Monks et al. (2018) is quite similar to ODD, as STRESS (STrengthening the Reporting of Empirical Simulation Studies) also lists a number of questions to be answered about the purpose of the model, about the underlying logic and the algorithms used, about components such as environment, agent types and interactions, about data sources and about experimentation and implementation details. Whereas ODD is "highly relevant to modellers in ABS", (Monks et al. 2018, p. 58) qualify it as having "limited applicability elsewhere" (which makes ODD the optimal choice in the context of this paper).
- Rahmandad and Sterman (2012) discuss visualisation aspects in their recommendations before they list a number of model reporting requirements which start with the description of the algorithms (functions, equations) as minimum requirements and continue with preferred requirements (sources of data, definition of all variables, source code), but as they exemplify their recommendations with a system dynamics model (the Bass diffusion model), it is questionable whether this method can easily be extended to other simulation approaches such as agent-based models.
- Likewise, the approach documented in Waltemath et al. (2011) and applicable mainly to models in biology is more or less restricted to "to simulation descriptions of biological systems that could be (but are not necessarily) written with ordinary and partial differential equations" [p. 4]. This does not exclude that "MIASE compliance will be directly applicable to a wider range of simulation experiments, such as the ones performed in computational neuroscience or ecological modeling", but it does not seem to have been applied in ecological modelling so far. (Monks et al. 2018, p. 58) call MIASE "simple, high-level and generalisable across simulation approaches".
- Yilmaz and Ören (2013) and Lutz et al. (2016) promote an even more sophisticated approach than ODD and STRESS. Both advocate an XML description (SED-ML, the Simulation Experiment Description Markup Language in Yilmaz and Ören (2013) and SRML,[3] the Simulation Reference Markup Language in Lutz et al. (2016)) of simulation models for domains "such as transportation, logistics, communication, marketing, physics, etc." (Lutz et al. 2016, p. 13)—hence obviously not in the context of computational social science, although the approach can certainly be generalised in this direction. SED-ML "was originally devised for the Systems Biology domain" (Yilmaz and Ören 2013, p. 212)—as in the case of MIASE—and is said to provide "a sound basis for further extensions

[3]Not to be confused with SRML, a Service Modeling Language, https://www.cs.le.ac.uk/srml/, accessed 13/03/2019, nor with the Semantic Rule Meta Language, Kálmán et al. (2006).

to cover a broad class of models", but so far both, SED-ML and SRML have not found many applications, and apparently none in computational social science.

As the approaches listed above are less appropriate for agent-based modelling and simulation, many papers, theses and books use plain UML instead, particularly class diagrams and sequence diagrams—see Figs. 1 and 2—to specify agent classes and agent behaviour. Class diagrams can sometimes be generated with a method first presented in Polhill (2015): Here a method has been developed which derives an ontology directly from a NetLogo model (see Fig. 3), currently showing the base sets and the relations defined on these in the model. But for a complete understanding of a model other kinds of diagrams are also necessary.

3 The 'Non-statement View' of Structuralism

The "non-statement view"[4] of structuralism is an approach at formalising theories introduced by Sneed (1979) and elaborated upon by Balzer et al. (1987) (see also Balzer and Brendel 2019) mainly for theories in physics which has rarely been used in the context of simulation (Balzer 2009; Troitzsch 1992, 1994, 1996, 2008, 2013, 2017, 2019) and also rarely been applied to theories from psychology (Westmeyer 1992), economics (Alparslan and Zelewski 2004; Balzer 2009) and the social sciences (Abreu 2014, 2019; Avendaño de Aliaga and Horenstein 1997; Balzer 1992; Balzer and Dreier 1999; Dreier 2000; Druwe 1985; Troitzsch 1992, 1994, 1996, 2008, 2013, 2017, 2019), to cite only a few authors who went this way.

In short, the "non-statement view" of structuralism defines a theory element as a mathematical structure consisting of a core and a set of intended applications. Whereas the set of intended applications in the current case of a community referendum **CR** is just a multi-wave survey of a community referendum on some political issue (not necessarily about drinking water fluoridation), the core of the theory element **CR** is again a mathematical structure consisting (at least) of three sets of *models*.[5] The set of *potential models*—$\mathbf{M}_p(\mathbf{CR})$—is described by all terms which are necessary to speak about the underlying theory whether these terms (so-called **CR**-non-theoretic terms) are observable (usually with the help of other

[4]The term "non-statement view" points to the fact that this philosophy of science approach (Sneed 1979, p. xvii) does not see theories as systems of statements but as a unified mathematical structure consisting of theory elements consisting of classes of models, constraints and links among them (Stegmüller 1986, p. 468).

[5]Italics are used to show that the word *model* is used in the sense it has in the "non-statement view" and to distinguish it from other meanings of this word, particularly from agent-based simulation models, although the latter can also be conceived as *potential* or *even full models* of a theory, which in turn are defined as lists of terms the theory is about. These terms are either base sets, representing the terms needed to speak about the entities in question (which, for instance, represent citizens or TV channels) or functions defined on these base sets, representing the terms needed to speak about the properties of these entities and the methods of changing them—see Table 1.

theories) or whether **CR**-theoretic terms get a meaning only within the theory. Intended applications, i.e. empirical applications of **CR**, are usually only *partial potential models*—$\mathbf{M}_{pp}(\mathbf{CR})$—as they are restricted to the **CR**-non-theoretic terms. For the current purpose of reconstructing the set of *potential models* of **CR** only these are of interest, as a simulation model necessarily contains all classes, instance variables and even the axioms (which are only defined in the set of full models **M(CR)**) in order to be able to mimic their target systems. And these axioms define how the entities and their properties interact which other.

As the ODD description has shown, the model in Abelson and Bernstein (1963) was described in a way that allows for replication, albeit with some restrictions as the functions behind the rules were not sufficiently detailed—and this is also the reason why a *full model* of **CR** cannot be derived from Abelson and Bernstein (1963): not enough is known about the functions described in their assumptions A1–A22, B1–B27 and C1–C2.[6] Moreover one could argue that—obviously for reasons of programmability with the tools of the early 1960s— the series of events in the model are less realistic (or: empirically founded) than the rules as most of the latter have references to empirical research. The idea behind the original model (and, of course, also in the replication in Sect. 4) is that in each of the simulation time steps ("weeks") each citizen $c \in C$ is first exposed to a number of assertions $m \in \mathcal{M}$ issued by sources $s \in \mathcal{S}$ from public channels $\chi \in \mathcal{X}$, then processes them and only afterwards starts communicating with peers, exposes itself to their assertions, processes these and finally changes some of its attributes (position in the issue χ_p and interest in the issue χ_{i1}) and attributes of its relations to sources $C \cup \mathcal{S}$, channels $C \cup \mathcal{X}$ and peers $C \cup C$ (where these relations will have to be implemented as directed links in NetLogo). One can only guess that this chain or events were necessary due to the small memory of Abelson's and Bernstein's IBM 7090 (which they describe in the "Technical Note of the Computer Program" (Abelson and Bernstein 1963, 122)). A reimplementation does not need such restrictions, but for an accurate replication, this chain of events is also applied in the NetLogo model of Sect. 4.

Given all the information in Sect. 2.1 and some additional information from the original article (see also the class diagram, Fig. 1), this leads to the definition of the set of *potential models* $\mathbf{M}_p(\mathbf{CR})$ of the **c**ommunity **r**eferendum theory **CR**. Table 1 contains a synopsis between the $\mathbf{M}_p(\mathbf{CR})$ terms and the rules in Abelson and Bernstein (1963). The "kinds of entities" mentioned in Sect. 2.1 I.ii.a now show up as non-empty finite sets. The interactions mentioned in item II.vi are now relations (with lower-case names in Fig. 1). The "attributes (i.e. state variables and parameters)" by which "these entities [are] characterised" (item I.ii.b of the ODD) and which are listed in Fig. 1 above the dashed lines of the boxes describing the

[6]It would instead have been possible to describe all the functions implemented in the NetLogo model in terms of the structuralist reconstruction, but see below (equations 1–7) for a discussion of the functions sufficiently well defined in rules A5 and A12 to show that this reconstruction is also possible from the original article although in other functions some parameters must be left undefined.

classes and relations are now functions defined on the sets C, $\mathcal{S}$, $\mathcal{X}$ and $\mathcal{P}$ and also on the relations as the lower part of Table 1 shows.

With the information provided in the original article Abelson and Bernstein (1963), it is impossible to define the (full) model **M(CR)** as most functions—axioms in terms of the non-statement view—are only described as "direct function" (e.g. in rule A2), "inverse function" (e.g. in rule A3) or with something like "is greater ... if it is inconsistent with ... than if it is consistent" (e.g. rule B22). This makes it a little difficult to design a perfect replication of the original model, although "direct function" and "inverse function" might be understood as something like $y = kx$ and $x \cdot y = \text{const.}$[7]

But even after all these relations and functions are analysed, the constants are unknown, such that a replication will never be a true one without the code of the original FAP programme. The same is true for the initialisation of the citizen, source and channel agents where the data which were "input to the computer" are also lost without trace. All this becomes much clearer than in Abelson's and Bernstein's paper when one uses the non-statement view for translating the implicit theory behind the 1963 model into a replication—another hint at the relevance of this technique for the social sciences. Although the entities are well described there—see the ODD item I.ii.a—this is not the case for the rest.

Two examples for functions which are sufficiently well described are given in the following:

$c_m : C \times \mathcal{S} \to (0, 1)$ calculates the assertion match between a citizen and a source (according to rule B5, the latter might also be a citizen). According to A5, this

> is positive when s's assertions agree with those already accepted by i, and negative when they disagree. ... The 'assertion match' index between i and s is computed by scoring $+1$ for each single assertion agreement, -1 for each single assertion disagreement, and summing. (Abelson and Bernstein 1963, p. 100)

Informally described, this function goes through the list of memos in i's memory and counts the absolute differences between the o- and s-entries of the memos in this list (which are either 0 for agreement or 2 for disagreement, as o and s are coded $+1$ and -1 for both opinion and acceptance, respectively). This yields exactly what is described in rule A5 (in the NetLogo replication this is normalised to the set of real numbers $(0, 1)$ for further treatment). More formally this function yields

$$c_m(\gamma, \sigma) = \frac{\alpha(\gamma, \sigma) - \delta(\gamma, \sigma)}{2(\alpha(\gamma, \sigma) + \delta(\gamma, \sigma))} + \frac{1}{2} \tag{1}$$

where (2)

$$\alpha(\gamma, \sigma) = |\{m \in \mathcal{M}_\gamma(t) | S(m) = \sigma \wedge o(m) = s(m)\}| \tag{3}$$

[7] This is what Merriam-Webster defines for the terms "direct variation" and "inverse variation".

$$\delta(\gamma, \sigma) = |\{m \in \mathcal{M}_\gamma(t) | S(m) = \sigma \wedge abs(o(m) - s(m)) = 2\}| \quad (4)$$

and (5)

$$0 < c_m(\gamma, \sigma) < 1 \quad (6)$$

where $\gamma \in C$ and $\sigma \in \mathcal{S}$ denote the citizen and the source in question and S, o and s are functions yielding the source, opinion and state attributes of the memo or assertion in question (see Sect. 2.1 I.ii.a).

$\chi_{i1} : C \rightarrow (0, 1)$ changes a citizen's interest in the issue. According to A12 (A13 and A14 are not considered here as this would make this discussion even more complicated), a person's interest "increases as a direct function of the assertion match c_m between i and s" (Abelson and Bernstein 1963, p. 104) where i is the person in question and s is the source of the assertion (either another citizen or a source sending its assertion via a channel). Calling again the citizen γ and the source σ and γ's interest in the issue at time t $i_i(\gamma, t)$, then the changed interest $i_i(\gamma, t^+)$ in the issue is

$$i_i(\gamma, t^+) = i_i(\gamma, t)(1 + \rho(c_m(\gamma, \sigma)) - 0.5)(c_m(\gamma, \sigma) - 0.5) \quad (7)$$

where ρ is a function which rounds its argument ($\rho(x)$ is 0 for $0 < x < 0.5$ and 1 for $0.5 \leq x \leq 1$, such that $i_i(\gamma, t)$ is only changed when $\alpha(\gamma\sigma) > \delta(\gamma\sigma)$ as in rule A12. Abelson and Bernstein argue that "it is not clear how to view the potential effect on interest of disagreement (i.e. a negative assertion match) between self and source" (Abelson and Bernstein 1963, p. 104).

The next section will nevertheless show what comes out of an attempt to design all the other functions mentioned in Abelson's and Bernstein's rules in the simplest possible way.

4 Translation into a New Simulation Model

4.1 Designing Agent Types, Relations and Functions

The contents of Table 1 can now be used to define the breeds and functions for a NetLogo model which replicates Abelson's and Bernstein model as far as possible, given the absence of an exact parameterisation of most of the functions, as the rules often say only that there is a direct or inverse function between two or more instance variables of the citizen type. Hence Tables 2 and 3 link the base sets and functions identified in Table 1 to the breeds, links and function of a new NetLogo model.

This is—at least for the breeds and links—a straightforward process: base sets are represented as breeds and relations are represented with NetLogo's links. The instance variables of the classes and relations of Fig. 1, now defined as functions in the middle part of Table 1, are represented by variables belonging to ("owned by")

Table 2 Synopsis between $\mathbf{M}_p(\mathbf{CR})$ and NetLogo breeds

$\mathbf{M}_p(\mathbf{CR})$ Base sets and relations	NetLogo breeds
C	`citizens` keeping, among others, a memory for assertions (`memos`), a position on the issue, a predisposition, an interest in the issue and a vote (see Abelson 1964, p. 156)
$\mathcal{S}$	`sources` keeping an attitude position
$\mathcal{X}$	`channels` keeping a list of assertions (`memos`) to be transmitted and a bias
$\mathcal{P}$	`places` keeping a list of assertions (`memos`) posted from citizens and being read by other citizens
$\mathcal{E} \subset C \times \mathcal{S}$	`are-exposed-to` keeping the match of assertions between a citizen and a source, the latter's attitude to and satisfaction with the former and the receptivity the citizens bears of the source
$\mathcal{U} \subset \mathcal{X} \times \mathcal{S}$	`are-used-by` keeping the value of the attention the channel pays to the source
$\mathcal{A} \subset C \times \mathcal{X}$	`are-attracted-by` keeping the attraction the channel has for the citizen
$\mathcal{T} \subset C \times C$	`talk.to` keeping the match of assertions between the two citizens the listening citizen's attitude towards the posting citizen and the receptivity the former bears of the letter
$\mathcal{G} \subset C \times \mathcal{P}$	`go-to` without any instance variables
$\mathcal{M}$	memos (see Sect. 2.1, I.ii.a)

Table 3 Synopsis between $\mathbf{M}_p(\mathbf{CR})$ and NetLogo functions

$\mathbf{M}_p(\mathbf{CR})$ Functions[a]				NetLogo functions
Name	Caller	Arguments	Result	
g_m	$C \cup \mathcal{S}$	$\emptyset$	$\mathcal{M}$	`create-memo`
$a_{m\chi}$	$\mathcal{X}$	$\mathcal{M}^n, \mathcal{M}$	$\mathcal{M}$	`accept-distributed-memos`
g_m	C	$\mathcal{X}, \mathcal{M}$	$\mathcal{M}$	`get-memos-exposed-to`
r_u	C	$\mathcal{M}^n$	$\mathcal{M}^m$	`remove-memos-via-scorned-channels`
r_ℓ	C	$\mathcal{M}^n$	$\mathcal{M}^m$	`remove-memos-for-lack-of-interest`
c_m	$\mathcal{U} \cup \mathcal{T}$	$C, \mathcal{S}$	Q_0	`calc-memo-match`
c_r	C	$\mathcal{M}^n$	$\mathcal{M}$	`calc-receptivity-of-memos`
c_d	C	$\mathcal{M}^n$	$\mathcal{M}$	`upd-memos-from-disliked-sources`
χ_{i1}	C	$\mathcal{E}$	Q^0	`changed-interest-in-the-issue`
χ_{s1}	$\mathcal{E}$	$\emptyset$	Q^0	`changed-satisfaction-with-source`
χ_p	C	$\mathcal{E}$	Q^0	`changed-own-position`
χ_{as}	$\mathcal{E}$	$\emptyset$	Q^0	`changed-attitude-toward-source`
$\chi_{a\chi 1}$	$\mathcal{A}$	$\emptyset$	Q^0	`changed-attitude-toward-channel`
a_{mc}	$\mathcal{G}$	$\mathcal{M}^n, \mathcal{M}$	$\mathcal{M}$	`accept-posted-memos`
f	C	$\mathcal{M}^n$	$\mathcal{M}$	`forget-assertions`
χ_{s2}	C	Q_0^3	Q_0	`upd-attitude`
χ_{i2}	C	Q_0^2	Q_0	`upd-interest`
$\chi_{a\chi 2}$	$\mathcal{A}$	Q_0^2	Q_0	`upd-attraction`
v	C	$\emptyset$	$\{-1, 0, 1\}$	`vote-or-abstain`

[a]For short, arguments and results are denoted with the name of the set they belong to. For short, Q_0 is the set of all rational numbers x, $0 \leq x \leq 1$. Wherever m and n are used as superscripts, $m \leq n$

each of the breeds and links. The other methods listed in Fig. 1—below the dashed line of the class and relation boxes—were first (at least partly, to save space) defined as functions in the lower part of Table 1 and then translated to NetLogo code.

For the details of the functions not defined in equations 1–7, namely, how the rules of Abelson and Bernstein (1963) were translated into contemporary programme code—the NetLogo model is available from https://doi.org/10.25937/s10g-a461—as a complete formal description of these functions along the lines of equations 1–7 would certainly blow up the size of this chapter too much.

With the approach of Polhill (2015), an ontology can be generated from the NetLogo model which, at the same time, is very likely to be the ontology underlying Abelson's and Bernstein's original model (see Fig. 3).

Wherever Abelson and Bernstein (1963) specify rules and functions with terms such as "is a direct function of" (rules A1–A2, A4, A12–A13, A15, A19, A21–A22, B1, B3–B5, B25–B26, C2) or "is an inverse function of" (rule A3, B2) or wherever rules contain if clauses, the NetLogo model tries to find a way to define functions with results within the set Q_0 of the positive rational numbers ≤ 1 (this is mainly realised with the `increase` and `decrease` functions) since neither Abelson and Bernstein (1963) nor Abelson (1964) give more details about the form of these functions, not even in (Abelson 1964, p. 159) when Abelson writes:

> In order to effectuate the model, of course, it is necessary to quantify these propositions, inserting functional forms and assumed parameters. One advantage of the simulation approach over the mathematical models approach is that no restrictive simplicity considerations are imposed on the functional forms — linearity is not necessarily preferred. . . . there are many possible ways to make readjustments; when data is consistent with the model, it could be right for the wrong reasons.

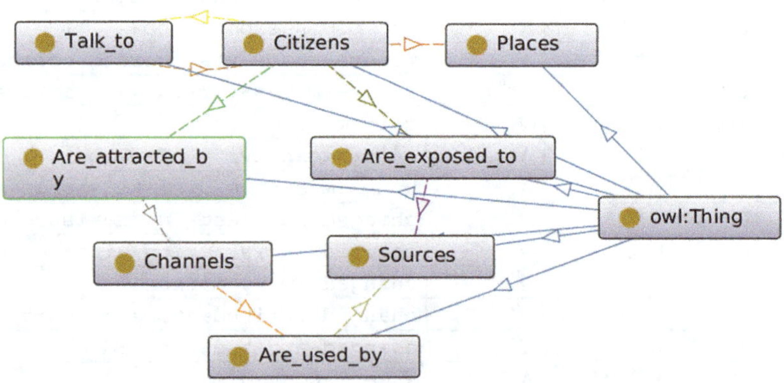

Fig. 3 Ontology generated by the NetLogo OWL extension of the reimplemented model

4.2 Results

Any verification of the NetLogo model against the original model of 1963 will fail because of the scarcity of information provided in Abelson and Bernstein (1963), and for the same reason, a sensitivity analysis is more or less in vain, too, although both could be over-parameterised, such that a sensitivity analysis could reveal which parameters are superfluous, but this would be beyond the scope of this chapter. What is possible is a comparison between some of the results of the original model and those of the NetLogo replication. This is done in this subsection.

One of the findings in Abelson and Bernstein (1963, p. 116, Table 1) is a "polarisation of attitude positions": the "average position scores" of those whose scores were positive and those whose scores were negative change monotonically over time from $+112 : -126$ at the beginning to $+193 : -145$ in the end in the original model, and something similar can be seen from a comparison of the histograms produced for the citizens' positions on the issue in the NetLogo model (see Fig. 4).

In the case of the interest in the issue, the replications yield results entirely different from the original—this is obviously because the initialisation of the interest in the issue attribute is random uniform in the NetLogo replication but unknown in the original. Whereas in the original (Abelson and Bernstein 1963, p. 118, Table 4) the distribution of the interest does not change drastically and the interest index of the citizen agents remain very low but with a slight increase for about 10 per cent of them, the replication shows that the distribution changes from uniform to a peaked distribution with a mode around 0.7 after 5 weeks and to a bimodal distribution with another peak at 1.0 in the end. It would be interesting to see what the history of this distribution would be like with another initialisation, but this is beyond the scope of this chapter.

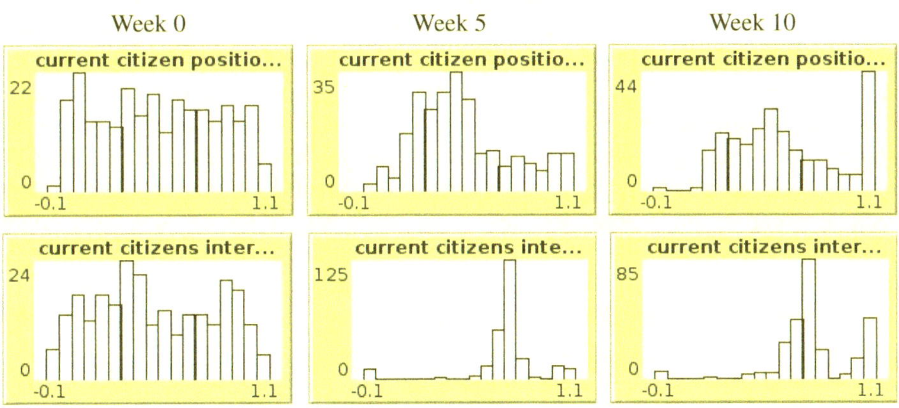

Fig. 4 History of citizens' positions on and interest in the issue

5 Conclusion and Outlook

The purpose of this chapter was to show how some different methods of formalisation are appropriate to make the process of building a simulation model from some theoretical considerations transparent. The same methods lend themselves to facilitate the replication of existent simulation models, as they allow not only to translate a computer programme from one language into another language—this would lead to preserving flaws from the old model into the new one—but to build a replication from a specification in a unified symbol system which is as precise and concise as possible.

The example used here was rather an informal description of a simulation model whose code has long been lost without trace, such that this description can also be considered as an informal design of a simulation model that has yet to be specified and implemented. This was done in three steps:

- The first step, the ODD+D description, sorted the information given by Abelson and Bernstein (1963) and Abelson (1964) according to this protocol.
- The second step, the UML class diagram presented as a desirable supplement of an ODD+D description (Grimm et al. 2006, p. 125, 2010, p. 2766).
- The third step was the transformation mainly of the class diagram into a definition of the *set of potential models* of Abelson's and Bernstein's theory, but without complementing it to a definition of a *full model* with the axioms of this theory (as the parameterisation is not sufficiently described in their two papers).

A fourth step could have been to formalise the whole theory net which Abelson and Bernstein span with numerous footnotes pointing to numerous theoretical and empirical papers which they use to substantiate most of their rules. This would certainly exceed the scope of this chapter but might be an interesting exercise for further work—here only a hint is in place that some of the theories mentioned by Abelson and Bernstein could also be translated into NetLogo modules plugged into the current model after a translation via the non-statement view reconstruction process, using intertheoretical links (Balzer et al. 1987, pp. 57–62) between, for instance, Thurstone's attitude measuring theory (Thurstone 1928) or Tannebaum's findings (Tannenbaum 1956) (see Sect. 2.1, II.i.a) and the current theory. This fourth step in reconstructing Abelson's and Bernstein's community referendum theory **CR** is not done, here but it would again show how helpful the non-statement view technique can be for a better and less ambiguous description of how short range theories can contribute to a more detailed understanding of a more complex target system.

The three steps taken in this chapter led to a new implementation of the 1963 model which shows some of the behavioural features of the original (history of the distribution of the citizen agents' attitude) but also shows differences (with respect to the history of the distribution of the interest in the issue), but with some more tuning of the initialisation and parameterisation of the model it might be possible to verify the replication against the results of the original model—but as these are only very sketchily reported in Abelson and Bernstein (1963), this is not a very promising exercise and dropped here. Abelson and Bernstein, when reporting their results, come to the conclusion that they "will not pursue this line of speculation further here, since [they] only mean to illustrate the wide assortment of variables and phenomena that can be addressed by means of the computer simulation model, and in any case a single dry run of an unvalidated model on hypothetical data does not constitute evidential proof" (Abelson and Bernstein 1963, p. 119).

Anyway, it is clear that the artificial world of the NetLogo agent-based model with its software agents representing citizens, newspapers, TV channels, politicians, etc. is a *full model* of the reconstructed community referendum theory of Abelson and Bernstein enlarged with a number of plausible assumptions on some specific parameters which could not be retrieved from the original FAP computer code which is unfortunately lost without traces some of whose results were reported in Abelson and Bernstein (1963) but which is also a *full model* of their theory. But it seems from a final remark in their paper that they were not entirely sure that the material they had collected in their surveys could claim to be an *intended application* of their theory, when they write: "Other adjustments in the model will be required as field evidence becomes available", obviously meaning that the design for the planned panel study would deliver more **CR**-non-theoretic, i.e. observable information to be compared to the output of their computer model. And then it would have been the model which inspired the new empirical research design. But it seems that the planned studies were never carried out. This raises at least three questions:

- Is the illustration of a wide assortment of variables and phenomena that can be addressed by means of computer simulation worthwhile at all?
- Can "a single dry run" or a large number of intelligently initialised and parameterised runs be used to validate the model?
- Can this, in the end, "constitute evidential proof?"

The answer to the third question is certainly a clear no. In the best case, after validation, one could preliminarily hope that the validated model for a certain initialisation and parameterisation displays a behaviour which "truly reflects the way in which the real system operates to produce this behaviour" (Zeigler 1985, 5). But this is certainly not a proof, but only a hint that the observed real system with its behaviour truly replicated by the model is a *full model* of the theory behind the simulation model.

The first question leads to the answer that certainly a sensitivity analysis is first necessary to find out whether the wide assortment of variables and phenomena put into the simulation is too wide or too narrow—too wide in the sense that the additional explanatory power of a certain attribute of the model or of one of its entities with respect to some interesting model output is very low such that its omission does not make a difference and too narrow in the sense that the overall explanatory power of all the model input is low with respect to some interesting model output such that the model output is only a result of the random effects used in the model.

This also leads to an answer to the second question: the output of a single run is prone to be only a result of random effects in a way that certain models can predict that the final value of the interesting variable is either about 0.9 or about 0.1 but nothing in between, and a single run would be misleading as it would only show one of the possible two outputs. Only a large number of runs can show the distribution of possible outcomes, and if the interesting output metric of the model is a non-theoretical term with respect to the theory (as in the case of **CR** certainly the voting turnout of a community referendum[8]), then the model can even be used to make (some of) the theoretical terms measurable with the help of the validated model.

This chapter applied two different approaches at describing how community referenda as a relevant object of social research into opinion and attitude formation have been treated for several decades. It showed how the notions about this object could be transformed into computer simulation models in a way more formal then the one taken by the first researchers attempting such a simulation. The two approaches analysed and described in more depth were ODD+D in Sect. 2.1 and a structuralist reconstruction according to the non-statement view in Sect. 3. The chapter should have shown that ODD+D is still verbose and less formal than the structuralist reconstruction. Both were not entirely complete as for reasons of space they had to leave out the description of the greater part of Abelson's and Bernstein's 50 rules and the functions and methods derived from these rules. Instead only a few extended examples could be given to illustrate the complete formalisation. The terms of the theory element definition as documented in Table 1 did not make any allusion to what they represent in the real world. Hence this kind of description is unambiguous much like the model coded in a computer programming language but, unlike the latter, does not need to contain functions and procedures for input, output, debugging and illustration. This might be the main advantage of going the stony way of the structuralist reconstruction: to have a recipe—a formal specification—for programming a computer which is reduced to the naked mechanism of the model representing the mechanism of the real target system without any allusion to the tools necessary to turn the recipe into executable computer code, strictly separating the functionality of the model from both the functionality of empirical research and of the programming environment.

The examples mentioned at the beginning of Sect. 3—and this chapter—show that the set-theoretic approach of the non-statement view is useful to describe the kernel of a theory as concise and precise as possible and that it is useful to apply additional formal design methods (see Sect. 2.1) to separate this kernel from what Ostrom called "bookkeeping information".

---

[8]But Abelson and Bernstein give a caveat (Abelson and Bernstein 1963, p. 113) when they formulate that "in practice, of course, voting turnout will depend on what other issues are on the ballot and upon other variables extraneous to the issue being simulated" which reduces the set of intended applications further, as only single-issue referenda in otherwise undisturbed circumstances can be analysed.

Alker, in his paper in which he analysed Abelson's and Bernstein's model, raised a number of questions related to the roles of Ostrom's three language systems, among others (Alker Jr. 1974, p. 153)

- What properties of ordinary language are most important for formal social science theory development?
- Among axiomatic language systems, which have the properties most essential for understanding social realities?

Unlike applications in physics or classical economics whose models are usually (and can be) equation-based (and mathematical equations are concise and precise enough), models in the social sciences typically need an agent-based approach to be able to represent large numbers of interacting human actors who are typically very inhomogeneous and who communicate in a much more complicated way than the entities dealt with in physics or biology and the *homines oeconomici*. Theoretical ideas about interacting and communicating human actors must—and can—be described with the same mathematical precision as theories in physics but usually have to be implemented in agent-based simulation models instead of being represented in systems of equations. To describe the way from first theoretical ideas to simulation models for the social sciences which can be calibrated and validated against empirical data, all three of Ostrom's symbol systems are necessary: the natural language used in the ODD description, the language of logic and mathematics of something like the set-theoretic approach of the non-statement view and finally an appropriate programming language to implement one's theoretical ideas in a manner that real-world mechanisms can be replicated.

References

R.P. Abelson, Mathematical models of the distribution of attitudes under controversy, in *Contributions to Mathematical Psychology*, ed. by N. Frederiksen, L.L. Thurstone, H. Gulliksen (Holt, Rinehart and Winston, Inc., New York, 1964), pp. 141–160

R.P. Abelson, A. Bernstein, A computer simulation model of community referendum controversies. Public Opin. Q. **27**(1), 93–122 (1963)

C. Abreu, Análisis estructuralista de la teoría de la anomia. Metatheoria **4**(2), 9–22 (2014)

C. Abreu, Análisis estructuralista de la teoría del etiquetamiento. Diánoia **64**(82), 31–59 (2019)

H. Alker Jr, Computer simulations: inelegant mathematics and worse social science? Int. J. Math. Educ. Sci. Technol. **5**, 139–155 (1974)

A. Alparslan, S. Zelewski Moral hazard in JIT production settings—a reconstruction from the structuralist point of view. Arbeitsbericht 21, Institut für Produktion und Industrielles Informationsmanagement Universität Duisburg-Essen (Campus: Essen) (2004)

M.D.C. Avendaño de Aliaga, N.S. Horenstein, Factibilidad de la reconstrucción metateórica de teorías en ciencias sociales. Centro de Investigaciones Jurídicas y Sociales III: 367–381. Accessed 19 Mar 2019 (1997)

W. Balzer, A theory of power in small groups, in *The Structuralist Program in Psychology: Foundations and Applications*, ed. by H. Westmeyer (Hogrefe & Huber, Seattle, 1992), pp. 192–210

W. Balzer, *Die Wissenschaft und ihre Methoden. Grundsätze der Wissenschaftstheorie*, 2nd edn. (Karl Alber, Freiburg/München, 2009)

W. Balzer, K.R. Brendel, *Theorie der Wissenschaften* (Springer, Wiesbaden, 2019)
W. Balzer, V. Dreier, The structure of the spatial theory of elections. Br. J. Philos. Sci. **50**(4), 613–638 (1999)
W. Balzer, D. Kurzawe, K. Manhart, *Künstliche Gesellschaften mit PROLOG: Grundlagen sozialer Simulation* (V&R unipress, Göttingen, 2014)
W. Balzer, C.U. Moulines, J.D. Sneed, *An Architectonic for Science. The Structuralist Program*. Synthese Library, vol. 186 (Reidel, Dordrecht, 1987)
M.D. Cohen, J.G. March, J.P. Olsen, A garbage can model of organizational choice. Adm. Sci. Q. **17**, 1–25 (1972)
M.D. Cohen, J.G. March, J.P. Olsen, Ein Papierkorb-Modell für organisatorisches Wahlverhalten, in *Entscheidung und Organisation. Kritische und konstruktive Beiträge, Entwicklungen und Perspektiven*, ed. by J.G. March (Gabler, Wiesbaden, 1990), pp. 329–372
V. Dreier Ein formales Basis-Modell zur Beschreibung und Rekonstruktion politischer Machtstrategien, in *Kontext, Akteur und strategische Interaktion: Untersuchungen zur Organisation politischen Handelns in modernen Gesellschaften*, ed. by U. Druwe, S. Kühnel, V. Kunz (VS Verlag für Sozialwissenschaften, Wiesbaden, 2000), pp. 189–211
U. Druwe, *Theoriendynamik und wissenschaftlicher Fortschritt in den Erfahrungswissenschaften. Erfahrung. Evolution und Struktur politischer Theorien* (Alber, Freiburg/München, 1985)
D.E. Ferguson, D.P. Moore, FORTRAN assembly program (FAP) for the IBM 709/7090. techreport, International Business Machines Corporation (IBM) (1961)
N. Gilbert (ed.), *Computational Social Science. Four-Volume Set* (Sage, Los Angeles, 2010)
V. Grimm, U. Berger, F. Bastiansen, S. Eliassen, V. Ginot, J. Giske, J. Goss-Custard, T. Grand, S.K. Heinz, G. Huse, A. Huth, J.U. Jepsen, C. Jørgensen, W.M. Mooij, B. Müller, G. Pe'er, C. Piou, S.F. Railsback, A.M. Robbins, M.M. Robbins, E. Rossmanith, N. Rüger, E. Strand, S. Souissi, R.A. Stillman, R. Vabø, U. Visser, D.L. DeAngelis, A standard protocol for describing individual-based and agent-based models. Ecol. Model. **198**, 115–126 (2006)
V. Grimm, U. Berger, D.L. DeAngelis, J.G. Polhill, J. Giske, S.F. Railsback, The ODD protocol: a review and first update. Ecol. Model. **221**, 2760–2768 (2010)
R. Hegselmann, T.C. Schelling, J.M. Sakoda, The intellectual, technical, and social history of a model. J. Artif. Soc. Soc. Simul. **20**(3), 15 (2017)
N. Inamizu, Garbage can code—mysteries in the original simulation model. Ann. Bus. Adm. Sci. **14**(1), 15–34 (2015)
M. Kálmán, F. Havasi, T. Gyimóthy, Compacting XML documents. Inf. Softw. Technol. **48**(2), 90–106 (2006)
R. Lutz, J. Bachman, C. Blais, J. Stevens, A. Lemmers, D. Ronnfeldt, G. Shanks, E. Stoudenmire, S. Weiss, Guide for Simulation Reference Markup Language—Primary Features. Techreport, Simulation Interoperability Standards Organization, Inc., Orlando, 2016
M. Möhring, Social science multilevel simulation with MIMOSE, in *Social Science Microsimulation*, chapter 6, ed. by K.G. Troitzsch, U. Mueller, N. Gilbert, J.E. Doran (Springer, Berlin, 1996), pp. 123–137
T. Monks, C.S.M. Currie, B.S. Onggo, S. Robinson, M. Kunc, S.J.E. Taylor, Strengthening the reporting of empirical simulation studies: introducing the STRESS guidelines. J. Simul. **13**(1), 55–67 (2018)
B. Müller, F. Bohn, G. Dressler, J. Groeneveld, C. Klassert, R. Matrtin, M. Schlüter, J. Schulze, H. Weise, N. Schwarz, Describing human decisions in agent-based models—ODD+D, an extension of the ODD protocol. Environ. Model. Softw. **48**, 37–48 (2013)
T. Ostrom, Computer simulation: the third symbol system. J. Exp. Soc. Psychol. **24**, 381–392 (1988)
G. Polhill, Extracting OWL ontologies from agent-based models: a NetLogo extension. J. Artif. Soc. Soc. Simul. **18**(2), 15 (2015)
H. Rahmandad, J.D. Sterman, Reporting guidelines for simulation-based research in social sciences. Syst. Dyn. Rev. **28**(4), 396–411 (2012)
T.C. Schelling, Dynamic models of segregation. J. Math. Sociol. **1**, 143–186 (1971)

J.D. Sneed, *The Logical Structure of Mathematical Physics*. Synthese Library, vol. 35, 2nd edn. (Reidel, Dordrecht, 1979)

W. Stegmüller, *Hauptströmungen der Gegenwartsphilosophie. Eine kritische Einführung, Band II.* Kröners Taschenausgabe, Bd. 309, 7 edn. (Kröner, Stuttgart, ,1986)

P. Suppes, Information processing and choice behavior, in *Problems in the Philosophy of Science*, ed. by I. Lakatos, A. Musgrave. Studies in Logic and the Foundations of Mathematics, vol. 49 (Elsevier, New York, 1968), pp. 278–304

P.H. Tannenbaum, Initial attitude toward source and concept as factors in attitude change through communication. Public Opin. Q. **20**(2), 413–425 (1956)

L.L. Thurstone, Attitudes can be measured. Am. J. Sociol. **33**(4), 529–554 (1928)

K.G. Troitzsch, Structuralist theory reconstruction and specification of simulation models in the social sciences, in *The Structuralist Program in Psychology: Foundations and Applications*, ed. by H. Westmeyer (Hogrefe & Huber, Seattle, 1992), pp. 71–86

K.G. Troitzsch, Modelling, simulation, and structuralism, in *Structuralism and Idealization*, ed. by M. Kuokkanen. Poznań Studies in the Philosophy of the Sciences and the Humanities (Editions Rodopi, Amsterdam, 1994), pp. 157–175

K.G. Troitzsch, Simulation and structuralism, in *Modelling and Simulation in the Social Sciences from a Philosophy of Science Point of View*, ed. by R. Hegselmann, U. Mueller, K.G. Troitzsch. Theory and Decision Library, Series A: Philosophy and Methodology of the Social Sciences (Kluwer, Dordrecht, 1996), pp. 183–208

K.G. Troitzsch, The garbage can model of organisational behaviour: a theoretical reconstruction of some of its variants. Simul. Model. Pract. Theory **16**, 218–230 (2008)

K.G. Troitzsch, Theory reconstruction of several versions of modern organization theories, in *Ontology, Epistemology, and Teleology for Modeling and Simulation*, ed. by A. Tolk. Intelligent Systems Reference Library, chapter 6, vol. 44 (Springer, Berlin/Heidelberg/New York/Dordrecht/London, 2013), pp. 121–140

K.G. Troitzsch, Axiomatic theory and simulation: a philosophy of science perspective on Schelling's segregation model. J. Artif. Soc. Soc. Simul. **20**(1), 10. http://jasss.soc.surrey.ac.uk/20/1/10.html. https://doi.org/10.18564/jasss.3372 (2017)

K.G. Troitzsch, Axiomatisation and simulation. Information **10**(2) 53. https://doi.org/10.3390/info10020053 (2019)

D. Waltemath, R. Adams, D.A. Beard, F.T. Bergmann, U.S. Bhalla, R. Britten, et al., Minimum information about a simulation experiment (MIASE). PLOS Comput. Biol. **7**(4). https://doi.org/10.1371/journal.pcbi.1001122 (2011)

H. Westmeyer (ed.), *The Structuralist Program in Psychology: Foundations and Applications* (Hogrefe & Huber, Seattle, 1992)

L. Yilmaz, T. Ören, Toward replicability-aware modeling and simulation: Changing the conduct of m&s in the information age, in *Ontology, Epistemology, and Teleology for Modeling and Simulation*, ed. by A. Tolk. Intelligent Systems Reference Library, chapter 11, vol. 44 (Springer, Berlin/Heidelberg/New York/Dordrecht/London, 2013), pages 207–226

B.P. Zeigler, *Theory of Modelling and Simulation* (Krieger, Malabar, 1985). Reprint, first published (Wiley, New York, 1976)

Part II
Methodological Toolsets

The Potential of Automated Text Analytics in Social Knowledge Building

Renáta Németh and Júlia Koltai

1 Introduction

In 2007 Savage and Burrows, in one of the most highly cited sociological paper of the decade, wrote about the coming crisis of empirical sociology. They predicted that a crisis would come if sociology, known for its innovative methodological resources, could not meet the challenges put forward by big data,[1] and thus would lose its leading role. This did not come to pass. Eight years later, the first member of the book series titled *Sociological Futures*, published by the British Sociological Association (Ryan and McKie 2015), referred to the end of the crisis in its title and saw important opportunities in big data research and also in automated text analysis.

RN's work was supported by the Higher Education Excellence Program of the Ministry of Human Capacities (ELTE-FKIP).

The work of Julia Koltai was funded by the Premium Postdoctoral Grant of the Hungarian Academy of Sciences.

[1] Although the former buzzword big data is seemingly going out of fashion, it does not have a better alternative. We use it in a general sense by referring to a vast amount of digital data which, in most cases, were created for some purpose other than our analysis (see also "found data").

R. Németh (✉)
Faculty of Social Sciences, Eötvös Loránd University, Budapest, Hungary
e-mail: nemeth.renata@tatk.elte.hu

J. Koltai
Centre for Social Sciences, Hungarian Academy of Sciences Centre of Excellence, Budapest, Hungary

Faculty of Social Sciences, Eötvös Loránd University, Budapest, Hungary
e-mail: koltai.julia@tk.mta.hu

T. Rudas, G. Péli (eds.), *Pathways Between Social Science and Computational Social Science*, Computational Social Sciences,
https://doi.org/10.1007/978-3-030-54936-7_3

In this paper, we review the possibilities that automated text analytics can provide for sociology. Automated text analytics[2] refers to the automated and computer-assisted discovery of knowledge from large text corpora. It lies at the intersection of linguistics, data science, and social science and uses many tools of natural language processing (NLP).

Our aim is to encourage sociologists to enter this field. We discuss the new methods based on the classic quantitative approach, using its concepts and terminology. We also address the question of how traditionally trained sociologists can acquire new skills. We are convinced that supporting this process is of crucial importance for the future of sociology because automated text analysis is going to be a standard tool of social research within a few years.

2 Challenges

The main challenges of applying automated text analytics in sociology are methodological. Some of them are well-known for traditionally trained sociologists, like the issue of *external validity, internal validity, and reliability*. The total survey error framework provides a conceptual structure to identify and quantify these challenges. Studies of large datasets can have the same shortcomings as surveys – that is why these questions were addressed by empirical social scientists (e.g., Hargittai 2015; Kreuter and Peng 2013). Indeed, when, according to a Twitter-based text analytical tool (hedonometer.org by the University of Vermont, Complex Systems Center), Louisiana is the least happy state in the USA, while a large-sample survey for the same period finds Louisiana to be the happiest one (Oswald and Wu 2011), we must examine the mechanisms generating our data. The explanation for this contradiction may lie in the difference between the coverage of the two samples.

Factors to be considered are external validity, coverage (who does use Twitter?), biased sample composition (those who tweet more are often over-represented in the population of tweets), and sampling procedure of Twitter's API. These factors also affect many other studies on digital texts: the digital divide is a decreasing but still existing problem; replacing texts for people as the unit of analysis may cause biases since more active people are more likely to appear in digital corpora. Finally, big data are often samples themselves, resulting from an unknown or hard-to-formalize sampling procedure. See, e.g., Common Crawl (http://commoncrawl.org), an open repository of textual web page data, widely used as a source representing language usage. However, what does it precisely represent? What is the probability of a given web page to get into the dataset? Another example is Google Ngram Viewer (https://books.google.com/ngrams). The viewer was created on the top of the largest digitized collection of books published between 1500 and 2000. The viewer gives the trend analysis of any n-words-long phrase (Ngram) using a yearly count found in the corpus. It is increasingly used to measure social and cultural

[2]Similar but not synonymous terms are text mining and computer-based content analysis.

trends by everyday users and researchers as well. However, results may be affected by many different potential representation and measurement errors, e.g., changes in the book publishing industry during the centuries or the bias caused by the various number of texts published already in digital formats in the different years.

Additionally, there are other big data-specific issues related to the quality of data, which are not present in surveys or interviews. These are, for example, the *presence of noise* (irrelevant data) or *fake data*. Another problem is the lack of demographic variables, which are routinely used for post-stratification in survey research. As most Internet data do not include these variables, researchers barely know anything about the social composition of users and, thus, about the external validity of data. For the same reason, post-stratification weighting is also not possible.

Such challenges contribute to the sociological skepticism toward big data-based, social-related findings that question the potential of this knowledge production and its contribution to the scientific discourse of sociology production (e.g., Kitchin 2014). It is revealing that, using Google Scholar search terms of "big data" and "sociology," the most highly cited paper found is a skeptical one (Boyd and Crawford 2012).

Additional difficulties are neither new nor specific to automated text analysis; they can be traced back to the age-old *quantitative-qualitative dichotomy*. One of them is the *close reading – distant reading* opposition (Moretti 2013) that attracted broad attention in the literature and cultural studies recently. According to the critiques, automated text analysis cannot produce deep insights and is overly reductionist. Automated text analysis methods cannot and do not aim to substitute human reading. They are capable of extracting information from texts but do not understand them. Hedonometer uses a bag-of-words model which is not capable of understanding the text but is appropriate to measure the average sentiment of tweets. Automated text analysis can construct models of language usage (see, e.g., topic models later), which – as an inherent specification of models – do not perfectly correspond to "reality" and can hardly detect sarcasm or latent intentions of speakers. Their importance is based on the fact that large text corpora are impossible to read and summarize by humans. The traditional quanti-quali debate can be cited again: we use surveys because we cannot conduct in-depth interviews with thousands of people, even if we know, for example, that self-reported answers do not fully correspond to reality. The consequence is not to refuse surveys (or automated text analytical tools) but to articulate epistemological issues and to incorporate the answers given for them in the process of data discovery, collection, preparation, analysis, and interpretation.

3 A New Methodological Basis of Sociology

In the present chapter, we review the methodological advances sociology can apply in the new field of digital textual data. The advances include the utilization of new data sources and new statistical models based on new research logic. For those interested in more details, we recommend Aggarwal and Zhai (2012).

There is a new step in the analysis process, which is not part of traditional research: pre-processing of texts. The other steps are not new; however, some of them have novel approaches.

3.1 New Data Sources

Before turning to new data sources, we have to mention that *traditional sociological textual data* like interviews, field notes, or open-ended survey questions can be analyzed by automated tools as well potentially providing new and inspiring insights into old questions. Traditional surveys with open-ended questions can also be analyzed with automated text analysis resulting in a hybrid approach.

Advances in NLP technologies allow the extension of the scope of open-ended questions. This hybrid technique increases the depth and the internal validity of surveys by moving them into the qualitative direction.

One of the most important of the new textual sources is social media. A significant and continuously increasing part of the population uses social media in their daily lives. We refer to social media as the use of technologies that turn communication into an interactive, two-way dialogue. Besides Twitter, Facebook, and Instagram, YouTube, blogs, and online forums also belong to this category. With the application of new technologies, it is possible to convert images, video, etc. to textual data and analyze the visual and textual content together (Aggarwal and Zhai (2012) give a good review).

Significant advantages of using social media for research are the low cost of data collection, an enormous amount of data, and rich metadata like geolocation, author info, exact time, and friends/followers. In addition, the network nature of the data is highly important as communication pathways can be traced through the links, which connect people. Typical analytical approaches are, e.g., social network analysis, sentiment analysis, and identification of influential/antisocial users. However, there are limitations to social media research. Social interactions within social media are mediated interactions, where individuals imagine an audience, and they build their self-representation accordingly (see, e.g., Marwick and boyd 2011). Also, the algorithms used by social media companies, which generate sampling procedures of the data, are not transparent. Therefore, researchers do not know if the data collected through APIs is a representative sample of all posts or just a biased portion of them. The question arises as to what extent the information extracted from social media can be identified with "reality."

Further textual sources, which represent more traditional, one-way communication, are *web pages and online editorial media*, which can contain multimedia contents as well.

There is a tendency for the digitalization of originally not digital textual contents. *Digitized archives* generally contain texts that originally were not produced for the public. These documents can be historical and contemporary public administration documents (birth and death certificates, healthcare patient records, property records, military and police records), private letters, diaries, journals and newspapers, books, etc. A huge project by Google Books Library (Michel et al. 2011) is the digitalization of books from the 1500s. A good example of the analysis of such data sources is Grimmer (2010), who measured how senators explain their work in Washington to constituents based on a collection of over 24,000 press releases. Another historical analysis is the project, named "Mining the Dispatch" by Robert K. Nelson (McClurken 2012).[3] Nelson and his team analyzed the dramatic social and political changes of the Civil War through the articles of the newspaper *Richmond Daily Dispatch.*

3.2 A Brief Overview of NLP Methods

3.2.1 Pre-processing

Raw textual digital data is often unstructured, "noisy," and full of surplus information; thus, it is fundamentally different from the well-structured data sources of classical sociological research, which mainly contain the relevant information for the analysis. Therefore, it is not recommended to use the corpus itself per se, but to *pre-process the text before the analysis*. Several methods are available to prepare corpora for analysis – and as these corpora are frequently quite large, these pre-processing methods are algorithmized.

The pretreatment process usually starts with "clearing" the corpora in order to have only the relevant information in it. In practice, it means that all the punctuation marks (e.g., ., !, ?), articles (e.g., the, an), and conjunctions (e.g., of, by) are deleted. The use of stopwords (the deletion of content, where the given stop word is present) can help filter out content, which is not relevant at all for the analysis. It can be especially important if data is collected by polysemic words. Another main task is to reduce inflectional forms of words to a common base form. Stemming and lemmatization can be useful processes for this (Manning et al. 2008). Depending on the goal of the research, another step can be the detection of word classes and other linguistic or syntactic categories within the text. These tags can also help the lemmatization process. Besides these methods, the handling of multiple words/terms is also important. One such group is geographical or proper names, which usually contain several words (such as East Germany or Barack Hussein Obama), which nevertheless have to be treated together. The detection of proper names ("named entity recognition") is especially important in the case of opinion mining and sentiment analysis, where the extraction of entities has exceptional importance – considering, for example, the analysis of political texts and the politicians, who appear in them. Another such group consists of expressions, which belong together (like bus driver or carpe diem). These problems are usually solved

[3]http://dsl.richmond.edu/dispatch/pages/about

by the use of dictionaries, with which these names or expressions with multiple words can be identified and then handled together. The selection and way of application of different pre-processing steps depend on the nature of the research. Most studies fail to emphasize that these consecutively taken steps depend on each other. This dependency implies that the selection and the order of the steps have profound effects on the results (see, e.g., Denny and Spirling 2018). Accurate planning of this phase is immensely important.

3.2.2 Bag of Words and Beyond

Basic methods of NLP treat these pre-processed corpora as separated words, without their syntactical linking (*bag-of-words* model); see, e.g., topic modeling later. Basic analytic methods use the simple distribution of roots of words in the pre-processed corpora with its absolute (number of occurrences) or relative (percentage) frequency, usually treated in a vector. Other methods weight the words by their relative occurrence, for example, like the number of different documents they appear in, divided by the number of all documents (Evans and Aceves 2016). These methods are especially useful for retrieving the relevant information from a large corpus, which would be impossible to achieve solely by humans.

Nevertheless, the more refined methods of NLP do use not only the frequency of words but also the structure of the text and sentences. Syntactical analysis of the sentences can give a deep understanding of the corpora; however, it needs massive computational power and rarely provides a relevant result for questions researched by social scientists. A simpler version of the analysis of structure is the examination of the co-presence of words. This method works very well, even in large corpora. In the analysis of the co-presence of words, researchers focus on small "windows" of the text, namely, the given number of words, that are next to each other (so-called n-grams; see Google Ngram Viewer) and scan the whole document by sliding this given width window through the text. (See Fig. 1, for example, on a sentence by Max Weber.) In the end, the database of the analysis contains cases, which include all these "windows" with the given number of words. The number of words, by which this "window" is filled, can be any positive integer greater than one (e.g., bigram, trigram, fourgram, and so on). However, if this number is too high, we can lose the focus of our interest, namely, the context of the words. With the analysis of the co-presence of words and their closeness-distances, associations latently present in the corpus can be examined.

Besides the analysis of words, a higher-level analysis is also possible. Co-presence of words can be examined not only by n-grams but also in a paragraph; the level of analysis does not have to be on the level of words but the level of sentences. The use of these higher-level analyses can help understand the context of a word, which can shed light on several higher-level associations and meanings (e.g., in the detection of dialects or subcultural language) (Evans and Aceves 2016). A good example of higher-level analysis is a paper by Demszky et al. (2019), where the authors used sentence embedding beside the analysis of words. An example of

Trigram1 "**Sociology is a** science which attempts the interpretive understanding of social action in order thereby to arrive at a causal explanation of its course and effects."

Trigram2 "Sociology **is a science** which attempts the interpretive understanding of social action in order thereby to arrive at a causal explanation of its course and effects."

Trigram3 "Sociology is **a science which** attempts the interpretive understanding of social action in order thereby to arrive at a causal explanation of its course and effects."

Trigram4 "Sociology is a **science which attempts** the interpretive understanding of social action in order thereby to arrive at a causal explanation of its course and effects."

Etc.

Fig. 1 An example of trigrams in a sentence

such an analytical tool is Stanford CoreNLP, which – besides other features – takes the grammatical and syntactical structure of the text into account (Hirschberg and Manning 2015).

3.3 The Goal of the Analysis and the Corresponding NLP Methods

The selection of the proper analytical NLP method is based on the same logic as in a "classical" sociological research. The choice needs to be based on the research question: (1) theory-driven approach for testing an already existing theory or (2) data-driven approach for exploring a not yet extensively studied topic and creating new theories. The group of supervised methods is the most suitable for the former, while the unsupervised methods are most suitable for the latter. These two approaches are more theoretical categories and are often used jointly in practice.

3.3.1 Supervised Methods

Supervised methods help researchers to (1) perform their analysis on larger datasets, than human capacity would be able to, or (2) expand their knowledge to external datasets they do not have background knowledge about, or (3) understand the mechanisms behind this expansion (see our discussion earlier about sociological relevance of the interpretation of "black box" parameters). A good example of the first one is the paper by Cheng et al. (2015), where the authors examined antisocial behavior in large online discussion communities. They employed humans to code the training part of the corpora, and based on these texts, they trained the computer to do the same on larger corpora. They also examined the importance of independent variables (features) of this classification, which shows the significance of the abovementioned aim (3). For the application of aim (2), Jelveh et al. (2014) provide an example. They examined the political ideology of economists based on their scientific papers. Originally, the authors were aware of the ideological attachment of a few of these economists (based on outer datasets like campaign contributions), but they expanded this knowledge by analyzing scientific papers written by the economists.

The method is supervised, as such supervision is needed to gain outside knowledge and for the training of the computer. The goal is to transmit this specific knowledge to the computer in order to be able to apply it to other datasets and texts. This outer knowledge has to be cautiously chosen by the researcher, as over-fitting of the model can result in bad classification later.

In practice, most frequently, a researcher (or a team of researchers) does the coding, the annotation of the text by hand, such as in the case of "classical" text analysis. Thus, a human-annotated text will be divided into a *train* part and a *validation* part. The "teaching" of the computer will be executed on the train part of the already coded text. The computer looks for patterns and tries to figure out the reasons why a researcher coded a sentence or word one way or another. Based on this training text, the computer creates the rules, by which it will code the text later on. To check the appropriateness of these rules and, thus, the computer coding itself, we use the other part of the already coded text, the so-called validation part. Not letting the computer know the result of the human coding, we make the computer do the annotation and compare its results with the human-coded results. This quality control usually happens by examining the area under the curve (AUC) of the receiver operator characteristic (ROC) curve or by other accuracy measures (e.g., precision or recall). If the results of the computer and the human are quite similar to each other, then the rules, which were based on the training text, are acceptable; if they are not similar, more train data is needed. If the training was acceptable, the best prediction model is used for the classification of new texts (which are not coded by human annotators) (Fig. 2).

When presented with complex annotations or codings, researchers may not have the capacity to provide enough training data. In these cases, crowdsourcing solutions, such as Amazon Mechanical Turk (Garg et al. 2018) or CrowdFlower (Kim et al. 2014), can be suitable to have enough human coded text as train or test data. Using crowdsourcing of an annotation or coding of a text can have hidden

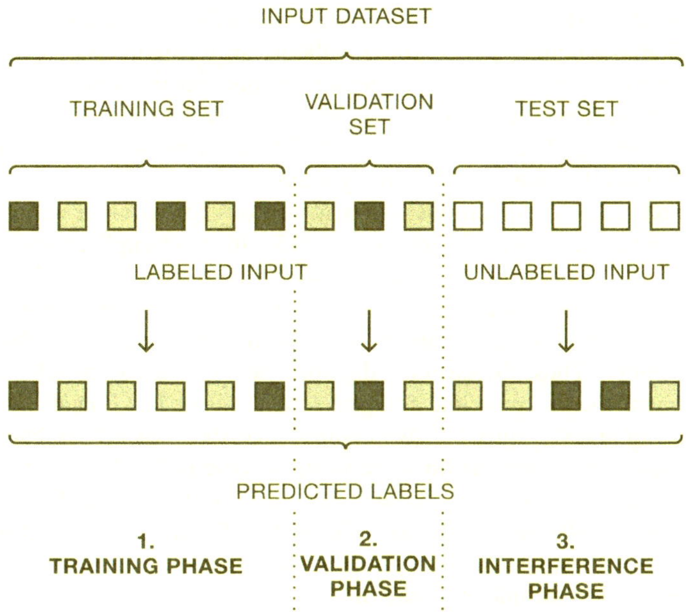

Fig. 2 The process of training, validation, and inference in supervised models

dangers, as human coders, who are not from the field of the specific research, can make different decisions from expert researchers of that field. In order to minimize this risk, it is advisable to take actions, which make the quality of the crowdsourced coding better. These actions may include the use of a very detailed codebook, using several coders for the same text and only accepting codings that are consistent across those coders, and/or continuous control of the coders with random checking. A good example of quality control can be found in the paper of Kim et al. (2014). For the coding phase, researchers can use not only their (or the crowdsource's) decision but external databases too, where the categories assigned to a text can be provided from other databases, as seen in the previously cited paper by Jelveh et al. (2014).

A special and frequently used type of supervised classification is *sentiment or emotion analysis*. The sentiment of a text is the attitude of the author toward an object (positive, negative, or neutral), while emotions are feelings from happiness to anger. A good example of its application is the Hedonometer mentioned above, where we can find the longitudinal "average happiness" for Twitter, measured with these techniques. Besides its scientific application, different classification methods can be used for the detection of sentiments and emotions of a sentence or a short text. Kharde and Sonawane (2016) introduce several techniques from the naïve Bayes method to the maximum entropy model with different data sources for the supervised part, like emoticons or external dictionaries. The goal is common: to classify the text to previously given sentiments or emotions and thus draw conclusions about the relationship of the content and the sentiment attached to it.

Yadollahi et al. (2017) give a detailed and well-defined taxonomy of sentiment analysis and emotion mining. According to them, this technique is quite widespread in business-related fields, such as customer satisfaction measurement or product recommendation. However, sentiment analysis can be easily adapted to sociological dimensions. See, e.g., Grimmer and Stewart (2013) discuss the placing of political actors on the liberal-conservative ideology scale.

Supervised methods contain well-known methods for social scientists, such as linear or logistic *regression* (Cheng et al. 2015). In these applications – contrary to the traditional theory-based approach – independent variables can be thousands of variables, which originate from the text (like a vector of word frequencies). As the number of these variables can be huge, dimension reduction techniques are frequently applied. These dimension reductions can be methods, which are familiar for social scientists, like principal component analysis or factor analysis or cluster analysis. Using sparse regression techniques, like the forward selection, is also common. Other supervised methods are more specific to NLP, such as *supervised topic modeling, support vector machine* (see, e.g., Bakshy et al. 2015), or decision tree. Regardless of the method we use, the goal is the same: to only choose those independent variables, which are important and have added value to our prediction or annotation, and to include all of them.

The application of supervised methods means that the theoretical considerations of the researcher influence the analysis, and thus, the interpretation of the predicted categories precedes the final results of the classification (Evans and Aceves 2016). As the annotation or coding is finalized, the dataset is ready to be analyzed by any statistical methods.

3.3.2 Unsupervised Methods

Unsupervised models are useful for discovering a field or topic, about which we do not have in-depth knowledge (see, e.g., Mohr et al. 2013). As the goal of these methods is exploration, there is no need for either background knowledge or prior annotation by the researcher. We allow the computer to find patterns in the data without theoretical supervision. Beyond discovering new topics, it can be useful for multigroup comparison, like comparing topics emerging from data of different cities (Nelson 2015) or countries (Marshall 2013). Unsupervised methods can also be useful as a supplementary method for supervised techniques. As we discussed above, researchers usually need data reduction techniques for the efficient application of unsupervised methods. These data reduction techniques are frequently unsupervised. Hereinafter we introduce the main unsupervised techniques, which are used in research dealing with large-scale corpora and suggest applied examples.

Some of the unsupervised methods can be familiar for social scientists, as they are frequently used in classical social research too. Widely used unsupervised classification research techniques dealing with large-scale data (Kozlowski et al. 2018; Kim et al. 2014) such as *cluster analysis, principal component analysis, and factor analysis* can be used according to the goal of the research. For example, cluster analysis can be applied to textual documents in a vector space, where the axes stand for the words, and the values of the vector space are these documents. The position of the documents in this vector space is defined by the frequencies of different words in a given document. Cluster analysis can be applied not only to documents but also to words based on their distances from each other.

The use of *network analysis* – which was also used by social scientists before the era of big data – is also an unsupervised and useful technique when working with large textual datasets. Nodes can be entities, topics, or users; edges can be associations or co-occurrences. The network figure of a text shows the relative topological positions, the importance, and centrality of entities; network indices inform the researchers about the density and structure of a document (see, e.g., Brummette et al. 2018).

Naturally, new methods also emerged specifically for large-scale textual data. The group of *unsupervised topic models* is one of them. The goal of this method is to find latent topics in a document or across documents. Compared to a cluster analysis, where the units of analysis are words, topic models are different in their outputs. While all the words are assigned to one cluster, and cannot be assigned to several clusters, in topic models, words are assigned to each topic but with different probabilities. The theoretical concept behind topic models assumes that there are a finite number of topics, which can describe a set of documents. The probability distribution of words defines these topics, e.g., there is a higher probability that a document about sport includes the word "winner" than the word "inflation," while a document about the economy has reverse probabilities for these two words. However, documents can contain a mixture of topics. For example, a document about building a stadium could contain 80 percent economic and 20 percent sports topics. See Fig. 3 regarding the process of topic modeling.

The possible application of topic models in social sciences is very broad. We can analyze and compare the attitudes and opinions of different groups (such as political party supporters, patients having a given disease, authors of a given scientific journal, etc.) or the changes of topics (e.g., in a newspaper) over time. An exciting historical application of topic modeling is the already mentioned "Mining the Dispatch" project (McClurken 2012), where topics of "fugitive slave" were identified by topic modeling among the advertisements of a newspaper published during the Civil War. After the identification, the trend of the appearance of this topic was examined, which showed the dramatic social changes that happened at that time.

It is possible to focus only on one long document and discover the topics in it, but it is also feasible to compare several documents according to their latent meanings. As this method is unsupervised, the labeling and interpretation of the topics is the task of the researcher, and – unlike in the case of supervised topic models – this interpretational phase has to happen after the classification by the researcher. Validation of the topics is not an easy task; Chuang et al. (2012) suggest some visualization techniques that can be useful in the validation process. A good introduction for (supervised and unsupervised) topic models for social researchers is

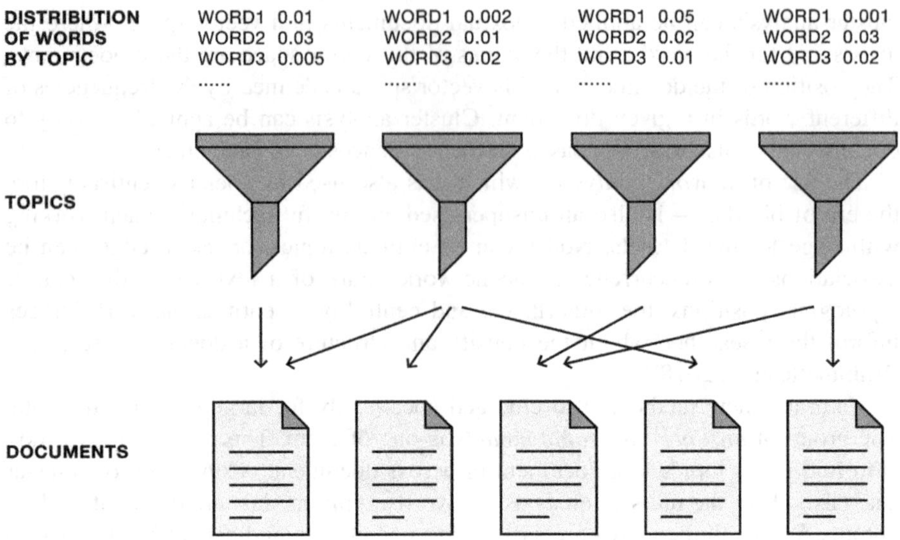

Fig. 3 The document generation process assumed by topic models

presented by Mohr and Bogdanov (2013) and Ramage et al. (2009). Wesslen (2018) presents good examples of the sociological application of topic modeling.

The more in-depth research of the semantic structure of texts became possible with *word embedding models*, which gained great popularity in the field of computational linguistics in recent years. The success of these models is based on their ability to help the understanding of large corpora. Earlier mentioned techniques do not model the meaning of words, while word embedding models do. The models identify the meanings with the contexts of the words. This distributional semantic approach has language philosophical antecedents: as its earliest representations, see the work of Wittgenstein from the 1930s and the paper by John R. Firth from 1957.

In word embedding models, words are embedded in a multidimensional semantic vector space, where words are positioned by their meanings, which is defined by their narrow textual environment. Two words will be positioned close or far from each other according to the similarity of their environment in the corpus. The multidimensional (usually a couple of hundred) vector space is created by the training of the words of a corpus by neural networks. These neural networks take into account the different semantical specifications of the corpus to reduce the space of words, where the semantical distances of words can be analyzed. Figure 4 presents a graphical illustration of a vector space, and as it shows, it is possible to grab semantic relationships (variations of greetings); partition the clusters of words with different meanings (sociologist and computational social scientist are separated from the other words); analogies can be created by vector differences (what Trump is for the USA, what can be for Russia?). However, the most inspiring possibility for sociologists is the identification of latent dimensions in the vector

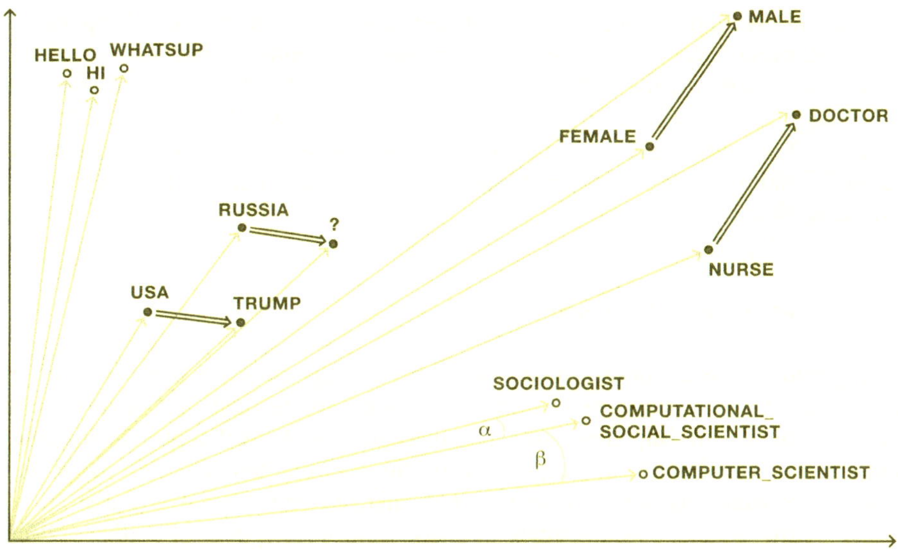

Fig. 4 An example for the vector space of a word embedding model

space, where dimensions represent some social difference. For example, if the vector from the nurse to the doctor is parallel with the vector from female to male, it refers to the male dominance of the profession of physicians. Using similar pairs to female-male (like girl-boy or women-men), one can identify the latent dimension of gender in this vector space. By projecting different expressions to this dimension, it is possible to grab the gender inequalities in different fields (e.g., differences in cultural consumption).

Word embedding methods can be used for several research-related questions, like sentiment analysis, classification of documents, or the detection of proper nouns. However, its most interesting feature is its ability to provide information about the associations between words, which can show the cultural analogies through texts. Kozlowski et al. (2018) provide one of the most interesting applications of this method. The authors identify latent dimensions (such as the gender dimension in Fig. 4) and show how cultural phenomena can be detected in various texts and how these phenomena change over time. Kulkarni et al. (2015) also show that longitudinal analysis can be efficiently processed with this method, which can help researchers detect and understand the change in the meaning of different expressions (and the cultural change behind it). The latest publications imply that word embedding models can detect such smooth distinctions as the identification of euphemistic coded words used in hate speeches (Magu and Luo 2018), or the detection of sarcasm (Joshi et al. 2016). Different computational methods, such as GloVe, fastText, and Word2vec, are available.

Unsupervised methods are different from supervised ones in the phase of interpretation of categories. While the interpretation of categories and theoretical considerations happen before the automatized classification in the case of supervised methods, these tasks have to be completed after the automatized analysis in the case of unsupervised methods. Nevertheless, it is true for both methods that the classical statistical analyses are conducted on the dataset, which is the result of these methods. The analysis goes the same way on this structured data as in classical social research; and the topics or groups of cases usually serve as independent variables in the analysis.

3.3.3 Which Method to Choose

One group of the methods contains tools for *classification for previously given categories*. Logistic regression or supervised topic models are part of this group of methods. It can help detect the frequency of the defined topics in different documents. These tools are especially useful for longitudinal analysis, where researchers would like to discover the topic changes over time. As we mentioned above, sentiment and emotion analysis is also part of this group of methods. With these latter techniques, the relationship between the content and the sentiment of the text can also be analyzed.

Another group of NLP methods is the *classification of text without previously given categories*. Unsupervised topic model and cluster analysis are typical examples of these methods. The goal is the exploration, where researchers do not have a priori knowledge about the examined field and would like to have a comprehensive picture of the topics that arise in the documents. In the case of text data from different sources, countries, or timepoints, unsupervised classification can help compare the different groups.

The third type of methods includes those techniques, which *show the associations, the latent relationships* that are in the corpus. Vector space-based word embedding models belong here. These methods are based on the phenomenon that the language mirrors the user's cultural frame of mind. Thus, the associations and relationships of words in a corpus show the associations and relations that are present in a culture, or in a society. The method allows researchers to explore the distances between different concepts or the detection of the interconnection of connotations. Word embedding methods are substantive methods, but can also be used for arranging other methods, such as cluster analysis or network analysis, which can be applied to the resulted vector space.

4 New Possibilities for Sociological Research

4.1 *How to Approach Automated Text Analysis as a Social Scientist*

As evidenced in the preceding chapters, the use of NLP methods requires more skills than conventional social science qualifications provide. Reasonable familiarity with programming, pre-processing, and data science analytical tools are important prerequisites. Thus, the *cost of entry is* relatively high for those who try to acquire the necessary skills. Therefore it is very common that social scientists lead *interdisciplinary research*, where computer scientists and/or computational linguists are co-authors of a paper, and they are the ones who conduct the data generation part of the research. The papers of Kim et al. (2014), Kozlowski et al. (2018), McFarland et al. (2013), Mohr et al. (2013), Niculae et al. (2015), Srivastava et al. (2018), and Tinati et al. (2014) are all good examples of this kind of collaborations.

Nevertheless, we would not like to dissuade anyone from getting deeper into this field. Possessing basic programming knowledge and with an openness for new methods, it is worth delving deeper into automated text analysis. For those who decide to start, we recommend the book by Ignatow and Mihalcea (2017), which summarizes the actual development of automated text analysis with actual data sources, program languages, software, and analytical tools, especially for social scientists. Aggarwal and Zhai (2012) provide a more technical but very detailed synthesis of NLP methods. These books present a broad overview and helpful start for the application of NLP methods.

Beyond engaging in interdisciplinary cooperation or learning new skills, *using software, which needs entry-level knowledge,* is another strategy of getting involved with NLP usage. One such software is the Stanford Topic Modeling Toolbox (Ramage et al. 2009). There are also examples for easy-to-use platforms, which deliver a free run of the results of deep and rich NLP analysis. These platforms allow traditional sociologists to use these NLP results for the analysis of their own research questions. One of these platforms is a word embedding demo of Turku NLP group,[4] where queries can be performed without the knowledge of programming. These queries produce information about the meaning of words in vector spaces trained by other researchers. Another platform enables interactive examination of gender bias through the expressions of different topics – also with the use of word embedding methods.[5] The earlier mentioned Google Ngram Viewer is also part of this software group. It offers researchers the opportunity to follow historical social and cultural trends embedded in the wording of books. The Viewer's advanced wildcard search features make expressively meaningful analysis possible. A good example is provided by Michel et al. (2011) or Ophir (2016), who discovered hidden

[4]http://bionlp-www.utu.fi/wv_demo/

[5]http://wordbias.umiacs.umd.edu

patterns of centuries-long conceptual trends, using the Viewer's wildcard search feature. The terms "truth," "love," and "justice" were analyzed, and the results were paralleled with Sorokin's (1937) findings of the connection between "systems of truth" and "ethics of love." These examples prove that the use of these entry-level platforms can also be an avenue for sociologists to utilize the advantages of NLP in their research.

4.2 Combining the New with the Traditional: Mixed Approaches

The new techniques and methods we presented above are, in some sense, just by-products of information science and business analytics; they were not meant for supporting social research. Therefore it is an under-examined and open question which of the recently developed methods can be applied to sociological problems outside the scope of business applications (excellent examples are Evans and Aceves (2016) and Ignatow and Mihalcea (2017)). In the present chapter, we discuss how new methods and new textual data sources can be used jointly with the traditional ones, which are the sociological questions not yet studied in this area, and how the new approach can offer new insight into old research questions.

The most natural and widely used example for mixed methods in automated text analysis is the case of supervised learning when *models are trained on human-coded data*. Human coding requires close reading, and hence practically, it is a kind of qualitative text analysis. Such approaches directly extend qualitative text analysis to investigating large corpora. The problem of inter-rater reliability is of great importance here, as it profoundly affects not only label quality but predictive power as well. Two examples using crowdsourcing platforms for human coding are Iyyer et al. (2014), who aimed to identify political ideology from congressional debates, books, and magazine articles, and Cheng et al. (2015), who tried to detect trolls in online discussion communities.

Another field to mention is *qualitative content analysis*, a traditional and well-elaborated approach in social sciences that has been utilizing quantitative tools for a long time. Hence automated text analytics naturally emerges in this field, and its use could produce a more generalizable but interpretative approach. Ignatow and Mihalcea (2017) present a good summary of integrating NLP methods into classic sociological research. This approach could support the knowledge-driven applications of automated analytical tools, but it has some methodological challenges as well. Chen et al. (2016) provide a review of the challenges of applying automated analysis in qualitative coding. A theoretically elaborated example is Murthy (2016), who studied Twitter to rely on established grounded theory to give input to a computational Twitter analysis. According to Murthy's conclusion, mixed methods can offer new ontologies and epistemologies, an entirely new knowledge.

Grounded theory is the most elaborated qualitative content analysis method which has great potential in automated text analytics. Nelson (2017) proposes integration as a three-step methodological framework called *computational grounded theory*. The first step is a computational, unsupervised pattern detection step. The second step is a pattern refinement step that involves close reading. The third, pattern confirmation step, uses NLP techniques again to access the inductively identified patterns. Nelson (2015) provides a great example of her theory, a historical study of women's organizations in Chicago and New York City through the documents they left behind (e.g., internal memos and newsletters).

Other qualitative approaches present in the analysis of large textual data are ethnography of online communities, e.g., Facebook groups. An example is given by Baym (1999) with an ethnographic study of an online soap opera fan group. "*Netnography*"[6] adapts the traditional participant observation technique to the study of digital communications. It could be effectively integrated into automated text analysis (or vice versa) to get an exciting and balanced means of analysis. Di Giammaria and Faggiano (2017) gave an excellent example of the mixed analysis of this kind. They analyzed the Roman Five Star Movement Blog, by combining Facebook quantitative text analysis, ethnographic analysis of online conversations, social network analysis, and semiotic analysis of visual materials.

Additionally, integration of different approaches may be realized by *combining different data sources* like digital text corpora with survey data and/or census data. An example of combining data sources is Jelveh et al. (2014), who extracted political ideology from economists' papers, where ideologies in the training set were determined from datasets of political campaign contributions and petition signing activities. Another example is Garg et al. (2018), where word embedding models were trained on textual data (news and books) spanning a 100-year period in order to detect changes in stereotypes and attitudes in the USA. Human annotators were recruited on Amazon Mechanical Turk to assign gender labels to occupations. External validation was conducted by comparing the trends with shifts in US Census data, as well as with changes in historical surveys of stereotypes. They used digitized textual data, census data, survey data, and qualitative, human-coded data simultaneously.

4.3 What the Approach Can Offer to Classic Sociological Questions

The most important epistemological advantage of digital data is that it provides *observed* instead of self-reported *behavior*. This type of data offers real-time observation with continuous follow-up. The new approach offers access to *new data sources* (social media, digitized archives) and *new text analytic methods* (e.g., topic

[6]The term comes from Kozinets (1998).

modeling, word embedding modeling) not known before even by quantitative text analysis in sociology. NLP *technologies* developed for industry can be *transformed to answer sociological questions*.

Large digital corpora are spread over time, space, and topic. Therefore, (1) they provide the opportunity to conduct studies, which are otherwise impossible or at least hard-to-conduct within the traditional approach, like *longitudinal studies* by utilizing the dynamic flow of data. (2) They make possible *cross-country or cross-region comparisons* without conducting costly data collection far from the place of research. Finally, (3) given their size, they make possible the investigation of *small subpopulations* of societies (like soap opera fans) or subpopulations, which are *hard-to-contact otherwise* (e.g., drug users, members of illegal movements).

By definition, social sciences are less about individuals than interactions. When having textual data, *social phenomena* like identity, norm, conformity, deviance, status, conflict, or cooperation *emerge from communicative interactions*, not from separate individual statements. Large digital corpora present an outstanding opportunity for this analysis. Good examples are Lindgren and Lundström (2011) on rules of a global movement denoted by the #WikiLeaks hashtag on Twitter, Danescu et al. on power relations among Wikipedia editors (2012), Cheng et al. (2015) on antisocial behavior in online communities, and Srivastava et al. (2018) on adaptation to organizational culture analyzing internal employee emails.

The level of analysis (network/group or individual) can be freely selected when analyzing digital data. Sociological theories operating either at the *macro* (large-scale social processes) *or the micro* (face-to-face interactions or individual-level values, attitudes, and acts) *level* can be approached through digital data, and the syntheses of the two levels can also be realized. Sociology had an important role here since, as Resnyansky (2019) highlights, standard social media research concentrates on processes that are manifested at the micro level while forgets about structure. Researchers should not treat the text as standing alone in a vacuum, but they should pay attention to its social construction by taking the users' context and network as a starting point.

5 Summary

The analysis of a vast amount of digital textual data offers sociologists a broad perspective. Classic questions may be tested on new empirical bases; new insights and theories may also be generated. This new approach should be regarded as a complement and not a replacement for the traditional methods of sociology. Therefore, all of these data sources together may present sociology a flexible and broad enough empirical base.

Beyond encouraging interdisciplinary collaborations and revision of university training, further methodological developments are needed to decrease the currently high entry cost. Well-elaborated and easy-to-follow guidelines and rule of thumbs would make sociologists' task easier in data collection, preparation, and model training. Analytical platforms that do not require advanced programming skills also support the transformation.

Fast development of NLP is expected in the next decade. Sociology will exploit the potential of this development if it is able to update its research culture while preserving its critical reflections. A renewed discipline like this will be hopefully able to understand better the profound changes in our contemporary society.

References

C. C. Aggarwal, C. Zhai (eds.), *Mining Text Data* (Springer, New York, 2012)

E. Bakshy, S. Messing, L.A. Adamic, Exposure to ideologically diverse news and opinion on Facebook. Science **348**(6239), 1130–1132 (2015). https://doi.org/10.1126/science.aaa1160

N.K. Baym, *Tune in, log on: soaps, fandom, and online community*, 1st edn. (SAGE Publications, Inc., Thousand Oaks, 1999)

D. Boyd, K. Crawford, Critical questions for big data: provocations for a cultural, technological, and scholarly phenomenon. Inf. Commun. Soc. **15**(5), 662–679 (2012). https://doi.org/10.1080/1369118X.2012.678878

J. Brummette, M. DiStaso, M. Vafeiadis, M. Messner, Read all about it: the politicization of "fake news" on twitter. J. Mass Commun. Q. **95**(2), 497–517 (2018). https://doi.org/10.1177/1077699018769906

N.-C. Chen, R. Kocielnik, M. Drouhard, V. Peña-Araya, J. Suh, K. Cen, et al. *Challenges of Applying Machine Learning to Qualitative Coding*. Presented at the ACM SIGCHI workshop on human-centered machine learning, 2016

J. Cheng, C. Danescu-Niculescu-Mizil, J. Leskovec, Antisocial behavior in online discussion communities (2015). arXiv:1504.00680 [cs, stat]. http://arxiv.org/abs/1504.00680. Accessed 30 Oct 2018

J. Chuang, D. Ramage, C. Manning, J. Heer, Interpretation and trust: designing model-driven visualizations for text analysis, in *Proceedings of the 2012 ACM Annual Conference on Human Factors in Computing Systems – CHI'12*. Presented at the 2012 ACM Annual Conference (ACM Press, Austin, 2012), p. 443. https://doi.org/10.1145/2207676.2207738

C. Danescu-Niculescu-Mizil, L. Lee, B. Pang, J. Kleinberg, Echoes of power: language effects and power differences in social interaction, in *Proceedings of the 21st International Conference on World Wide Web – WWW'12*. Presented at the 21st international conference (ACM Press, Lyon, 2012), p. 699. https://doi.org/10.1145/2187836.2187931

D. Demszky, N. Garg, R. Voigt, J. Zou, M. Gentzkow, J. Shapiro, D. Jurafsky, Analyzing polarization in social media: method and application to tweets on 21 mass shootings (2019). arXiv:1904.01596 [cs]. http://arxiv.org/abs/1904.01596. Accessed 4 Apr 2019

M.J. Denny, A. Spirling, Text preprocessing for unsupervised learning: why it matters, when it misleads, and what to do about it. Polit. Anal. **26**(2) (2018). https://doi.org/10.1017/pan.2017.44

L. Di Giammaria, M.P. Faggiano, Big text corpora & mixed methods – the roman five star movement blog. Bull. Sociol. Methodol./Bulletin de Méthodologie Sociologique **133**(1), 46–64 (2017). https://doi.org/10.1177/0759106316681088

J.A. Evans, P. Aceves, Machine translation: mining text for social theory. Annu. Rev. Sociol. **42**(1), 21–50 (2016). https://doi.org/10.1146/annurev-soc-081715-074206

J.R. Firth, *A Synopsis of Linguistic Theory. Studies in Linguistic Analysis* (Blackwell, Oxford, 1957)

N. Garg, L. Schiebinger, D. Jurafsky, J. Zou, Word embeddings quantify 100 years of gender and ethnic stereotypes. Proc. Natl. Acad. Sci. **115**(16), E3635–E3644 (2018). https://doi.org/10.1073/pnas.1720347115

J. Grimmer, A Bayesian hierarchical topic model for political texts: measuring expressed agendas in senate press releases. Polit. Anal. **18**(1), 1–35 (2010). https://doi.org/10.1093/pan/mpp034
J. Grimmer, B.M. Stewart, Text as data: the promise and pitfalls of automatic content analysis methods for political texts. Polit. Anal. **21**(3), 267–297 (2013)
E. Hargittai, Is bigger always better? Potential biases of big data derived from social network sites. Ann. Am. Acad. Pol. Soc. Sci. **659**(1), 63–76 (2015). https://doi.org/10.1177/0002716215570866
J. Hirschberg, C.D. Manning, Advances in natural language processing. Science **349**(6245), 261–266 (2015). https://doi.org/10.1126/science.aaa8685
G. Ignatow, R.F. Mihalcea, *An Introduction to Text Mining: Research Design, Data Collection, and Analysis*, 1st edn. (SAGE Publications, Inc., Los Angeles, 2017)
M. Iyyer, P. Enns, J. Boyd-Graber, P. Resnik, Political ideology detection using recursive neural networks, in *Proceedings of the 52nd Annual Meeting of the Association for Computational Linguistics (Volume 1: Long Papers)* (Association for Computational Linguistics, Baltimore, 2014), pp. 1113–1122. http://www.aclweb.org/anthology/P14-1105. Accessed 30 Oct 2018
Z. Jelveh, B. Kogut, S. Naidu, Detecting latent ideology in expert text: evidence from academic papers in economics, in *Proceedings of the 2014 Conference on Empirical Methods in Natural Language Processing (EMNLP)* (Association for Computational Linguistics, Doha, 2014), pp. 1804–1809. http://www.aclweb.org/anthology/D14-1191. Accessed 30 Oct 2018
A. Joshi, V. Tripathi, K. Patel, P. Bhattacharyya, M. Carman, Are Word Embedding-Based Features Useful for Sarcasm Detection? Presented at the conference on empirical methods in natural language processing, 2016
A.V. Kharde, S.S. Sonawane, Sentiment analysis of twitter data: a survey of techniques. Int. J. Comput. Appl. **139**(11), 5–15 (2016). https://doi.org/10.5120/ijca2016908625
A. Kim, J. Murphy, J. Richards, A. Hansen, J. Murphy, R. Haney, Can tweets replace polls? A U.S. health-care reform case study, in *Social Media, Sociality, and Survey Research,* ed. by C.A. Hill, E. Dean, J. Murphy (Wiley, Hoboken, 2014), pp. 61–86. https://www.rti.org/publication/can-tweets-replace-polls-us-health-care-reform-case-study. Accessed 1 Nov 2018
R. Kitchin, Big data, new epistemologies and paradigm shifts. Big Data Soc. **1**(1), 2053951714528481 (2014). https://doi.org/10.1177/2053951714528481
R.V. Kozinets, On Netnography: initial reflections on consumer research investigations of cyberculture. ACR North Am. Adv. **NA-25** (1998) http://acrwebsite.org/volumes/8180/volumes/v25/NA-25. Accessed 30 Mar 2019
A.C. Kozlowski, M. Taddy, J.A. Evans, The Geometry of Culture: Analyzing Meaning Through Word Embeddings (2018). arXiv:1803.09288 [cs]. http://arxiv.org/abs/1803.09288. Accessed 30 Oct 2018
F. Kreuter, R. Peng, Extracting information from big data: issues of measurement, inference and linkage, in *Privacy, Big Data, and the Public Good: Frameworks for Engagement* (2013), pp. 257–275. https://doi.org/10.1017/CBO9781107590205.016
V. Kulkarni, R. Al-Rfou, B. Perozzi, S. Skiena, Statistically significant detection of linguistic change, in *Proceedings of the 24th International Conference on World Wide Web – WWW'15.* Presented at the 24th International Conference (ACM Press, Florence, 2015), pp. 625–635. https://doi.org/10.1145/2736277.2741627
S. Lindgren, R. Lundström, Pirate culture and hacktivist mobilization: the cultural and social protocols of #WikiLeaks on twitter. New Media Soc. **13**(6), 999–1018 (2011). https://doi.org/10.1177/1461444811414833
R. Magu, J. Luo, Determining code words in euphemistic hate speech using word embedding networks, in *Proceedings of the Second Workshop on Abusive Language Online*, Brussels, 2018, pp. 93–100
C.D. Manning, P. Raghavan, H. Schütze, *Introduction to Information Retrieval*, 1st edn. (Cambridge University Press, New York, 2008)
E.A. Marshall, Defining population problems: using topic models for cross-national comparison of disciplinary development. Poetics **41**(6), 701–724 (2013). https://doi.org/10.1016/j.poetic.2013.08.001

A.E. Marwick, D. Boyd, I tweet honestly, I tweet passionately: Twitter users, context collapse, and the imagined audience. New Media Soc. **13**(1), 114–133 (2011). https://doi.org/10.1177/1461444810365313

J.W. McClurken, Richmond daily dispatch, 1860–1865 and mining the dispatch. J. Am. Hist. **99**(1), 386–388 (2012). https://doi.org/10.1093/jahist/jas157

D.A. McFarland, D. Ramage, J. Chuang, J. Heer, C.D. Manning, D. Jurafsky, Differentiating language usage through topic models. Poetics **41**(6), 607–625 (2013). https://doi.org/10.1016/j.poetic.2013.06.004

J.-B. Michel, Y.K. Shen, A.P. Aiden, A. Veres, M.K. Gray, Google Books Team, et al., Quantitative analysis of culture using millions of digitized books. Science (New York, N.Y.) **331**(6014), 176–182 (2011). https://doi.org/10.1126/science.1199644

J.W. Mohr, P. Bogdanov, Introduction—topic models: what they are and why they matter. Poetics **41**(6), 545–569 (2013). https://doi.org/10.1016/j.poetic.2013.10.001

J.W. Mohr, R. Wagner-Pacifici, R.L. Breiger, P. Bogdanov, Graphing the grammar of motives in National Security Strategies: cultural interpretation, automated text analysis and the drama of global politics. Poetics **41**(6), 670–700 (2013). https://doi.org/10.1016/j.poetic.2013.08.003

F. Moretti, *Distant Reading* (Verso, London, 2013)

D. Murthy, The ontology of tweets: mixed methods approaches to the study of twitter, in *The SAGE Handbook of Social Media Research Methods*, ed. by L. Sloan, A. Quan-Haase (SAGE, London, 2016), pp. 559–572

L. Nelson, *Political Logics as Cultural Memory: Cognitive Structures, Local Continuities, and Women's Organizations in Chicago and New York City* (2015). https://www.academia.edu/10250788/Political_Logics_as_Cultural_Memory_Cognitive_Structures_Local_Continuities_and_Womens_Organizations_in_Chicago_and_New_York_City. Accessed 31 Oct 2018

L. Nelson, *Computational Grounded Theory: A Methodological Framework* (2017). https://doi.org/10.1177/0049124117729703. Accessed 30 Mar 2019

V. Niculae, S. Kumar, J. Boyd-Graber, C. Danescu-Niculescu-Mizil, Linguistic harbingers of betrayal: a case study on an online strategy game (2015). arXiv:1506.04744 [physics, stat]. http://arxiv.org/abs/1506.04744. Accessed 31 Oct 2018

S. Ophir, Big data for the humanities using Google Ngrams: discovering hidden patterns of conceptual trends. First Monday **21**(7) (2016). https://doi.org/10.5210/fm.v21i7.5567

A.J. Oswald, S. Wu, Well-Being Across America (2011). https://doi.org/10.1162/REST_a_00133

D. Ramage, E. Rosen, J. Chuang, C.D. Manning, D.A. McFarland, *Topic modeling for the social sciences*. Presented at the workshop on applications for topic models, neural information processing system, Stanford Computer Science (2009)

L. Resnyansky, Conceptual frameworks for social and cultural big data analytics: answering the epistemological challenge. Big Data Soc. **6**(1), 2053951718823815 (2019). https://doi.org/10.1177/2053951718823815

L. Ryan, L. McKie (eds.), *An End to the Crisis of Empirical Sociology? Trends and Challenges in Social Research* (Routledge, London, 2015)

M. Savage, R. Burrows, The coming crisis of empirical sociology. Sociol. J. British Sociol. Assoc. **41**, 885–899 (2007)

P.A. Sorokin, *Fluctuation of Systems of Truth, Ethics, and Law*, vol 2 (American Book Co., New York, 1937)

S.B. Srivastava, A. Goldberg, V.G. Manian, C. Potts, Enculturation trajectories: language, cultural adaptation, and individual outcomes in organizations. Manag. Sci. **64**(3), 1348–1364 (2018). https://doi.org/10.1287/mnsc.2016.2671

R. Tinati, S. Halford, L. Carr, C. Pope, Big data: methodological challenges and approaches for sociological analysis. Sociology **48**(4), 663–681 (2014). https://doi.org/10.1177/0038038513511561

R. Wesslen, Computer-assisted text analysis for social science: topic models and beyond. arXiv:1803.11045 [cs] (2018). http://arxiv.org/abs/1803.11045. Accessed 17 Feb 2019

A. Yadollahi, A.G. Shahraki, O.R. Zaïane, Current state of text sentiment analysis from opinion to emotion mining. ACM Comput. Surv. **50**, 25–25 (2017). https://doi.org/10.1145/3057270

Combining Scientific and Non-scientific Surveys to Improve Estimation and Reduce Costs

Joseph W. Sakshaug, Arkadiusz Wiśniowski, Diego Andres Perez Ruiz, and Annelies G. Blom

1 Introduction

For more than a decade, the survey research industry has witnessed increasing friction between two distinct sampling paradigms: probability and nonprobability sampling. Probability sampling is characterized by the process of drawing samples using random selection with every element having a known (or knowable) nonzero chance of being selected into the sample. Nonprobability sampling, on the other hand, involves some form of arbitrary selection of elements into the sample for which probabilities of selection cannot be accurately determined. Both sampling paradigms have strengths and weaknesses. The primary appeal of probability sam-

Electronic Supplementary Material: The online version of this chapter (https://doi.org/10.1007/10.1007/978-3-030-54936-7_4) contains supplementary material, which is available to authorized users.

J. W. Sakshaug (✉)
University of Mannheim, Ludwig Maximilian University of Munich, and Institute for Employment Research, Nuremberg, Germany
e-mail: joe.sakshaug@iab.de

A. Wiśniowski
Department of Social Statistics, University of Manchester, Manchester, UK
e-mail: a.wisniowski@manchester.ac.uk

D. A. Perez Ruiz
School of Mathematics, University of Manchester, Manchester, UK
e-mail: diego.perezruiz@manchester.ac.uk

A. G. Blom
School of Social Sciences, University of Mannheim, Mannheim, Germany
e-mail: blom@uni-mannheim.de

T. Rudas, G. Péli (eds.), *Pathways Between Social Science and Computational Social Science*, Computational Social Sciences,
https://doi.org/10.1007/978-3-030-54936-7_4

pling is that it is based on a formal theory of design-based inference which makes the sampling mechanism ignorable, permitting unbiased estimation of the population mean along with measurable sampling error. However, achieving an ignorable sampling mechanism through probability sampling is proving difficult in today's survey environment with response rates hovering around very low levels. Another limitation of probability sampling is the need to obtain reasonably large samples in order to produce robust population estimates with small standard errors, which can be problematic for survey firms working with a small- to medium-sized budget.

The primary advantage of nonprobability sampling is cost. Nonprobability samples can be drawn and fielded in a number of ways, such as through volunteer web panels, that are relatively inexpensive compared to probability samples. However, nonprobability sampling is also limited in a number of ways. The lack of an underlying theory akin to design-based inference raises concerns over the accuracy of estimates derived from nonprobability samples (Baker et al. 2013). Specifically, several benchmark studies have shown that nonprobability samples tend to be less accurate than probability samples for descriptive population estimates (Blom et al. 2017; Chang and Krosnick 2009; Cornesse et al. 2020; Dutwin and Buskirk 2017; Malhotra and Krosnick 2007; Pennay et al. 2016; Yeager et al. 2011). Studies of multivariate estimates, however, show fewer discrepancies between probability and nonprobability samples (Ansolabehere and Rivers 2013; Pasek 2016). To maximize accuracy of nonprobability samples, a variety of quota sampling, sample matching, and weighting adjustment procedures have been proposed that attempt to conform the composition of the nonprobability sample with that of a reference probability sample or other population benchmark source (Ansolabehere and Rivers 2013; Lee 2006; Lee and Valliant 2009; Rivers 2007; Rivers and Bailey 2009; Valliant and Dever 2011). However, such procedures have their own limitations, as they assume that the compositional adjustment variables – typically limited in scope to demographic characteristics – fully explain the selection mechanism that led to inclusion in the nonprobability sample. A potentially strong assumption if the target variable of interest affected by selection bias is only weakly related to the adjustment variables. Moreover, these adjustment procedures do not solve the issue of quantifying uncertainty in the estimates, which is why measures of variability are rarely presented alongside point estimates derived from nonprobability samples.

Given the pros and cons of probability and nonprobability sampling, it would be advantageous to devise a method that combines both paradigms in a way that exploits each of their strengths to compensate for their weaknesses. From the perspective of unbiased point estimation and variability assessment, probability sampling offers an attractive framework which would be imprudent to forfeit if impediments such as cost and convenience were no issue. However, the reality is that collecting large probability samples is prohibitively expensive for most survey budgets. Thus, if probability sampling and its benefits were to be retained, then sample sizes would have to be reduced considerably. But the robustness of the resulting estimates would suffer and the variances of the estimates may become too large to draw reliable conclusions. The situation resembles small area applications in which sample sizes within target domains are too sparse to derive reliable estimates (Rao 2003). In such applications, auxiliary information collected from other sources (e.g., census data) are commonly incorporated into the estimation process using a model-based or fully Bayesian approach in order to reduce variability and improve the robustness of the survey estimates (Briggs et al. 2007; Marchetti et al. 2016; Porter et al. 2014; Schmertmann et al. 2013).

In a similar vein, nonprobability samples could potentially be used to supplement small probability samples and improve the robustness of the resulting probability-based estimates. In this way, the probability sampling paradigm and its properties could be retained, but at a potentially lower cost, if the combination of probability and nonprobability cases is cheaper than a probability-only sample that achieves equivalent robustness. We investigated this idea in two prior studies using data from multiple probability surveys and eight nonprobability surveys of the general population fielded simultaneously with overlapping questionnaires (Sakshaug et al. 2019; Wiśniowski et al. 2020). Using a standard Bayesian approach and considering different sample sizes for the probability survey (e.g., n = 50, 100, etc.), we showed that supplementing a probability survey with a nonprobability survey can yield estimates of two well-known personality indices with substantially lower mean-squared errors compared to estimates derived from small probability-only samples. Moreover, the estimates derived under the combined approach for the smallest probability samples were just as efficient as estimates derived from much larger probability-only samples. The MSE efficiency gains were primarily driven by reductions in variability which offset increases in bias and resulted in some considerable cost efficiencies (up to 54% based on actual and hypothetical cost data) compared to a probability-only setting.

In the present article, we build on this prior work by investigating the cost and error properties of the proposed combining approach for a different set of commonly used target variables: body height and weight. Using real data from the aforementioned surveys, we examine whether similar efficiency gains and cost savings can be achieved. The article is organized as follows: in Sect. 2 we motivate why the Bayesian approach offers a suitable framework for combining probability and nonprobability samples. In Sect. 3 we describe the methodology and modeling approaches. The data sources and model evaluation procedures are presented in Sect. 4. The model results are presented in Sect. 5. Final conclusions and limitations of the method are discussed in Sect. 6.

2 Why Bayesian?

The Bayesian paradigm is well-suited for applications involving the combination of various data sources. Applications from such fields as health, ecology, and migration have utilized Bayesian methods to incorporate multiple information streams into a single inferential framework (Berbaum et al. 2002; Murphy and Daan 1984; O'Hagan et al. 2006; Qian and Reckhow 1998; Renooij and Witteman 1999; van der Gaag et al. 2002; Vescio and Thompson 2001). Bayesian inference makes use of Bayes theorem (Bayes 1763):

$$p(\theta|obs) \propto p(obs|\theta) \times p(\theta),$$

which states that the posterior distribution of some unknown parameter, θ, conditional on the observed data, $p(\theta|obs)$, is proportional to the probability distribution of the observed data under an assumed model $p(obs|\theta)$, known as the likelihood, and a probability distribution $p(\theta)$ that characterizes the investigator's prior beliefs about the unknown parameter, known as the prior distribution (or the prior, for short). The prior can be specified in various ways according to the investigator's subjective knowledge about the unknown parameter, which may come in the form of focus groups, expert opinion, record systems, or other potentially subjective information that isn't already accounted for in the likelihood.

In population studies it is common for prior distributions to be constructed from less reliable data sources (in comparison to the likelihood). For example, in their estimation of the Kurdish population, Daponte et al. (1997) combine data from various sources with varying levels of quality, including reliable survey data and lower-quality census data and construct priors for several migration parameters based on approximations from the literature. Raymer et al. (2013) combine multiple sources of migration data from sending and receiving countries in Europe and supplement these data with expert-based priors to estimate migration flows. If no subjective knowledge about the unknown parameter is available or assumed, then "weakly informative" priors can be used to produce a purely data-driven inference, one that is completely dominated by the likelihood and akin to frequentist inference (Gelman et al. 2013). However, regardless of how the prior distribution is specified, its influence on the posterior gradually deteriorates as the number of observations represented in the likelihood increases. This is an important feature of Bayesian inference as it is principally structured to give greater influence to the observed data and less influence to the prior distribution as more observations are collected.

In the present application, we treat data coming from a probability survey as the likelihood and use data coming from a single, simultaneously collected nonprobability survey to construct priors for the unknown parameters of interest. In this approach the system of inference incentivizes the collection of probability-based observations over the nonprobability-based ones, which is consistent with the notion that probability-based sampling continues to be the preferred sampling paradigm from a quality perspective. However, in the case of very small probability samples – which are the primary focus of our study – the nonprobability-based priors are expected to dominate the inference. For priors constructed from biased nonprobability data, this has the potential to introduce bias in the posterior inference, as compared to a posterior inference that relies on flat or "weakly informative" priors (i.e., priors that do not make use of the nonprobability data). Hence, for small probability samples, we purposely skew the posterior estimations in the direction of the nonprobability data in order to reduce variability in the posterior estimates. We note this is a form of reversal on the common adjustment strategy that skews nonprobability sample data in the direction of a large reference probability sample or other "gold standard" data source. We do not discount these strategies and actually recommend that they be applied on the nonprobability data before constructing the priors.

A major advantage that the Bayesian framework offers in this context is the borrowing of information from the strongly informative, nonprobability-based priors to compensate for the sparse information available in the small probability sample. This "borrowing of strength" is expected to produce posterior estimations that are more efficient (i.e., have less variability) compared to estimations based on vague or weakly informative priors, as we showed in our previous analysis of two well-known personality measures (Sakshaug et al. 2019). A key question of interest is whether the expected reduction in variance is dramatic enough to tolerate an increase in bias for other variables of interest. In the forthcoming analysis, we evaluate the bias-variance trade-off inherent to the proposed Bayesian method and determine whether the method reduces the overall error in survey estimates of height and weight from a mean-squared error (MSE) perspective. We also examine the potential cost implications of adopting the method. We demonstrate the method and accomplish these aims using a national probability-based web survey and eight nonprobability web surveys conducted simultaneously by different survey vendors in Germany.

3 Methodology and Modeling Approach

To produce the estimates of the quantities of interest, we employ a linear regression model. Let us denote the $n \times 1$ vector of the response variable by $\mathbf{y} = (y_1, \ldots, y_n)^T$ and an $n \times p$ design matrix $X = [\mathbf{x}_1, \ldots, \mathbf{x}_p]$ containing the variables that were used in the sample design and/or help explain the selection mechanism. The sampling model (likelihood) can be written as:

$$\mathbf{y} \sim N(X\boldsymbol{\beta}, \sigma^2 I)$$

where $N()$ is a normal distribution, $\boldsymbol{\beta} = (\beta_1, \ldots, \beta_p)'$ is a column vector of length p, σ^2 is a variance parameter, and I is the $n \times n$ identity matrix.

A conjugate prior distribution for a single regression coefficient, β_j, for $j = 1, \ldots, p$ is

$$\beta_j \sim N(\beta_{j0}, \sigma^2_{\beta_j 0}), \tag{1}$$

with fixed location and variance hyperparameters, β_{j0} and $\sigma^2_{\beta_j 0}$, respectively. We consider three specifications of these parameters as presented in Table 1. The first specification (denoted as **Model 1**) serves as a reference as we employ weakly informative priors that lead to the probability sample being the only information to shape the posterior. The means of the priors are centered at zero with a very large variance ($\sigma^2_{\beta_j 0} = 10^6$).

In **Model 2** we utilize information from a given nonprobability sample, which we assume has been fielded simultaneously with a parallel probability sample.

Table 1 List of proposed models

Model	Parameters	Location	Variance
Model 1	$\beta_j \sim N(\beta_{j0}, \sigma^2_{\beta_j 0})$	$\beta_{j0} = 0$	$\sigma^2_{\beta_j 0} = 10^6$
Model 2	$\beta_j \sim N(\beta_{j0}, \sigma^2_{\beta_j 0})$	$\beta_{j0} = \hat{\beta}_j^{NP}$	$\sigma^2_{\beta_j 0} = d^2(\hat{\beta}_j^P, \hat{\beta}_j^{NP})$
Model 3	$\beta_j \sim N(\beta_{j0}, \sigma^2_{\beta_j 0})$	$\beta_{j0} = \hat{\beta}_j^{NP}$	$\sigma^2_{\beta_j 0} = \left(\hat{\sigma}^{BNP}_{\beta_j 0}\right)^2$

First, we apply linear regression to both probability (P) and nonprobability (NP) samples using the maximum likelihood (ML) method (which yields estimators of the parameters that are equivalent to the posterior means under non-informative priors). The ML estimators are denoted by $\hat{\beta}_j^P$ and $\hat{\beta}_j^{NP}$ for the P and NP data, respectively. Next, we assume that the hyperparameter β_{j0} is equal to $\hat{\beta}_j^{NP}$, whereas the variance hyperparameter $\sigma^2_{\beta_j 0}$ is set to be the Euclidean distance between the ML estimators from the probability and nonprobability surveys:

$$\sigma^2_{\beta_j 0} = d^2(\hat{\beta}_j^P, \hat{\beta}_j^{NP}) = (\hat{\beta}_j^P - \hat{\beta}_j^{NP})^2, \quad \forall j.$$

This method of setting the hyperparameter for the regression coefficient implies that the standard deviation, $\sigma_{\beta_j 0}$, is equal to the difference between the probability- and nonprobability-based estimates. This approach ensures some variability around the mean and keeps the uncertainty at a relatively low level. The obvious limitation of this method is that if the probability and nonprobability estimates are very close to each other, then the variance hyperparameter will be extremely small. This could lead to severe underestimation of the uncertainty in the location parameter β_j and, thus, a false sense of accuracy, especially when the probability sample size is very small.

In the third approach (denoted by **Model 3**), we again set the mean of the hyperparameter $\beta_{j0} = \hat{\beta}_j^{NP}$ as in Model 2. For the variance hyperparameter in (1), we apply a bootstrap procedure[1] based on the NP data. We first draw 1,000 bootstrap samples from the nonprobability survey data, estimate the regression coefficient in each of them, and calculate the variance $\left(\hat{\sigma}^{BNP}_{\beta_j 0}\right)^2$ of all regression coefficients. We then set the variance hyperparameter in (1) to the estimated variance, which yields the following prior distribution for the regression coefficient:

$$\beta_j \sim N\left(\hat{\beta}_{j0}^{NP}, \left(\hat{\sigma}^{BNP}_{\beta_j 0}\right)^2\right).$$

[1]Bootstrap methods have been used in many contexts and were originally proposed by Efron (1979). The general approach is to randomly draw subsamples with replacement from the full sample a large number of times and estimate the statistic of interest in each subsample before combining them using a bootstrap estimator.

The method ensures that the variance is always positive. However, the hyperparameter relies on the bootstrapped nonprobability sample which may propagate its unrepresentativeness, especially for very large sample sizes. This effect, again, can be reduced if the size of the probability sample is increased or by correcting for the bias in the NP sample before the bootstrapping procedure.

We finish the specification of the prior distribution by setting the prior for the variance of the regression model, σ^2. We utilize a Gamma distribution assumed for the *precision* (i.e., inverse variance, σ^{-2}):

$$\sigma^{-2} \sim \Gamma(r, m)$$

where $\Gamma(\cdot, \cdot)$ denotes a Gamma distribution with hyperparameters r being a shape and m being a rate. The precision is used for numerical convenience as the Gamma distribution is a standard probability distribution in most statistical packages, whereas the conjugate prior for the variance is a so-called inverse-Gamma.

In our application we set these hyperparameters to be $r = m = 10^{-3}$. This specification is approximately non-informative and gives preference to the data. It remains the same for Models 1 through 3 which facilitates comparisons of the results between the models.

4 Application to the German Internet Panel

4.1 Probability and Nonprobability Data and Target Variables

To demonstrate the method and evaluate the different models, we make use of the German Internet Panel (GIP). The GIP is a probability-based longitudinal survey designed to represent the population of Germany aged 16–75. The survey is funded by the German Research Foundation (DFG) as part of the Collaborative Research Center 884 "Political Economy of Reforms" based at the University of Mannheim. Sample members are selected through a multi-stage stratified area probability sample. At the first stage, geographic districts (i.e., primary sampling units; PSUs) are selected from a database of 52,947 districts in Germany, each containing roughly the same number of households. Listers are instructed to record every household within the selected PSUs using a predefined random route. A random sample of listed households is then drawn, and all age-eligible members of the sampled households are invited to join the panel (Blom et al. 2015). The GIP is designed to cover both the online and offline population and provides Internet service and/or Internet-enabled devices to facilitate participation among the offline population (Blom et al. 2017). The first GIP recruitment process occurred in 2012, achieving a 18.5% recruitment rate (based on Response Rate 2; AAPOR 2016). A second recruitment effort occurred in 2014 and achieved a recruitment rate of 20.5% (AAPOR Response Rate 2). Panelists receive a survey invitation every 2 months, which consists of an online interview of approximately 20–25 min. The interview covers a range of social, economic, and political topics. The questionnaire module we utilize was administered in March 2015 coinciding with wave 16 of the GIP. During this month, completed interviews were obtained from 3,426 out of 4,989 (or 68.7% of) panelists.

As part of a separate methodological project studying the accuracy of nonprobability surveys (Blom et al. 2017), the GIP team sponsored several nonprobability web survey data collections that were conducted in parallel with the March 2015 GIP. The nonprobability web surveys were carried out by different commercial vendors in response to a call for tender. The key stipulation of the call was that the vendor should recruit a sample of 1,000 respondents that is representative of the general population aged 18–70 years living in Germany. The call for tender did not specify how representativeness should be achieved. A total of 17 bids were received of which 7 were selected based on technical and budgetary criteria. Upon learning about the specific goals of the study, an eighth commercial vendor voluntarily agreed to take part in the study without remuneration. Further details of the eight nonprobability surveys can be found in Table 2, including costs and quota sampling variables. For confidentiality reasons, we refer to the individual nonprobability surveys as Survey 1, Survey 2, and so on, ordered from least expensive (pro bono) to most expensive (10,676.44 EUR). Cost information for the GIP survey is unavailable.

Our interest lies in two continuous outcome variables: body height (in centimeters) and weight (in kilograms). These variables were collected in the GIP and all nonprobability surveys. Due to the bimodal nature of these variables, we perform the analysis separately for males and females. Histograms of the height and weight variables produced from the GIP data are provided in Appendix A. Unconditional (design-based) sample means of height derived from each probability and nonprobability survey are provided in Appendix B. As shown in Sect. 3, covariates for modeling the height and weight outcomes are incorporated into the three modeling approaches. In this application we use seven standard covariates:

Table 2 List of probability and nonprobability surveys

Survey	No. respondents	Quota variables	Total cost (in Euros)	Average cost per respondent (in Euros)
GIP	3,426	N/A	Unavailable	Unavailable
1	1,012	Age, gender, region, education	0 (pro bono)	N/A
2	1,000	Age, gender, region	5,392.97	5.40
3	999	Age, gender, region	5,618.57	5.63
4	1,000	Age, region	7,061.11	7.07
5	994	Age, gender, region	7,411.00	7.46
6	1,002	Age, gender, region, education	7,636.22	7.62
7	1,000	Age, gender, region	8,380.46	8.39
8	1,038	Age, gender, region	10,676.44	10.29

regularly employed (binary), education (four categories), marital status (binary), region (four categories), self-reported health status (binary), age (continuous), and a weighting variable produced by the GIP team which includes a raking adjustment to population benchmarks. These variables are available and are used in each probability and nonprobability survey. Ordinary least squares regression coefficients for each survey are plotted for the outcome variables height and weight in Appendix C.

4.2 Model Evaluation

To evaluate the performance of the three proposed models (Sect. 3), we divide the probability survey data (GIP) into two data sets: a training set (denoted by $\mathbf{y}$) and a test set (denoted by $\tilde{\mathbf{y}}$). We then utilize the training set to fit the three models using Bayesian inference. Next, based on the fitted models, we predict the outcome variables in the test set $\tilde{\mathbf{y}}$.

The result of the procedure is the so-called posterior predictive distributions for each data point. They depict the probability of obtaining a given observation under the assumed model. The mean of the posterior predictive distribution can be treated as a point prediction of the given observation. We denote these means by $\bar{\tilde{\mathbf{y}}}$. The next step is to evaluate the error properties of the predictions with the three models for the test data set. Here, we consider three measures: **bias**, **variance**, and the **mean-squared error (MSE)** of posterior predictive means for $\tilde{\mathbf{y}}$.

The MSE is defined as:

$$\mathrm{MSE}(\bar{\tilde{\mathbf{y}}}) = \mathbb{E}\left[(\bar{\tilde{\mathbf{y}}} - \bar{\mathbf{y}}^P_{\text{test}})^2\right],$$

where $\bar{\mathbf{y}}^P_{\text{test}}$ are the model-adjusted predictions in the test set of the probability survey. The MSE can be decomposed into variance and bias:

$$\mathrm{MSE}(\bar{\tilde{\mathbf{y}}}) = \mathrm{Bias}^2(\bar{\tilde{\mathbf{y}}}) + \mathrm{Var}(\bar{\tilde{\mathbf{y}}}).$$

We compute the bias as the difference between the mean of the posterior means, $\bar{\tilde{\mathbf{y}}}$, and the mean of the model-adjusted predictions $\bar{\mathbf{y}}^P_{\text{test}}$, i.e.,

$$\mathrm{Bias}(\bar{\tilde{\mathbf{y}}}) = \frac{1}{n}\sum \bar{\tilde{\mathbf{y}}} - \frac{1}{n}\sum \bar{\mathbf{y}}^P_{\text{test}}$$

whereas $\mathrm{Var}(\bar{\tilde{\mathbf{y}}})$ is the unbiased estimator of the variance of $\bar{\tilde{\mathbf{y}}}$.

Since we employ a model which is built in an ad-hoc manner rather than justified by any theoretical framework, it may be misspecified. For example, the regression model may suffer from omitted variables and confounders; thus, comparing the predictions with raw data may lead to bias which is due exclusively to the model misspecification. To avoid this problem, we apply the same model to the test data, predict the outcome variable, and use these predictions as model-adjusted predictions $\bar{\mathbf{y}}^{P}_{\text{test}}$.

We calculate the bias, variance, and MSE of the posterior predictive means for the three models described in Sect. 3 under different probability sample size scenarios. To accomplish this, we estimate coefficients of the three models with training sets ranging in size from 50 to 600 cases with intervals of 50 and from 600 to 1,000 with intervals of 100. The samples are constructed cumulatively so that the same cases used in the smaller samples are included in the larger samples. The samples are also sorted by the GIP timestamp variable to account for early and late respondents.

After excluding cases with missing data and assigning 1,000 cases from the probability survey to the training set, the remaining cases are assigned to the test set. For the height and weight variables, the test set includes 304 cases each. The entire procedure of splitting the probability data randomly into training and test sets was conducted 100 times to produce 100 estimates of variance, bias, and MSE for each probability sample size. The forthcoming results report the averages of these 100 repetitions.

The posterior characteristics were computed using five MCMC chains with samples of 10,000 and a 1,000 iteration burn-in sample. This ensured convergence of all chains used for creating the posterior distributions. The analysis was implemented in JAGS (Just Another Gibbs Sampler) and R (R Core Team 2016) using the library *rjags*.

4.3 Assessing Model Efficiency

To assess the efficiency of the two models informed by the nonprobability data (Models 2 and 3) relative to the reference model (Model 1), we calculate the ratio of the MSEs and variances of the posterior predictive means.

This measure indicates that if the ratio takes a value less than 1, then the informative model is more efficient than the reference model. Conversely, if the ratio is equal to or greater than 1, then the informative models do not yield gains in efficiency over the reference model. The ratio of the variances of the posterior predictive means is given by:

$$\epsilon(\text{Var}(\bar{\bar{\mathbf{y}}}_{\text{Model1}}), \text{Var}(\bar{\bar{\mathbf{y}}}_{\text{Model2}})) = \frac{\text{Var}(\bar{\bar{\mathbf{y}}}_{\text{Model2}})}{\text{Var}(\bar{\bar{\mathbf{y}}}_{\text{Model1}})},$$

$$\epsilon(\text{Var}(\bar{\bar{\mathbf{y}}}_{\text{Model1}}), \text{Var}(\bar{\bar{\mathbf{y}}}_{\text{Model3}})) = \frac{\text{Var}(\bar{\bar{\mathbf{y}}}_{\text{Model3}})}{\text{Var}(\bar{\bar{\mathbf{y}}}_{\text{Model1}})}$$

And the ratio of the MSEs of the posterior predictive means is given by:

$$\epsilon(\text{MSE}(\bar{\bar{y}}_{\text{Model1}}), \text{MSE}(\bar{\bar{y}}_{\text{Model2}})) = \frac{\text{MSE}(\bar{\bar{y}}_{\text{Model2}})}{\text{MSE}(\bar{\bar{y}}_{\text{Model1}})},$$

$$\epsilon(\text{MSE}(\bar{\bar{y}}_{\text{Model1}}), \text{MSE}(\bar{\bar{y}}_{\text{Model3}})) = \frac{\text{MSE}(\bar{\bar{y}}_{\text{Model3}})}{\text{MSE}(\bar{\bar{y}}_{\text{Model1}})}$$

5 Results

We now report the results of the three modeling approaches applied to the GIP survey. As a reminder, Model 1 is the reference model and does not make use of any nonprobability data in the prior specification. In contrast, Models 2 and 3 both use nonprobability data to construct informative priors for the unknown model parameters. The outputs of these models are posterior predictive distributions of height and weight, which we summarize in the form of means. For brevity, we refer to the posterior predictive means simply as means when discussing the results. We also show the results for only one of the nonprobability surveys used in the prior distribution, the middle-priced survey (NP = 5). Similar results were found when the other nonprobability surveys were used (results available upon request). The variance, bias, and MSE (as described above) of the mean estimates of height and weight are presented separately for males and females below.

5.1 *Variability in Mean Estimates of Height and Weight*

Figure 1 (top row) shows the variance for mean estimates of height for males and females when using nonprobability survey 5 in the prior distribution. The figure shows that combining probability and nonprobability samples under Models 2 and 3 yields smaller variances than those produced under the probability-only sample (Model 1), particularly for the smallest probability sample sizes (50 or 100 cases), where the largest efficiency gains occur. For example, each plot shows that the variance produced under Model 1 is at least three times larger than the corresponding variance produced under Models 2 and 3 for a sample size of 50 cases and at least two times larger for a sample size of 100. Efficiency gains under Models 2 and 3 continue at a decreasing rate until a probability sample size of about 500 cases is reached; at this point the variances under all three models converge. What is also worth noting from this figure is that the variances produced under Models 2 and 3 for the smallest probability sample sizes are roughly equivalent to the variances produced under Model 1 for the largest probability sample sizes (about 500 cases or more). The above pattern tends to occur regardless of which nonprobability sample

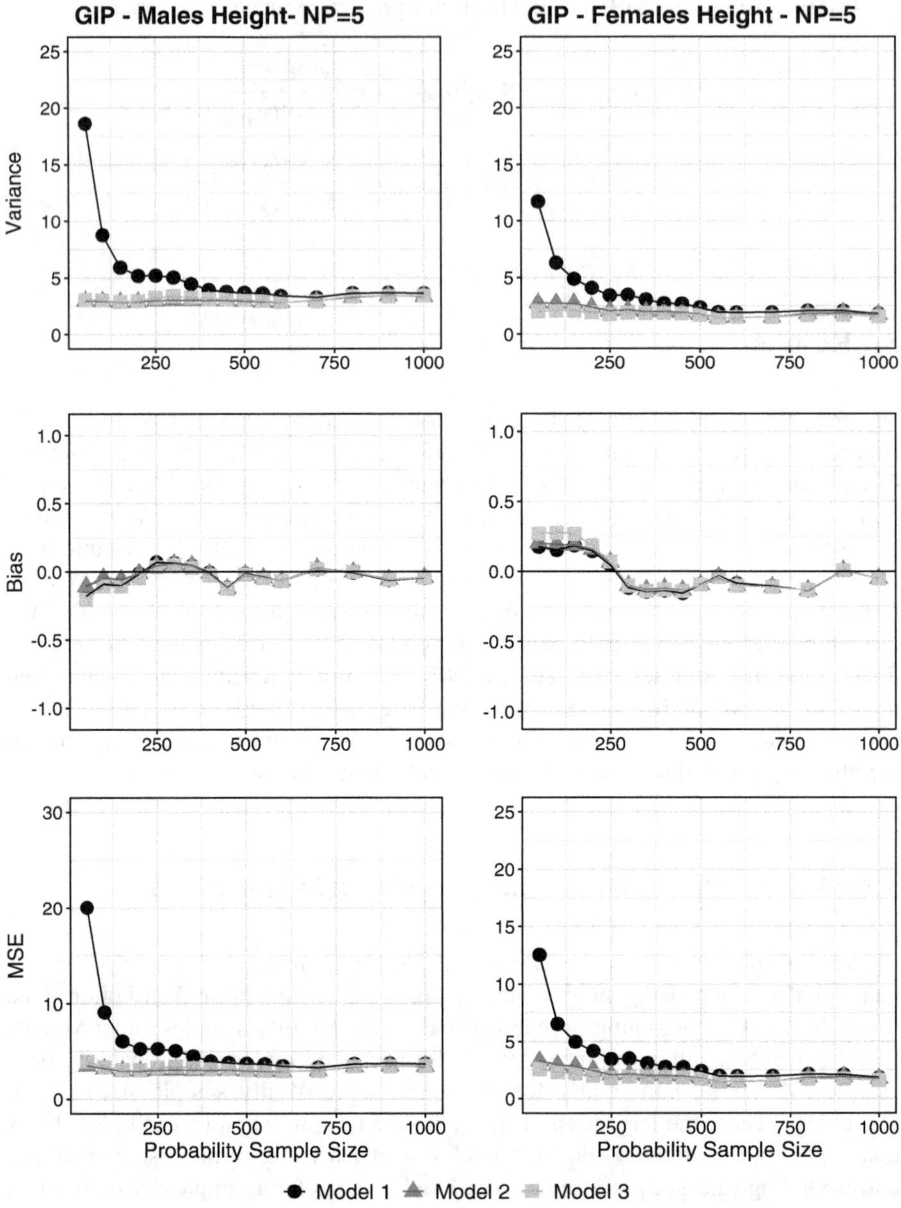

Fig. 1 Variance, bias, and mean-squared error (MSE) of mean estimates of body height among males (left panel) and females (right panel) in the German Internet Panel. Probability sample sizes are denoted on the horizontal axis

and modeling approach is used (results not shown). In fact, Models 2 and 3 are virtually indistinguishable in the plots, suggesting that the observed efficiency gains are robust to the specification of the variance hyperparameter in Eq. 1.

The variances for mean estimates of weight for males and females (depicted in Fig. 2, top row) show similar results, namely, that Models 2 and 3 yield substantially smaller variance estimates compared to Model 1 for the smallest probability sample sizes. Further, and as observed in Fig. 1, the variances under Models 2 and 3 for the smallest sample sizes are similar to the variances under Model 1 for the largest probability sample sizes.

5.2 Bias in Mean Estimates of Height and Weight

In this section we examine the impact of bias when probability and nonprobability samples are combined under Models 2 and 3. Here, we tacitly assume that the mean estimates of height and weight from the probability-only test set, which serves as the benchmark data source, are unbiased. Figure 1 (middle row) shows the bias in mean estimates of height for males and females. The figures show relatively small bias across the three models with a maximum bias of about 0.20 units (centimeters) for males and 0.30 units for females. Differences in bias between the three models are most apparent for the small probability sample sizes where the influence of the nonprobability-based priors in Models 2 and 3 is at their peak. However, there is no consistent pattern of bias across all of the nonprobability surveys used in Models 2 and 3 (results not shown). All of the biases are rather small and are observed only for the smaller probability sample sizes; beyond a probability sample size of around 500 (and, in some cases, less), biases in all three models reach equivalency as the impact of the likelihood dominates the prior.

Similar bias results are observed for the weight variable (illustrated in Fig. 2, middle row). That is, bias in each of the three models is rather small (roughly less than 1.5 and 1.0 kg for males and females, respectively), with only subtle differences between the three models for the smaller probability sample sizes. In summary, it is apparent for both the height and weight variables that combining small probability samples with larger nonprobability samples (under Models 2 and 3) increases bias only slightly (if at all) relative to the probability-only samples (under Model 1).

5.3 Mean-Squared Error (MSE) in Mean Estimates of Height and Weight

Next, we examine the joint impact of bias and variance under each modeling approach in the form of mean-squared errors (MSEs). MSEs for mean estimates of height among males and females are shown in Fig. 1 (bottom row). MSEs for

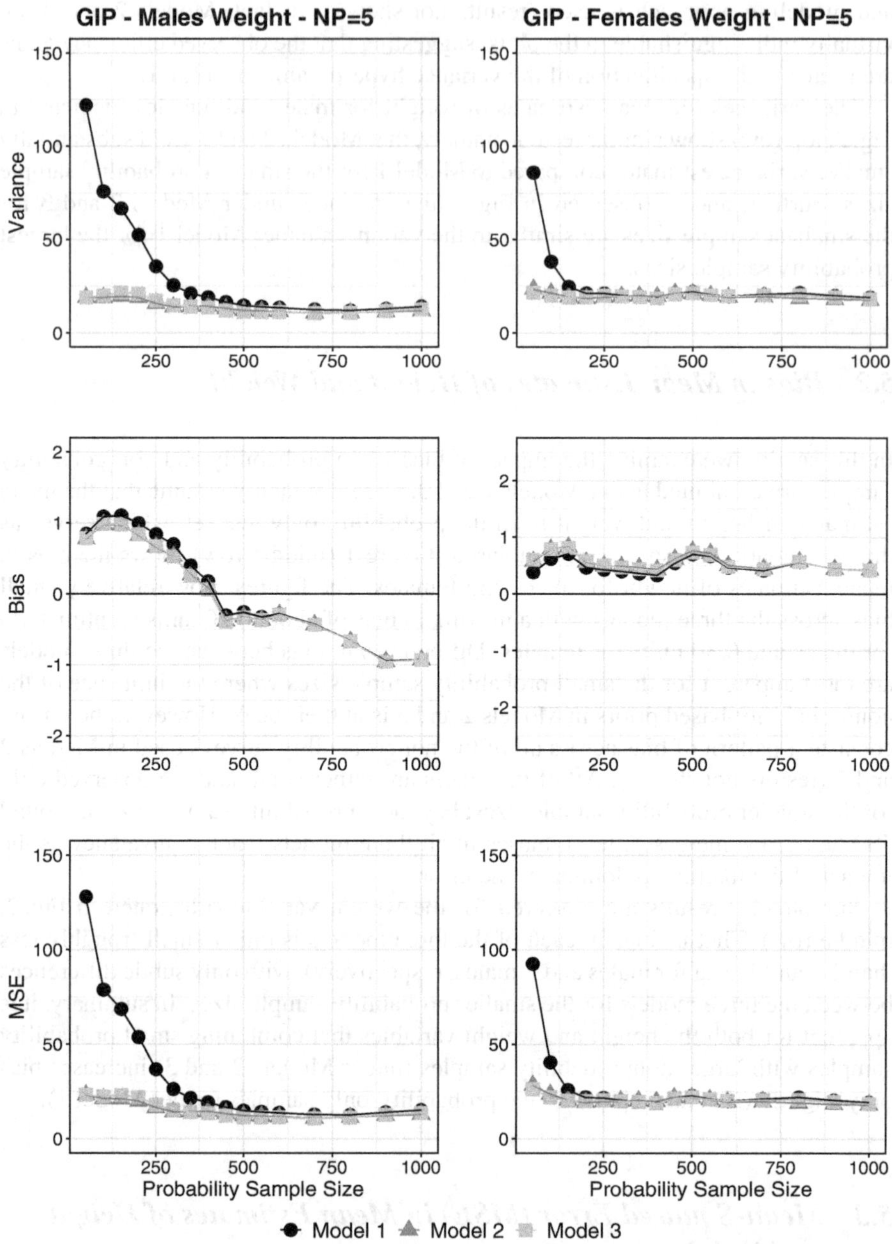

Fig. 2 Variance, bias, and mean-squared error (MSE) of mean estimates of body weight among males (left panel) and females (right panel) in the German Internet Panel. Probability sample sizes are denoted on the horizontal axis

the weight variable among males and females are presented in Fig. 2 (bottom row). Collectively, the MSE results share a similar pattern with the variance results presented in Sect. 5.1. That is, the MSEs are considerably smaller when small probability samples are combined with the nonprobability samples (Models 2 and 3) as compared to the situation where only probability samples are used (Model 1). The MSE gap closes at an increasing rate when a modest probability sample size of generally 500 or so cases is achieved. Similar to the variance results, Models 2 and 3 produce MSEs for the smaller probability sample sizes (250 cases or less) which are similar to the MSEs produced under Model 1 for the largest sample sizes.

5.4 *Efficiency Ratios for MSE and Variance*

To summarize the MSE and variance efficiency gains presented in the previous sections, we now report efficiency ratios by comparing the MSE and variances achieved under the combined modeling approaches (Models 2 and 3) with those of the probability-only reference model (Model 1). Figure 3 shows efficiency ratios for the height variable for females (top row) and males (bottom row). Instead of reporting individual efficiency ratios for the eight nonprobability surveys used, we

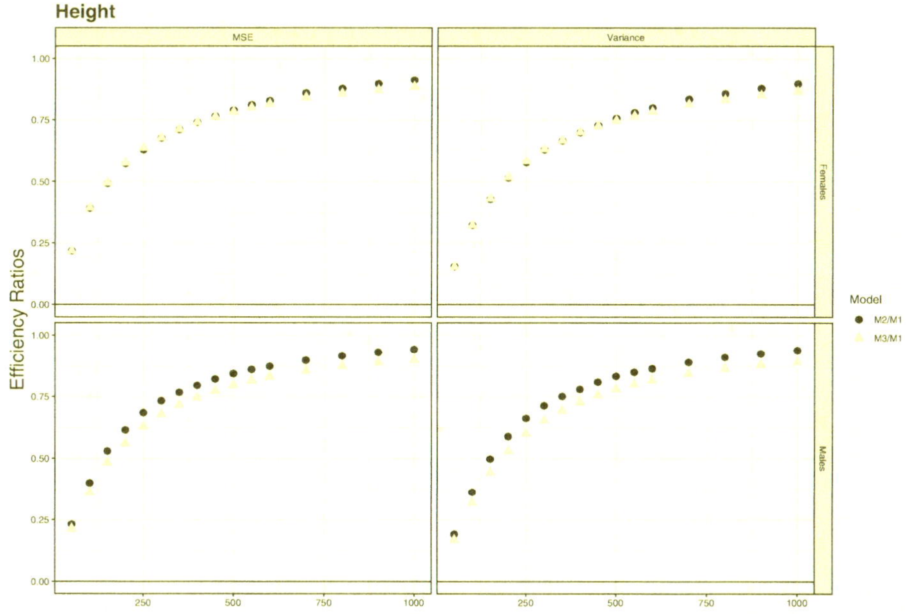

Fig. 3 Efficiency ratios of mean-squared error (MSE) and variance for mean estimates of body height among males and females. Ratios are averaged across all nonprobability surveys. Probability sample sizes are denoted on the horizontal axis

simply report the average efficiency ratios across all nonprobability surveys. These results show that, on average, Models 2 and 3 reduce variances by at least 80% and reduce MSEs by at least 77% compared to Model 1 for the smallest probability sample size of 50 cases. For a sample size of 100 probability cases, the combined modeling approaches reduce variances by at least 60% and MSEs by at least 58%, on average, compared to the probability-only model. Even for modest sample sizes of about 250 probability cases, the variance and MSE reductions for the height variable are considerable – around 40% or higher, respectively.

Concerning the weight variable, Models 2 and 3 for males and females (shown in Fig. 4) also show evidence of large variance and MSE efficiencies, although these efficiencies are somewhat less considerable compared to the height variable. For example, Models 2 and 3 yield variance efficiencies of at least 72% and MSE efficiencies of about 70% compared to Model 1 for the smallest probability sample size of 50 cases. For a sample size of 100 cases, variance efficiencies for Models 2 and 3 range between 49 and 58%, and for MSE efficiencies, the range is between 46 and 56%, depending on the sex. The slightly larger efficiencies for variances compared to MSEs suggests that bias has only a very minor effect on the error of the estimates.

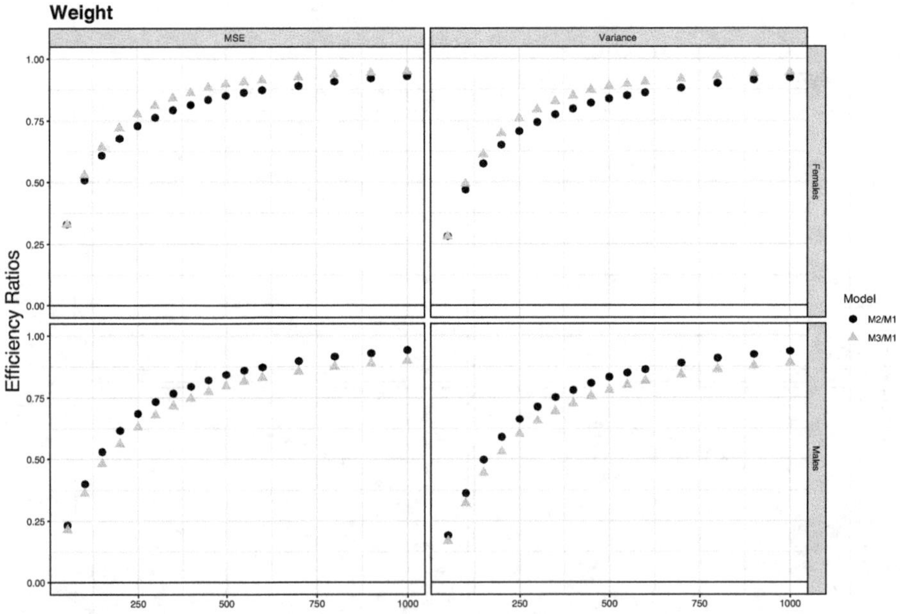

Fig. 4 Efficiency ratios of mean-squared error (MSE) and variance for mean estimates of body weight among males and females. Ratios are averaged across all nonprobability surveys. Probability sample sizes are denoted on the horizontal axis

5.5 Potential Cost Savings for a Fixed MSE

The final analysis examines the cost implications of combining a small probability sample with a larger nonprobability sample in a way that achieves a similar MSE as would be achieved under a probability-only sample. To do this, we fit a crude cost model of assumed per unit costs for the GIP sample sizes on the actual MSEs achieved for these sample sizes under Model 1; for reference, a data table used in the model fitting process is provided in Appendix D. We do not have precise cost values for the GIP survey, so we can only assume a reasonable per unit cost. In this case, we assume 22 EUR per GIP unit, which is roughly between 2 and 4 times the cost of a completed case in the 7 remunerated nonprobability surveys (see Table 2).[2] The fitted cost models are shown here: height (males), $Log(Costs_{M1}) = 11.85 - MSE_{M1} \cdot 0.74 + MSE^2_{M1} \cdot 0.03$ ($R^2 = 0.90$); height (females), $Log(Costs_{M1}) = 10.99 - MSE_{M1} \cdot 0.78 + MSE^2_{M1} \cdot 0.04$ ($R^2 = 0.96$); weight (males), $Log(Costs_{M1}) = 10.06 - MSE_{M1} \cdot 0.04 + MSE^2_{M1} \cdot 0.00$ ($R^2 = 0.92$); and weight (females), $Log(Costs_{M1}) = 11.90 - MSE_{M1} \cdot 0.15 + MSE^2_{M1} \cdot 0.00$ ($R^2 = 0.70$). Using these fitted cost models, we then plug in the actual MSE values obtained from Models 2 and 3 to estimate the Model 1 costs of achieving the same MSE under a probability-only sample. Next, we compare the estimated probability-only (i.e., Model 1) costs to the costs of combining probability and nonprobability samples (i.e., Models 2 and 3). The costs of combining both samples are derived using the actual per unit costs for the nonprobability cases (shown in Table 2) and the assumed per unit cost (22 EUR) of the probability cases.

Detailed cost estimates and estimated cost differences for the three modeling approaches are shown for the five smallest probability sample size groups (i.e., n = 50, 100, 150, 200, and 250) in Appendix E. Here we summarize these results in terms of percent cost savings for the combined probability/nonprobability modeling approaches (Models 2 and 3) relative to the probability-only scenario (Model 1). Percent cost savings for the height variable (by sex) are presented in Table 3. The cost results are shown for Models 2 and 3 and for each nonprobability survey used in the modeling process. Positive values denote estimated cost savings, while negative values denote no cost savings. Cost savings for the height variable vary considerably across the seven nonprobability surveys. For example, among the 50 males analyzed for their height under Model 2, the results range from no cost savings (−21.98%; NP4) to a cost savings of 43.99% (NP3) with an average cost savings of 7.70%. The average cost savings among males under Model 2 increases considerably for the larger sample sizes (n = 100, 150, 200, and 250), ranging from 23.95 to 43.12%; similar cost savings can be seen for males under Model 3. Smaller (albeit still

[2]We assume the GIP per unit cost is higher than the per unit costs of the nonprobability surveys due to the interviewer-administered recruitment and setup costs of equipping the offline population. Further, we reason that, in practice, a high response rate would be desired for the small probability sample to minimize the risk of nonresponse bias in the sparse sample, for which extensive recruitment efforts may be needed.

Table 3 Percent cost savings for the height variable, by model and sex

	Model 2 – Height – Males								
GIP sample size	NP2	NP3	NP4	NP5	NP6	NP7	NP8	Avg.	Std. dev.
50	−10.41	43.99	−21.98	41.79	4.06	4.80	−8.38	7.70	25.72
100	29.03	60.61	32.03	44.42	39.44	32.39	−1.78	33.73	18.96
150	51.00	66.40	49.05	45.82	42.96	42.97	3.66	43.12	19.17
200	52.84	61.92	38.59	40.75	38.9	41.21	8.97	40.45	16.39
250	33.28	51.14	15.58	31.34	17.20	19.04	0.06	23.95	16.28
	Model 3 – Height – Males								
GIP sample size	NP2	NP3	NP4	NP5	NP6	NP7	NP8	Avg.	Std. dev.
50	27.01	41.19	−25.66	24.97	−2.23	20.49	22.85	15.52	22.25
100	35.45	56.82	18.58	38.83	38.50	31.70	30.52	35.77	11.54
150	45.89	61.25	28.98	43.64	44.52	39.33	32.45	42.29	10.50
200	46.75	57.12	25.21	37.71	43.40	38.38	30.48	39.86	10.55
250	34.87	46.77	15.49	20.19	28.62	33.51	19.18	28.37	10.99
	Model 2 – Height – Females								
GIP sample size	NP2	NP3	NP4	NP5	NP6	NP7	NP8	Avg.	Std. dev.
50	19.53	9.81	5.55	−23.64	−23.35	41.68	−4.12	3.64	23.36
100	17.47	26.67	24.99	−16.22	1.23	38.14	14.43	15.24	17.99
150	13.52	25.54	21.23	−19.67	7.24	28.65	14.18	12.96	16.18
200	15.11	26.20	34.29	−7.66	21.82	29.86	21.97	20.23	13.76
250	16.62	28.78	42.01	6.83	29.69	30.88	24.86	25.67	11.24
	Model 3 – Height – Females								
GIP sample size	NP2	NP3	NP4	NP5	NP6	NP7	NP8	Avg.	Std. dev.
50	33.34	1.08	6.56	17.10	−14.36	33.94	−0.25	11.06	18.03
100	37.79	15.89	24.76	18.92	8.40	36.56	12.46	22.11	11.49
150	35.73	15.56	24.43	16.94	13.99	31.67	9.96	21.18	9.65
200	31.16	17.96	33.73	18.58	25.72	29.59	20.09	25.26	6.46
250	30.41	20.96	40.39	22.49	33.04	30.60	23.95	28.83	6.87

modest) cost savings can be seen for female height: across the sample size spectrum, the average cost savings range between 3.64 and 25.67% and between 11.06 and 28.83% under Models 2 and 3, respectively. However, similar to male height, there is considerable variation in cost savings for female height across the nonprobability surveys.

Table 4 Percent cost savings for the weight variable, by model and sex

	Model 2 – Weight – Males								
GIP sample size	NP2	NP3	NP4	NP5	NP6	NP7	NP8	Avg.	Std. dev.
50	−10.16	−57.95	−40.42	13.14	−101.66	−50.34	−68.30	−45.10	37.78
100	−22.34	−60.69	−50.78	8.44	−92.26	−56.45	−45.79	−45.70	31.65
150	−37.44	−73.91	−59.68	−1.29	−100.60	−76.46	−46.62	−56.57	32.06
200	−40.29	−70.75	−53.73	−7.42	−83.88	−74.75	−47.39	−54.03	25.84
250	−23.71	−47.34	−32.45	−3.76	−54.39	−52.91	−35.32	−35.70	18.05
	Model 3 – Weight – Males								
GIP sample size	NP2	NP3	NP4	NP5	NP6	NP7	NP8	Avg.	Std. dev.
50	4.86	−42.30	−33.39	13.26	−71.36	−27.13	−128.25	−40.62	48.02
100	−7.79	−53.53	−41.39	5.08	−74.25	−33.63	−102.64	−44.02	37.15
150	−19.84	−69.43	−51.87	−8.65	−82.92	−52.27	−96.34	−54.47	31.88
200	−27.57	−70.58	−55.75	−15.15	−80.85	−56.03	−89.43	−56.48	27.15
250	−18.59	−52.16	−45.04	−9.70	−65.34	−44.61	−66.01	−43.06	21.69
	Model 2 – Weight – Females								
GIP sample size	NP2	NP3	NP4	NP5	NP6	NP7	NP8	Avg.	Std. dev.
50	−41.81	23.09	−54.31	−73.10	1.57	−198.33	−113.60	−65.21	74.22
100	11.93	50.83	−3.02	−25.21	23.86	−125.90	−28.11	−13.66	56.68
150	33.12	56.16	13.56	−0.79	33.84	−77.25	−4.68	7.71	43.11
200	42.31	46.14	16.67	2.32	26.62	−41.29	−5.75	12.43	30.46
250	35.95	32.08	14.14	−10.69	16.61	−31.88	−13.69	6.07	25.36
	Model 3 – Weight – Females								
GIP sample size	NP2	NP3	NP4	NP5	NP6	NP7	NP8	Avg.	Std. dev.
50	−2.25	29.90	−7.98	−42.71	−29.60	−126.47	−97.39	−39.50	55.11
100	24.01	53.99	24.46	−7.40	−10.48	−73.54	−37.10	−3.72	42.70
150	34.33	59.37	33.82	2.47	2.42	−30.66	−18.67	11.87	32.05
200	28.64	50.08	23.36	−4.92	−1.83	−29.54	−33.36	4.63	30.92
250	19.43	39.68	15.44	−21.39	−15.37	−35.62	−52.42	−7.18	33.02

Percent cost savings for the weight variable (by sex) are presented in Table 4. For males, the overwhelming majority of cost estimates are negative, suggesting no cost benefit of combining the probability and nonprobability samples. In other words, for the male weight variable, a probability-only sample under Model 1 can be expected to achieve the same MSE as would be achieved under Models 2 and 3 at a cheaper cost. However, for the female weight variable, larger expected cost savings can be seen, particularly for nonprobability surveys 2, 3, and 4. When these surveys are incorporated into the modeling process, expected cost savings occur for nearly all of the sample size groups, ranging from a modest 15.44% to a considerable 59.37% under Model 3 and 11.93 to 56.16% under Model 2.

6 Discussion

In this study we evaluated an alternative method of utilizing nonprobability samples for the analysis of survey data. Instead of the traditional approach of adjusting a nonprobability sample towards a supplementary probability sample (or other high-quality data source), we considered the reverse approach of using nonprobability data to supplement probability-based estimations when only small probability samples are available. We demonstrate that constructing prior distributions based on nonprobability data to inform small-sample probability-based estimates of body height and weight reduces the variability in the estimates considerably compared to the corresponding estimates derived from the small probability samples alone. Moreover, the reduced variances are comparable to variances obtained from much larger probability-only samples. Using nonprobability data to inform traditional probability-based estimations comes with a risk of introducing bias in the estimates, which is why nonprobability samples are typically adjusted towards a reference probability sample, and not the other way around. However, in our case study, this concern was not well-founded as the increased bias, if any, was minor in comparison with the reduction in variability. Taken together, the joint impact of these errors yields a smaller MSE under the combined sample approach relative to the probability-only inference. In many cases, the expected cost savings from these MSE efficiencies was substantial, reaching as high as 66% for a fixed MSE. Cost savings were particularly evident for the body height variable but less so for the weight variable which, in some cases, did not yield any cost savings.

At a time when cost-cutting measures in the survey research industry are often viewed as having harmful effects on data quality, it is useful to know that methods can be implemented to reduce both survey costs and errors simultaneously. In light of our findings, the proposed method of supplementing small probability samples with cheaper nonprobability samples to reduce survey errors and costs shows promise and may be considered a relevant addition to the survey practitioner's toolbox. The method is appealing for several reasons, a key one being that it is based on a system of estimation that gives increasing weight to probability-based observations while simultaneously decreasing the weight of the nonprobability-based observations, by using Bayesian inference. We believe this should be the preferred way of handling nonprobability samples, at least at the present time where probability surveys still appear to hold a quality edge over their nonprobability counterparts (Blom et al. 2017; Chang and Krosnick 2009; Cornesse et al. 2020; Dutwin and Buskirk 2017; Malhotra and Krosnick 2007; Pennay et al. 2016; Yeager et al. 2011). A further advantage of the Bayesian framework is the capability to estimate measures of uncertainty which is a topic largely ignored in the analysis of nonprobability samples. From a computational viewpoint, the methodology is easily implemented using freely available software, and no noticeable speed differences were encountered for the alternative constructions of the nonprobability-based priors.

The method, nevertheless, does have limitations. For one, it is restricted to continuous outcome variables. Extending the method to handle other variable types (e.g., ordinal, nominal, count) would be a worthy next step in this line of research. A second area ripe for further work is the specification of the prior distribution. The two (informative) prior specifications we considered both have limitations which could, under certain circumstances, make the prior too strong. We purposely kept the prior construction as simple as possible, but more sophisticated approaches could be explored here. Lastly, while demonstrating the method over several nonprobability surveys collected by different commercial vendors is a strength of this study, it did highlight some variability in the results that warrants further investigation. Future work ought to investigate the properties of nonprobability surveys that make them potentially good (or bad) candidates for prior distributions.

Despite these limitations, the proposed method is appealing from both a cost and error perspective. Any method that succeeds in reducing both is a welcome contribution to survey practice where data collection costs are ever increasing and concerns over data quality are widespread. Regarding increasing costs, it is understandable that many researchers are embracing nonprobability surveys as a way to cut costs, but abandoning probability sampling altogether is rather extreme in light of the empirical comparative evidence. Combining both sampling schemes under a framework that exploits their advantages seems a sensible way forward.

References

AAPOR, *Standard Definitions: Final Dispositions of Case Codes and Outcome Rates for Surveys*, 9th edn. (American Association for Public Opinion Research, 2016)

S. Ansolabehere, D. Rivers, Cooperative survey research. Ann. Rev. Polit. Sci. **16**, 307–329 (2013)

R. Baker, J.M. Brick, N.A. Bates, M. Battaglia, M.P. Couper, J.A. Dever, K.J. Gile, R. Tourangeau, Summary report of the AAPOR task force on non-probability sampling. J. Surv. Stat. Methodol. **1**(2), 90–143 (2013)

T. Bayes, An essay towards solving a problem in the doctrine of chances. Philos. Trans. **53**, 370–418 (1763)

K.S. Berbaum, D.D. Dorfman, E.A. Franken, R.T. Caldwell, An empirical comparison of discrete ratings and subjective probability ratings. Acad. Radiol. **9**(7), 756–763 (2002)

A.G. Blom, C. Gathmann, U. Krieger, Setting up an online panel representative of the general population: the German internet panel. Field Methods **27**(4), 391–408 (2015)

A.G. Blom, J.M.E. Herzing, C. Cornesse, J.W. Sakshaug, U. Krieger, D. Bossert, Does the recruitment of offline households increase the sample representativeness of probability-based online panels? evidence from the German internet panel. Soc. Sci. Comput. Rev. **35**(4), 498–520 (2017)

A.G. Blom, D. Ackermann-Piek, S.C. Helmschrott, C. Cornesse, J.W. Sakshaug, The representativeness of online panels: coverage, sampling and weighting, in *Paper Presented at the General Online Research Conference* (2017)

D. Briggs, D. Fecht, K. De Hoogh, Census data issues for epidemiology and health risk assessment: experiences from the small area health statistics unit. J. R. Stat. Soc. Ser. A (Stat. Soc.) **170**(2), 355–378 (2007)

L. Chang, J.A. Krosnick, National surveys via RDD telephone interviewing versus the internet comparing sample representativeness and response quality. Public Opin. Q. **73**(4), 641–678 (2009)

C. Cornesse, A.G Blom, D. Dutwin, J.A. Krosnick, E.D. De Leeuw, S. Legleye, J. Pasek, D. Pennay, B. Phillips, J. W. Sakshaug, B. Struminskaya, A. Wenz, A Review of Conceptual Approaches and Empirical Evidence on Probability and Nonprobability Sample Survey Research. J. Surv. Stat. Methodol. **8**(1), 4–36 (2020)

B.O. Daponte, J.B. Kadane, L.J. Wolfson, Bayesian demography: projecting the Iraqi Kurdish population, 1977–1990. J. Am. Stat. Assoc. **92**(440), 1256–1267 (1997)

D. Dutwin, T.D. Buskirk, Apples to oranges or gala versus golden delicious? comparing data quality of nonprobability internet samples to low response rate probability samples. Public Opin. Q. **81**(S1), 213–239 (2017)

B. Efron, Bootstrap methods: another look at the jackknife. Ann. Stat. **7**, 1–26 (1979)

A. Gelman, J.B. Carlin, H.S. Stern, D.B. Rubin, *Bayesian Data Analysis*, Vol. 3 (Chapman & Hall/CRC, Boca Raton, 2013)

S. Lee, Propensity score adjustment as a weighting scheme for volunteer panel web surveys. J. Off. Stat. **22**(2), 329 (2006)

S. Lee, R. Valliant, Estimation for volunteer panel web surveys using propensity score adjustment and calibration adjustment. Sociol. Methods Res. **37**(3), 319–343 (2009)

N. Malhotra, J.A. Krosnick, The effect of survey mode and sampling on inferences about political attitudes and behavior: comparing the 2000 and 2004 ANES to internet surveys with nonprobability samples. Polit. Anal. **15**, 286–323 (2007)

S. Marchetti, C. Giusti, M. Pratesi, The use of twitter data to improve small area estimates of households? share of food consumption expenditure in italy. AStA Wirtschafts-und Sozial-statistisches Archiv **10**(2–3), 79–93 (2016)

A.H. Murphy, H. Daan, Impacts of feedback and experience on the quality of subjective probability forecasts. Comparison of results from the first and second years of the Zierikzee experiment. Mon. Weather Rev. **112**(3), 413–423 (1984)

A. O'Hagan, C.E. Buck, A. Daneshkhah, J.R. Eiser, P.H. Garthwaite, D.J. Jenkinson, J.E. Oakley, T. Rakow, *Uncertain Judgments Eliciting Expert's Probabilities* (Wiley, Chichester, 2006)

J. Pasek, When will nonprobability surveys mirror probability surveys? considering types of inference and weighting strategies as criteria for correspondence. Int. J. Public Opin. Res. **28**(2), 269–291 (2016)

D.W. Pennay, D. Neiger, P.J. Lavrakas, K.A. Borg, S. Mission, N. Honey, Australian online panels benchmarking study, in *Presented at the 69th Annual Conference of the World Association for Public Opinion Research, Austin, May* (2016)

A.T. Porter, S.H. Holan, C.K. Wikle, N. Cressie, Spatial fay–herriot models for small area estimation with functional covariates. Spatial Stat. **10**, 27–42 (2014)

S.S. Qian, K.H. Reckhow, Modeling phosphorus trapping in wetlands using nonparametric Bayesian regression. Water Res. Res. **34**(7), 1745–1754 (1998)

R Core Team, *R: A Language and Environment for Statistical Computing* (R Foundation for Statistical Computing, Vienna, 2016)

J.N. Rao, *Small-Area Estimation.* (Wiley Online Library, Hoboken, 2003)

J. Raymer, A. Wiśniowski, J.J. Forster, P.W. Smith, J. Bijak, Integrated modeling of European migration. J. Am. Stat. Assoc. **108**(503), 801–819 (2013)

S. Renooij, C. Witteman, Talking probabilities: communicating probabilistic information with words and numbers. Int. J. Approx. Reason. **22**(3), 169–194 (1999)

D. Rivers, Sampling for web surveys, in *Joint Statistical Meetings* (2007)

D. Rivers, D. Bailey, Inference from matched samples in the 2008 us national elections, in *Proceedings of the Joint Statistical Meetings*, Vol. 1, pp. 627–39 (YouGov/Polimetrix Palo Alto, 2009)

J.W. Sakshaug, A. Wiśniowski, D.A. Perez-Ruiz, A.G. Blom, Supplementing small probability samples with nonprobability samples: a Bayesian approach. J. Off. Stat. **35**(3), 653–681 (2019)

C.P. Schmertmann, S.M. Cavenaghi, R.M. Assunção, J.E. Potter, Bayes plus brass: estimating total fertility for many small areas from sparse census data. Popul. Stud. **67**(3), 255–273 (2013)

R. Valliant, J.A. Dever, Estimating propensity adjustments for volunteer web surveys. Sociol. Methods Res. **40**(1), 105–137 (2011)

L.C. van der Gaag, S. Renooij, C.L.M. Witteman, B.M.P. Aleman, B.G. Taal, Probabilities for a probabilistic network: a case study in oesophageal cancer. Artif. Intell. Med. **25**(2), 123–148 (2002)

M.D. Vescio, R.L. Thompson, Forecaster?s forum: subjective tornado probability forecasts in severe weather watches. Weather Forecast **16**(1), 192–195 (2001)

A. Wiśniowski, J.W. Sakshaug, D.A. Perez-Ruiz, A.G. Blom, Integrating probability and nonprobability samples for survey inference. J. Surv. Stat. Methodol. **8**, 120–147 (2020)

D.S. Yeager, J.A. Krosnick, L. Chang, H.S. Javitz, M.S. Levendusky, A. Simpser, R. Wang, Comparing the accuracy of RDD telephone surveys and internet surveys conducted with probability and non-probability samples. Public Opin. Q. nfr020 (2011)

Harnessing the Power of Data Science to Grasp Insights About Human Behaviour, Thinking, and Feeling from Social Media Images

Diana Paula Dudău

1 Introduction

Although much of the online social sharing is visual and that the popularity of image-sharing platforms such as Instagram and Flickr has increased considerably, there is a need for social science research aimed at extracting meaning from images. So far, many successful efforts have been made to analyse linguistic contents, network characteristics, or specific online behaviours such as Facebook likes, along with metadata. With the growth of deep learning, it is now possible to better tackle also more unstructured data such as images. However, the main tasks embedded in image analysis – e.g. obtaining quantitative features from the raw visual data, detecting the most relevant area(s) of an image, summarizing image collections, or handling weakly annotated data – remain essential challenges. Also, clear guidelines on how to exploit social media images – from the scrapping phase to the data analysis phase – are still scarce.

Several pioneering studies have used social media images to build predictive models for psychological features, based on face detection software, pixel analysis, metadata, and posting behaviours, and relied on the human annotation to train the proposed algorithms. Such results prove that the transition from working in traditional labs, which usually assures reasonable variable control with the cost of losing external validity, to setting sophisticated virtual labs, which allows access to large samples of authentic behaviours, not only has begun but has started to move beyondexploiting user-generated linguistic contents and metadata. A picture taken

D. P. Dudău (✉)
Department of Psychology, West University of Timisoara, Timișoara, Romania
e-mail: diana.dudau@e-uvt.ro; diana.paula.dudau@gmail.com

T. Rudas, G. Péli (eds.), *Pathways Between Social Science and Computational Social Science*, Computational Social Sciences,
https://doi.org/10.1007/978-3-030-54936-7_5

in the real life can capture and express more than any words and metadata can say about that particular social context and how the involved person(s) were feeling and acting.

However, training algorithms to efficiently decode social, emotional, and behavioural cues from images posted on social media is a difficult job that has only recently become possible. Also, using humans to label complex inputs such as images can be very time-consuming, expensive, and even imprecise.

The current chapter builds on the need to assemble a repertoire of image analysis methods and techniques intended to guide researchers interested in interdisciplinary approaches to the emerging field of computational social science. The chapter contains two main parts. The first one highlights the connection between traditional social science and computational social science and presents the methodology and the results of a literature review on the link between social media images and various psychological constructs. The remaining part introduces the reader to the basics of image processing and analysis. A particular subsection is devoted to pixel-level features, another one to face detection, and the third one to convolutional neural networks. Further key points will emerge as the existing research located at the interplay between computer vision applied to social media images, Big Data, and psychology will be discussed from a theoretical, topical, and methodological point of view.

2 Social Media Images: Pathways Between Traditional Social Science and Computational Social Science

By definition, psychology involves finding solutions to a challenging problem: the impossibility of directly observing or assessing constructs that are sealed deep into the human mind. Such constructs are emotions, beliefs, values, motivation, personality, intelligence, and many other aspects that make up the object of study of this field. They belong to the so-called black box and cannot be measured straightforward as it happens with arterial pressure, glucose level, or other objective indicators.

Three main approaches to this problem stand out. The first one, which is also the most common and intuitive, is to ask individuals for explicit reports of how they think about themselves and others, how they feel, and how they act in various situations. The answers can be recorded either in a closed-ended fashion (on a Likert or dichotomous scale) or in an open-ended fashion (everyday speech obtained in writing or during interviews or focus groups). Either way, the people's statements are treated as accurate inputs for psychological assessment or research, which is not necessarily an issue. On the contrary, this approach is advantageous for two main reasons: (1) each individual is the one with the highest access to own experiences and self-content; and (2) self-report questionnaires are cheap, easy to score, and time-saving (e.g. Paulhus and Vazire 2007). However, self-report method also carries several threats that are hard to detect and control and can distort the results and interpretations: (1) impression management; (2) self-deception or self-serving biases; (3) memory errors (e.g. Goffin and Boyd 2009; Gosling et al. 1998; Paulhus and Vazire 2007; Robins and Beer 2001; Rothstein and Goffin 2006; Tourangeau 2000).

Hence, implicit measures – i.e. the implicit association test, its variants, and affect misattribution procedure – came as an alternative solution to the problem of measuring psychological constructs (e.g. Gawronski and Houwer 2014; Sava et al. 2012). However, not much is known about their underlying mechanisms and whether they provide contextually malleable outcomes (Goodall 2011; Han et al. 2010).

The third approach, which relies on digital traces found in social media, emerged naturally in the last decade, as part of the never-ending endeavour towards scientific advancement, which defines psychology as any other science. The need to find new ways to measure psychological constructs to solve real-world or theoretical problems or to gain new perspectives on various topics in psychology and related fields met the compelling technological and computational developments that allowed researchers to track and exploit countless amounts of linguistic and non-linguistic data.

These data are generated freely by users worldwide in sync with their real lives, are triggered and recorded in a natural setting, and remain unaltered over the years. However, most importantly, they contain not only overt information, which can be subjected to impression management, but also hidden cues about one's thoughts, feelings, and preferences. These latent cues, such as linguistic style or the hue, saturation, and brightness of the shared images, can be uncovered with the aid of machine learning. The same is true for the overt details accumulated over long periods, within huge piles of data. All this information is subtle and difficult to control by the user but usually can be framed by some psychological theory.

The bottom line is that within this state-of-the-art approach based on digital traces, a three-way connection between traditional social science and computational social science has already been established and needs to be strengthened. This pathway stands on three pillars: theoretical, topical, and methodological.

The following subsections of the current chapter are intended to showcase how these pillars are supported in the area of social media image analysis. Most past research has focused on other types of digital traces than images. With the spread of smart mobile devices, which are equipped with cameras and can instantly connect to the Internet, the popularity of social media visual sharing has also increased considerably, giving rise to an emerging line of research situated at the interplay between psychology/social science and computer vision.

To highlight the three-fold pathway between traditional social science and computational social science, I will present the results of a literature review on the link between social media images and various psychological constructs. The identified papers revolved around three main themes: (1) personality and individual differences; (2) depression and mental health; and (3) emotions and sentiment analysis.

Throughout, to argue the theoretical pathway, I will discuss from a psychological point of view why it is reasonable to think that social media images can contain footprints of personality and mental health status and to what extent such theoretical expectations were sustained in the literature so far. As far as the topical pathway is concerned, I will point out several ardent issues that can be tackled by interpreting social media images. Along the way, some methodological approaches that make possible the connection between traditional social science and computational social science will come up.

2.1 The Relationship Between Social Media Images and Personality, Depression, Emotions, and Other Psychological Constructs: A Literature Review

2.1.1 Search Strategy, Eligibility Criteria, and Results

Images are different from speech/language because they can display all at once many elements: objects, faces, poses, the presence of others, pets, fashion style, locations, sceneries, actions, contexts, etc. Images are static scenes of real happenings. Also, they can include cartoons or other forms of animations or drawings, which still incorporate various characteristics (e.g. colour-level features) and some representation regardless of how abstract they are.

Social media images, like any other image, capture various items, details, and stories. Besides, they are shared deliberately as social currency and vehicles for self-presentation (Deeb-Swihart et al. 2017; Kim et al. 2010; Rainie et al. 2012; Wei and Stillwell 2017). In this line of thought, the profile picture, which is a clear statement for what represents us best in the social network, stands out as a unique communication tool (Liu et al. 2016). Even expressing appreciation for some pictures and not for others is a way to interact with other people and to signal something about ourselves (Guntuku et al. 2018). Moreover, social media images can expose not only intentional but also unintentional self-disclosures – for instance, only the face is a rich, well-known source of cues for psychological interpretations. A fine-grained analysis, performed with the aid of cutting-edge computational methods, on large samples of participants and images can reveal more than naked eyes can tell about personality traits, mood, emotions, values, preferences, and other personal characteristics.

To obtain a thorough depiction of the ways social media images have been leveraged so far in the field of computational social science as input to measure or predict various psychological constructs, I carried out a systematic literature search in the following databases: Web of Science; PsycINFO; Scopus; ScienceDirect; Wiley, Taylor & Francis; SAGE; IEEE Xplore Digital Library; the library of the Association for the Advancement of Artificial Intelligence (AAAI); arXiv; and the website of the World Well-Being Project (WWBP).

The search was made in March 2019, using a formula comprised of four types of keywords: (1) from the area of mental health ("depress*", "anx*", "suicid*", "well-being") and personality and individual differences ("personality", "intelligen*", "emotion*", "behav*"), to set the research topic; (2) from the area of visualization to narrow the search to image digital traces ("image*", "photo*", "picture*", "pixel*", "face*", "facial", "physiognom*"); (3) from the area of social media, to specify the source of digital traces ("twitt*", "facebook", "Instagram", "social media", "social network*"); and (4) from the area of data science to obtain a more restrictive filter in the databases that are not focused exclusively on computer science ("computer vision", "visual comput*", "computational aesthetics", "machine learning", "deep learning", "neural network*", "CNN*", "data mining").

Overall, the search generated 4,476 results. These results were subjected to a stepwise selection, according to several inclusion/exclusion criteria. The eligible papers had to harness the power of machine learning/deep learning to estimate/predict variables specific to psychology, using social media images (not photos taken in the lab or another type of images). Also, the features embedded in images had to be extracted with the aid of computer vision algorithms or deep learning – all papers that used only human annotation or self-report questionnaires were excluded.

In the first step, by reading the titles and, occasionally, the abstract, I identified 289 papers as being potentially relevant for the current review. Then, I read all the abstracts thoroughly, and in some cases, I also scanned the full texts. This step ended with the retention of 162 papers. In the next step, I eliminated 25 duplicates and analysed the full text of 133 papers (I did not have full access to 4 papers).

At the end of this systematic selection process, 85 papers proved suitable for the scope of the current analysis. To these, I added five more relevant papers that stood out from the literature or other sources. The 90 papers address the following topics: personality, 19 papers; intelligence, 1 paper; gender differences, 7 papers; personal interests, 2 papers; diversity and other social topics, 8 papers; depression and mental health, 9 papers; and emotions, 44 papers. In the remaining of this chapter, I will use some of these papers to emphasize the theoretical, topical, and methodological pathway between traditional social science and computational social science. I will allocate a stand-alone subsection for personality and individual differences. Another subsection will be devoted to depression and mental health. The papers on emotions and sentiment analysis will be used selectively throughout the section about image processing and analysis, to illustrate the discussed technical knowledge.

2.1.2 Personality and Individual Differences

In the twentieth century, personality science made possible the realization of one of the most important achievements in psychology – establishing a robust and widely accepted structure of personality, namely, the Big Five model (Wright 2014). This model posits that personality is composed of five main traits: openness to experience, conscientiousness, extraversion, agreeableness, and neuroticism (e.g.

McCrae and Costa 2008). Personality defined within this framework has demonstrated predictive value across multiple areas and outcomes: academic performance (Chamorro-Premuzic and Furnham 2003), longitudinal course of depression (Kim et al. 2016), mortality risk (Graham et al. 2017), political preferences (Ekstrom and Federico 2018), unemployment (Viinikainen and Kokko 2012), loneliness (Abdellaoui et al. 2018), well-being (Soto 2015), self-regulation, and health (Hampson et al. 2016), to name a few.

In the twenty-first century, personality science remains a robust field, ready to face challenging topics concerning human behaviour. However, it is also a field that needs to take steps forwards towards "a more refined and pluralistic data acquisition and modelling" (Wright 2014).

Computational personality analysis, which is a machine learning approach to personality assessment (Ilmini and Fernando 2017), can be the answer to at least some of the current problems and debates within personality psychology. For instance, the desideratum to move the field towards building ideographic models of personality based on data streams of various types, in order to catch the whole person as life unfolds (Wright 2014), can be addressed by leveraging the digital footprints found in social media. The thriving research conducted so far at the interplay between data science/computer science and psychology, having as input the contents generated in social media, sustains the idea that this is a promising start. Also, notably, although this line of research is state-of-the-art personality science, it tends to maintain a theoretical and methodological connection with the "traditional" approach. Computer vision, when used as a way to perform computational personality analysis, strengthens these ideas. The following studies are good examples in this regard.

Personality assessment based on social media images usually requires the following steps: (1) choosing a theoretical model to define personality; (2) obtaining benchmarks for the accuracy of assessment; (3) extracting features from images; and (4) using those features as predictor variables in a statistical learning model.

Most of the studies yielded by the systematic literature search described in the previous section rely on the Big Five model (e.g. Guntuku et al. 2018; Torfason et al. 2017; Xiong et al. 2016). One exception is the study of Nie et al. (2016), which defines personality within the Keirsey temperament model. Another notable exception is the study of Preoțiuc-Pietro et al. (2016), which focused on predicting the dark personality triad (narcissism, psychopathy, and Machiavellianism) from Twitter linguistic posts and images (aesthetic features – greyscale, brightness, blur, etc. – and general facial features, number of faces, face ratio, the 3D face posture, the yaw angle of the face, and the degree of smiling).

In terms of assessment criteria, three approaches have been usually applied. One of them is to ask participants to fill a self-report questionnaire. For example, Celli et al. (2014) recruited 112 Facebook users who agreed to answer two self-report questionnaires (for personality and interpersonal style). The participants' answers were used to label their Facebook profile picture for the machine learning phase. Although it might seem infeasible to gather self-reports from large samples of users, this approach is quite common – 13 out of 19 studies selected for the current review used this method (3 studies involved the same Flickr database: Segalin et al. 2016, 2017b; Samani et al. 2018). This approach suggests a vital pathway between traditional and computational social science.

A second approach is to operationalize personality using others' perceptions. This approach consists of asking observers to rate someone else's personality, based on digital traces/images (Guntuku et al. 2018). Several studies have implemented this approach either uniquely or in combination with the self-report approach (e.g. Biel et al. 2011, 2012; Teijeiro-Mosquera et al. 2015; Segalin et al. 2017b).

An alternative to both self-report assessment and others' assessment is to grasp personality from non-visual digital traces automatically. For example, Dhall and Hoey (2016) predicted the Big Five traits from various facial and scene features of Twitter profile images, using as benchmarks the personality estimates obtained via a text analysis of the most recent 3,200 tweets of each user. Also, in three of the studies included in the current review, a text-based prediction approach was applied to infer the assessment criteria for personality. These criteria were further used to validate the prediction model based on Twitter images (Bhatti et al. 2017; Guntuku et al. 2017; Liu et al. 2016).

A variety of features from different types of social media images (e.g. profile picture, liked pictures, or posts) crawled from various platforms (Facebook, Instagram, Twitter, Flickr, YouTube) were used successfully as estimates of personality. Kim and Kim (2018) studied three types of features extracted from 25,394 Instagram photos belonging to 179 students: content features, facial features (number of faces; facial expression – 8 emotions), and pixel features (RGB values, hue, brightness, saturation, pleasure, arousal, dominance). The results revealed that personality could relate to such features in a way consistent with what would be expected based on the Big Five model (e.g. positive correlation between extraversion and number of faces in the photos). Liu et al. (2016) also managed to detect visual futures that could reflect personality traits on a sample of 66,000 Twitter users, in a way consistent with the Big Five model. They investigated features related to colour (e.g. greyscale, RGB values, brightness, contrast, saturation, hue, colourfulness, naturalness, sharpness, blur), image composition (e.g. average rule of thirds, edge distribution), facial presentation (e.g. type of glasses; pitch, roll, and yaw angle), and facial expressions (e.g. basic emotions). For feature extraction, Kim and Kim (2018) and Liu et al. (2016) relied on computer vision software such as Face^{++} and EmoVu. Other authors harnessed more directly the power of deep learning and convolutional neural networks, in particular, along with other computer vision algorithms or methods (e.g. Guntuku et al. 2017; Segalin et al. 2016, 2017a). More recently, using a deep learning framework, Farnadi et al. (2018) managed not only to predict Big Five personality traits based on Facebook images but also to integrate multiple modalities (textual, visual, and relational data). Also, notably, Samani et al. (2018) conducted a cross-modal (posts, likes, and profile images) and cross-platform (Twitter and Flickr) study, showing that merging data from two social networks improves the prediction performance.

The bottom line is that the development of machine learning and computer vision enabled promising results not only in terms of personality prediction but also in terms of estimating other psychological constructs from the area of individual differences or social psychology.

Thus, for example, Wei and Stillwell (2017) revealed that intelligent people tend to have some colour, texture, and content preferences for their Facebook profile picture. Also, Deeb-Swihart et al. (2017) detected signals of identity statements such as wealth, health, and physical attractiveness in the selfies posted on Instagram, whereas Cinar et al. (2015) managed to infer the interests of the Pinterest users applying a multimodal approach based on images and text. Several authors have also investigated gender differences. For example, You et al. (2014) predicted gender depending on the images posted by the same users on different social networks. Another noteworthy example is the study of Haseena et al. (2018), in which a deep convolutional neural network achieved good accuracy in classifying gender based on facial features.

In the area of social psychology, Singh et al. (2017) studied diversity in interaction by computing the mix of race, age, and gender in the Instagram photos of users from New York City. Also, You et al. (2017) grasped the cultural diffusion among Facebook users, based on classifying their pictures using a convolutional neural network. In one of the oldest studies included in this review, Mavridis et al. (2010) predicted the friendship network of a sample of Facebook users, based on the photos containing their faces. Bakhshi and Gilbert (2015) revealed the importance of colours in image diffusion by analysing a sample of one million images from Pinterest. Also, other authors showed that social media images could be used to address more pressing social problems such as understanding the crisis events (Dewan et al. 2017) or detecting protest activity and estimating the level of perceived violence (Won et al. 2017).

2.1.3 Depression and Mental Health

Nowadays, it is well-known that mental and substance use disorders have become a concerning problem worldwide.

On the one hand, the personal, social, and economic burden caused by these conditions is disturbing. The Global Burden of Disease Study from 2010 revealed that the mental and substance use disorders were the leading cause of YLDs. The most damaging mental health conditions in terms of DALYs were depression and anxiety (Murray et al. 2012; Whiteford et al. 2013). Currently, the leading cause of disability worldwide is depression (World Health Organization 2018). Also, it has been estimated that mental illness will be the leading problem among non-communicable diseases in terms of lost economic output between 2010 and 2030 and also that the direct and indirect costs of mental health care will increase considerably in the same period (Bloom et al. 2011; World Bank Group and World Health Organization 2016).

On the other hand, mental disorders, especially depression and anxiety, are under-diagnosed and under-treated conditions (Kasper 2006; Kessler and Bromet 2013; Kroenke et al. 2007; Sheehan 2004). This fact worsens the burden imposed by these conditions. Clinical diagnosis requires a detailed assessment of patient's symptoms, life events, stressors, etc. and can be performed only by mental health professionals. However, due to several social, personal, and financial factors, people generally hesitate to seek psychological or psychiatric help (Corrigan et al. 2014; Jennings et al. 2015; Wang et al. 2007).

Therefore, finding new screening tools for mental health conditions is an ardent topic in clinical psychology and health. Digital traces on social media can provide an avenue towards a cost-effective solution to this multidimensional problem. In the literature so far, there can be noticed a few but promising attempts to use social media images as cues for depression and other mental health issues.

Manikonda and De Choudhury (2017) showed that the visual contents of Instagram posts carried different self-disclosure needs related to mental health than the ones expressed in textual form – cries for help, disclosure of emotional distress, and a clear exposure of personal vulnerability. They analysed a corpus of 2,757,044 posts tagged for 10 mental health disorders. The tags were selected with an association rule mining approach and were categorized by a psychiatrist. With the aid of computer vision, three attributes were extracted from the images: (1) visual features; (2) visual themes; and (3) emotions. The Open Source Computer Vision Library (OpenCV; Bradski 2000) was used to obtain the greyscale histograms, the visual saliency, and the speeded up robust features of the mental health images.

Other researchers managed to distinguish between depressed and nondepressed individuals using social media images. Two approaches stand out: (1) analysing visual features as primary markers of depression and (2) integrating multiple features from different modalities.

Analysing the visual features of Instagram photos, Reece and Danforth (2017) developed several machine learning classifiers that outperformed human ratings and general practitioners' unassisted diagnostic accuracy. The best-performing algorithm for detecting depressed individuals was a 100-tree Random Forests classifier. The features extracted from photos were the total number of faces in each photo (a face detection software was used in this regard), the use of Instagram-provided filters, as well as the pixel-level averages for hue, saturation, and value/brightness. The benchmark for the machine learning problem was established by asking the participants to answer a self-report screening questionnaire (i.e. Center for Epidemiologic Studies Depression Scale), which is a method specific to traditional social science. These results were consistent with what is known about depression, revealing that depressed individuals were more likely to post bluer, greyer, and darker images and to receive fewer likes. Also, they tended to post more photos containing faces but fewer faces per photo, which was a potential indicator for low social interaction (Reece and Danforth 2017).

Kang et al. (2016) proposed a multimodal method to identify depressed Twitter users. They combined three single-modal analyses of daily tweets: a learning-based text analysis intended to infer mood from language, a word-based emoticon analysis performed with a lexicon, and a support vector machine-based image analysis comprised of visual feature extraction and mood prediction. The first part of the image analysis consisted of extracting the colour composition and the scale-invariant feature transform (SIFT) descriptors for shapes. SIFT algorithm converts the visual content into local feature coordinates and provides a good representation of the image regions even when the characteristics of the images, such as illumination, scale, or rotation, vary (Dey 2018). Next, a support vector machine with a radial basis function kernel was used to classify users' moods based on images. The obtained multimodal method was tested on 45 users and proved to be more accurate than the baseline.

Shen et al. (2017) found a multimodal dictionary learning approach to discriminate between depressed and nondepressed Twitter users. The solution involved six groups of features: social network behaviours, profile information, visual features, emotional features from text and emoji, topic-level features, references to antidepressants, and depression symptoms. The profile and the home page pictures were used for the visual features. These features were colour combinations, brightness, saturation, cold colour ratio, and clear colour ratio.

More recently, Yazdavar et al. (2019) also managed to blend multimodal features (visual, textual, and ego network) in order to detect depressed individuals from Twitter data. For the visual component, they extracted the aesthetic features from posted images (colour analysis, hue variance, sharpness, brightness, blurriness, naturalness) and used the profile picture to infer gender, age, and facial expression.

Facial expression was defined with Face++ API, according to Ekman's model of six emotions. The results sustained the superiority of this multimodal framework over existing approaches.

Promising results supporting the predictive value of social media images for other outputs related to mental health have also started to emerge due to the development of deep learning. Lin et al. (2014) estimated the psychological stress of users from various microblogging platforms, using a deep neural network model that incorporated multiple features, including visual attributes. Peng et al. (2017) analysed the facial cues of 10,480 Instagram users and managed to measure the sleep-deprivation fatigue by weekday, over different age, gender, and ethnic groups.

Yazdani and Manovich (2015) showed that the contents of nonphotographic Twitter images aggregated per city could relate to self-reported social well-being responses from Gallup surveys. Also, notable progress has been made towards the automatic detection of alcohol and substance use, based on multimodal social media data (Pang et al. 2015; Roy et al. 2017).

3 An Intuitive Introduction to Image Processing and Analysis

At the heart of grasping psychological insights from social media images lie algorithms and scripts able to perform three types of processing: (1) extracting basic information from pixel values, namely, attributes regarding the colour, texture, shape, or spatial location embedded in images; (2) detecting faces and objects; and

72	73	73	66	65	65	68	69	69
72	73	74	65	66	70	73	71	66
72	75	67	74	69	79	74	66	70
65	72	70	76	70	74	73	70	68
61	73	74	71	70	66	71	73	65
62	72	72	62	69	62	68	71	62
63	70	67	61	73	65	64	65	61
65	72	69	67	76	71	63	61	63
66	74	71	68	68	72	66	62	67

Fig. 1 The grid of pixel intensities for a small region of a greyscale image. (The image of the guitar was retrieved from https://pixabay.com/)

(3) learning the required functionality using the extracted visual features (e.g. emotion recognition, personality prediction), which usually implies the understanding of the visual contents at a high level of abstraction.

An image is a digital representation of a real-life object, in the form of a two-dimensional grid of numeric values, which are called pixels (Datta 2016; Singh 2019). Each pixel stores a piece of information. An image is composed of thousands and sometimes millions of such tiny carriers of information, although the human eye perceives a continuous, coherent pattern (Datta 2016). In other words, an image can be conceptualized as a function $f(x,y)$ that assigns coordinate pairs to integer values. Each integer value defines the intensity or colour of the corresponding point (Dey 2018). Thus, an image is nothing more than a numeric map of an object (Fig. 1).

Extracting knowledge from images is a challenging task. There are several problems that can distort or impede the results, including (Villán 2019):

1. The ambiguity of images caused by changes in perspective – the same object from different perspectives can lead to significant changes in the visual appearance of an image.
2. The fact that images can be affected by many factors such as illumination, weather, reflections, etc.
3. The overlaps between objects – the occluded objects can be difficult to detect or classify.

Also, there is a semantic gap between the features usually extracted by computers and the high-level concepts required to interpret and compare images in a way that enables outcomes similar to what humans are capable of (Liu et al. 2007). Thus, for instance, visual sentiment analysis is more complicated than textual sentiment analysis – the former requires a high level of abstraction, while the latter involves easy access to the semantic and context information (Fengjiao and Aono 2018).

According to Dey (2018), the image processing pipeline comprises the following steps: (1) image acquisition and storage; (2) preprocessing; (3) segmentation; (4) information extraction/representation; and (5) image understanding/interpretation.

In general, the first step is about capturing the images with a camera or other device and storing them on a hard disk or other storage space/device. In the case of social media image analysis, the data are mined from users' profiles by making API requests. For instance, to access Instagram API, an app must be created and registered using the Instagram Developer platform. Next, Python or other programming language is used for authenticating on the Instagram API, obtaining an access token, confirming access to the platform, and, finally, retrieving the images and their metadata. Similar steps are needed for other social media networks, too. A note to remember is that APIs are continually changing – implicitly, the current scripts may need adjustments at some point (Klassen and Russell 2019).

In the preprocessing phase, specific algorithms are used to run some transformations on images (e.g. greyscale conversion), improve the quality of images, and restore the images from noise degradation (Dey 2018).

Segmentation is the process of dividing the image into multiple sets of pixels (segments) by assigning labels to each pixel. In the end, pixels with the same characteristics should have the same label (Kovalevsky 2019). A proper segmentation is achieved if pixels in the same category have similar intensity values and form a connected region, while the neighbouring pixels belonging to different categories have dissimilar values (Dey 2018). Thus, the representation of an image is changed into something more meaningful and easier to analyse (Kovalevsky 2019). There are various approaches to image segmentation, including thresholding and Otsu's segmentation; edges-based/region-based segmentation techniques; Felzenszwalb, SLIC, QuickShift, and Compact Watershed algorithms; active contours, morphological snakes, and GrabCut algorithms; quantizing the colours; graph traversal algorithm; and equivalence classes (Dey 2018; Kovalevsky 2019). Overall, segmentation techniques can be either non-contextual (the pixels are grouped only by some global attributes, without regard to spatial relationships between features) or contextual (the spatial relationships are considered additionally; for example, spatially close pixels with similar grey levels are grouped) (Dey 2018).

In the fifth step of the image processing pipeline, which is information extraction/representation, some handcrafted feature descriptor can be computed from the image (e.g. HOG descriptors, with classical image processing), or some features can be learned from the image automatically (e.g. the weights and bias values learned in the hidden layers of a neural network). The goal is to obtain an alternative representation of the image (Dey 2018).

In the last step, this alternative representation is used for image classification or object recognition (Dey 2018).

Providing a detailed tutorial about all these steps exceeds the goal of the current chapter. The following subsections seek to outline some essential landmarks regarding the fourth phase – information extraction/representation – with an accent on colour features, face detection, and convolutional neural networks and referring to image sentiment analysis.

3.1 Pixel-Level Features

Although pixel-level features (low-level features) tend to be not only poorly connected to emotions but also challenging to interpret, they are useful as global descriptors of the overall image content (Li et al. 2018; Zhao et al. 2018a). Besides, they are part of the image processing evolution and have been used extensively in studies about other psychological constructs, as seen in the previous sections of this chapter. Therefore, pixel-level features cannot be excluded from a discussion regarding image processing and analysis. Some of them will be briefly explained in the current subsection.

The first step to gain psychological insights from visual contents found in social media is to understand the basics of images and their features. This understanding is the foundation that makes possible the development of scripts aimed at extracting information manageable by machine learning algorithms. These algorithms further enable the connection between those basic features and psychological constructs.

The digitization of images in the form of spatial coordinates and pixel intensities is mandatory for image processing and analysis. The range of values that can be used to represent the pixel intensities is imposed by the colour space/depth used to store the image (Datta 2016). "Colour space" or "colour depth" is analogous to "data type", which is a computer science concept met in the context of programming languages. In this context, a single bit can contain 2 values (0 and 1), 8 bits (1 byte) $2^8 = 256$ distinct values, 16 bits $2^{16} = 65{,}536$ distinct values, and so on (Datta 2016). In image processing, 8 bits per pixel is the simplest and widely used colour space, meaning that each pixel can hold any value between 0 and 255 (inclusive), where 0 is black, 255 is white, and the intermediate values between 0 and 255 represent different shades of grey (Datta 2016). Images encoded this way are greyscale images.

Coloured images cannot be represented by a single value per pixel, as greyscale images. They are made of multiple components or "channels"; in other words, each pixel at the *(x,y)* coordinate can be depicted as a tuple (Datta 2016; Dey 2018). This way, space is extended. There are many colour spaces; the most common ones are RGB, CIE L*a*b*, HSL and HSV, and YCbCr (Villán 2019). The RGB colour space contains three channels – red (R), green (G), and blue (B) – which means $8 \times 3 = 24$ bits per pixel (16 million colours). The combination of these three types of intensity values per each pixel is sufficient to cover the entire spectrum of colours (Datta 2016). The CIE L*a*b* colour space, which is also known as CIELAB or LAB, uses three values for each colour: L* is the lightness, a* the green-red channel, and b* the blue-yellow channel (Villán 2019). HSL (hue, saturation, lightness) and HSV (hue, saturation, value) are two colour spaces, where only one channel (H) is used to describe the colour. More specifically, hue refers to the actual colour (red, blue, etc.), saturation reflects the amount of grey that attenuates the hue/colour (desaturation means adding more grey), and value defines the darkening or the lightening of the colour, namely, the amount of black and white added to the hue (Malhoski and Rock 2018). YCbCr is a family of colour spaces used in video and digital photography systems (Villán 2019).

The texture is a feature not so clearly defined as the colour feature, but it provides crucial information, especially for image classification (Liu et al. 2007). Some authors defined texture as "the spatial arrangement of the grey levels of the pixels in a region of a digital image" (Bharati et al. 2004). Others saw texture more like a variability or pattern of brightness or colour (Russ and Neal 2016). Depending on method used for feature extraction, four main approaches to texture analysis can be defined (Bharati et al. 2004):

1. Statistical methods – the defining elements are the higher-order moments of the greyscale histograms of the texture regions.
2. Structural methods – the composition of some well-defined texture elements (e.g. regularly spaced parallel lines) is central.
3. Model-based methods – for each pixel, an empirical model from the weighted average of the pixel intensities in its neighbourhood is created.
4. Transform-based methods – the spatial frequency properties of the pixel intensity variations are used to convert the image into a new form (Bharati et al. 2004).

Some of the most commonly used texture features include spectral features, such as those obtained with Gabor filtering or wavelet transform, the six Tamura texture features (especially coarseness, directionality, and regularity), and Wold features, as defined by Liu et al. (2007).

Multiple transformations can be performed at the pixel level, with the aid of a programming language such as Python. But basically there are two types of transformation techniques: (1) ones involving some sort of computation for each pixel, resulting in images of the same size as the input image, where there is a one-to-one match between the pixel locations in the input and those in the output images (e.g. greyscale transformations, image filtering, thresholding, morphological operations), and (2) ones entailing computations that lead to other forms of representation for the input images (Datta 2016). The techniques in the first category form the prerequisites for more advanced computer vision procedures, being part of the preprocessing phase of the image processing pipeline, which will not be discussed in the current chapter. The techniques in the second category come into play in a more advanced phase of the pipeline and are meant to handle images with high spatial resolution, which is probably the case of most social media images. The performance of the algorithms used to process images may be affected when the size of the two-dimensional grid of pixel values, which is the default representation of an image, increases. A solution is to represent images as histograms. A histogram is an image descriptor that relies on the aggregation of the values associated with the pixels that make up an image (Datta 2016). The intensity values are plotted on colour histograms. Nevertheless, there are also other types of histograms, depending on the types of values associated with the pixels. For example, histogram of oriented gradients (HOG) is another popular feature descriptor; it was first used for human detection in static images, and it is more advanced than colour histograms, being used in computer vision (Villán 2019) (Fig. 2).

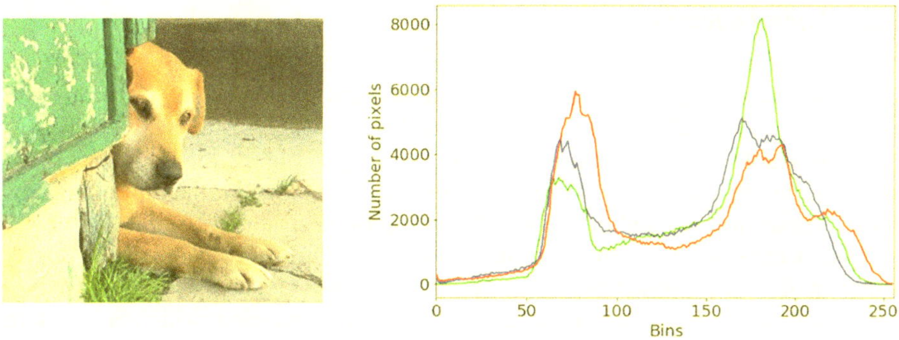

Fig. 2 The colour histograms for the blue, red, and green channels of the image in the left (The image of the dog is from the personal archive of the author)

In the area of visual sentiment analysis, few studies have included the extraction of colour and texture features. Liu et al. (2015) trained two support vector machine classifiers to detect the emotional valence (positive, negative, neutral) of Sina Weibo posts. The first classifier was based only on textual features while the second one on both textual and visual features. In terms of image processing, they extracted colour histograms for the RGB and HSV space and also computed the frequency of specific colours. The hybrid classifier was more accurate than the simple classifier. Also, using data from Sina Weibo, Tan et al. (2016) used textual information to classify images in terms of emotions, considering the mean and standard deviation values of hue, saturation, and brightness, the histogram of oriented gradients, the grey level co-occurrence matrix, three Tamura texture features (coarseness, contrast, directionality), and the pleasure, arousal, and dominance computed with brightness and saturation values. Zhang et al. (2015) extracted similar colour and texture features, along with colourfulness, to obtain a sentiment analysis model based on linguistic and visual contents on Sina Weibo. Yang et al. (2016) studied emotion contagion on Flickr, using a mix of colour and textural features (saturation contrast, bright contrast, cold colour ratio, figure-ground colour difference, figure-ground area difference, foreground texture complexity, background texture complexity) and network proprieties. The results showed that opinion leaders and structural hole spanners tend to have more influence than ordinary users in terms of positive emotion but are less influential in terms of negative emotion. Amencherla and Varshney (2017) crawled 2,580, 1,470, and 1,744 Instagram images for the hashtags *happy*, *awesome*, and *sad*, respectively, and extracted several colour features (hue, saturation, value, colourfulness, colour warmth). The association between these visual attributes and the emotional valence of hashtags was statistically significant in a way consistent with several psychological theories (e.g. happiness correlated with colourfulness).

3.2 Face Detection

Computer vision seeks to mimic what our nervous system manages to do with the images that enter through retina (Datta 2016). Hence, computer vision tasks are more challenging than those performed at the pixel level. In computer vision, although the inputs are still images (like in image processing), the output is some form of semantic information inferred by an algorithm (Datta 2016).

One of the most powerful computer vision algorithms remains the Viola-Jones, although it was proposed nearly 18 years ago. The Viola-Jones algorithm is a classical machine learning approach for visual object detection, which is based on Haar-like features and uses a cascade function trained on a mixed set of images (true positives and noise) (Viola and Jones 2001).

Haar-like features are blocks of rectangles with different pixel intensities that "capture the structural similarities between instances of an object class" (Papageorgiou et al. 1998). They are used to describe meaningful image subsections, depending on the pixel values difference between adjacent regions (Karim et al. 2018). There are three main types of Haar-like features: two-rectangle features (edge features), three-rectangle features (line features), and four-rectangle features (Dey 2018; Karim et al. 2018) (Fig. 3).

Typically, the Viola-Jones algorithm uses a frame with a specific size to systematically scan the entire image for significant features, by computing Haar-like features. The algorithm was proposed for face detection, but the same principles can apply to many other objects (Karim et al. 2018). Two types of shortcuts can lead to an increase in computation efficiency without losing robustness (calculating thousands of features for every patch would require much effort). The first one is to obtain an intermediate representation of the image, which allows a faster computation of rectangular features. It is called "integral image" and can be computed using the formula described in the paper written by Viola and Jones (2001). The second shortcut relies on the idea that most of the possible positions in an integral image will not contain a face; if a subarea does not contain a face, no more computational checks should be invested (Dey 2018). In this regard, a variant of AdaBoost learning algorithm is used to improve the classification performance (Viola and Jones 2001). The boost is obtained by combining a set of weak classification functions to form a stronger classifier. The strength of the final classifier is given by the ensemble of its weighted components, which is obtained through a sequence of learning problems

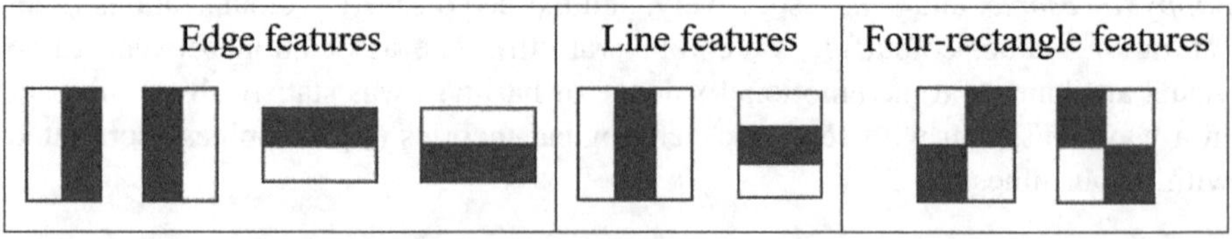

Fig. 3 The Haar-like features

(Tieu and Viola 2004). In the Viola-Jones algorithm, these weak classifiers are run one by one in a cascade; if the subarea under check fails at any step, then it is rejected (Dey 2018).

In the field of image sentiment/emotion analysis, several researchers have included the Viola-Jones algorithm for face detection (e.g. Abdullah et al. 2015; Dhall et al. 2015; Li and Deng 2018; Muhammad and Alhamid 2017).

The Viola-Jones algorithm is not only capable of delivering highly accurate results, but it can also work in real-time. However, it has a big drawback, among others: it requires much effort to gain the same detection efficacy with new objects (Karim et al. 2018). The low efficacy in this regard is one of the critical problems in computer vision, where the power of convolutional neural networks stands out.

3.3 Convolutional Neural Networks (CNNs)

Teaching a machine to identify the defining features of an object, and to recognize it, has been one of the most challenging tasks in computer science. Only several years ago, such a goal would have seemed unrealistic. Today, with the development of deep learning, an artificial neural network can learn by itself how to "understand" and classify objects. Although the traditional methods to extract handcrafted features are still useful, convolutional neural networks (CNNS), which are deep learning algorithms, have started to rule the field of image processing and analysis.

Deep learning means multilayered artificial neural networks (Chinnamgari 2019). An artificial neural network is a mathematical model for information processing, intended to resemble a biological neural network in the way it is organized. Thus, an artificial neural network has several particularities, as opposed to a classical machine learning algorithm. In an artificial neural network, the information is processed over a chain of interconnected neurons that carry signals one from another (Chinnamgari 2019; Spacagna et al. 2019). The neurons are displayed in layers – the outputs from a layer are the inputs for the next layer. There are three types of layers – the input layer, the middle layer(s), and the output layer (Chinnamgari 2019). Each neuron has an internal state, which is determined by the inputs it receives from other neurons, and also an activation function, which is calculated based on its state, setting the output signal (Spacagna et al. 2019). Examples of activation functions are the identity function, threshold activity function, logistic sigmoid, or rectifier function (rectified linear unit, ReLU). The connections between neurons vary in intensity, and this influences the way information is processed. Also, each neural network has a specific "architecture" – e.g. feedforward, recurrent, multi- or single-layered, etc. – which defines the connections between the neurons, the number of layers, and the number of neurons in each layer. The architecture is established by the researcher, usually through a cross-validation strategy (Chinnamgari 2019). Any neural network must go through a learning phase to achieve good performance. The most common way to train a neural network is with the gradient descent and backpropagation algorithms, but these are not the only available methods (Spacagna et al. 2019).

There are many types of artificial neural networks. The ones generally dedicated to computer vision are CNNs (LeCun et al. 1998). CNNs can find the probability of an image belonging to one of the known classes, by decomposing the image into small chunks of pixels and subjecting them to different computations based on specific filters. The first layers of the CNN are devoted to detecting shape (e.g. curves, rough edges), but the performance of recognizing objects is achieved after several convolutions. Initially, the filter values are randomly set. Therefore, in the beginning, the estimates of CNN are mostly wrong. However, with each iteration, by comparing its responses with the actual ones from labelled datasets, the filters are updated, and its performance improves (Chinnamgari 2019).

Using an artificial neural network to process images comes with an issue: the need to change the two-dimensional representation of the image into a one-dimensional array. The problem with this arrangement resides in the fact that the information given by how the pixels are grouped – assuming that the nearby pixels are closely related – might be lost (Spacagna et al. 2019). CNNs are powerful because they make the information coming from nearby neurons be more relevant than the information coming from distant neurons. In other words, in a CNN, the neurons know how to process information coming from neighbouring pixels (Spacagna et al. 2019). Notably, a CNN allows the input neurons to be displayed in a one-, two-, or three-dimensional space, respectively, producing corresponding outputs of the same type (Spacagna et al. 2019). Figure 4 depicts the architecture of a basic CNN.

Examples of common hidden layers of a CNN are (Chinnamgari 2019):

1. Convolution – a little matrix (usually 3×3) is used as a filter; every element of this matrix is multiplied by every pixel value of a 3×3 section of the image, and the results are summed; the final result is the convolution value for one pixel.
2. ReLU – is a non-linear activation function that replaces the negative components of the pixel matrix with zeros;
3. Max pooling – tells if a feature is present in the previous layer; only the maximum value from each subsection of the input matrix, as defined by the pooling layer, is kept; max pooling is useful to decrease the number of parameters, eliminate overfitting, and make the network see the whole picture.
4. Fully connected layer – establishes weights that connect every input with every output.
5. Softmax – is usually introduced as the last layer of the network, in a multiclass classification problem.
6. Sigmoid – is suitable as the last layer, in binary classification problems.

According to Li et al. (2018), there are two categories of deep learning approaches to visual sentiment analysis: (1) end-to-end methods which use CNNs in order to map directly the pixel representation with the visual sentiment orientation and (2) pipeline methods which use deep learning models to derive the sentiment

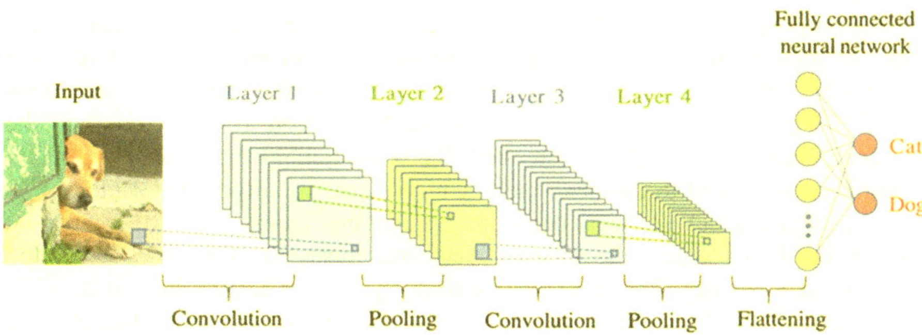

Fig. 4 The architecture of a basic CNN

orientation from the cognitive semantics obtained from visual contents (the mapping between image pixels and sentiment orientation is made via cognitive semantics).

There are multiple examples of how deep learning can be leveraged to extract emotions/sentiment from images. The literature review conducted for this chapter revealed 24 papers of such kind, addressing different challenges but mainly (1) combining visual traces with linguistic traces to improve the classification performance, which is the so-called multimodal approach (e.g. Huang et al. 2019; Lin et al. 2018); (2) labelling images for sentiment analysis, which is a difficult task given that each item usually can receive multiple labels and that several features, such as polarity or intensity, characterize emotions (e.g. Xiong et al. 2019; Yang et al. 2017); (3) transferring CNNs models trained on a large-scale dataset, to a new dataset for related tasks (e.g. Islam and Zhang 2016; Zhao et al. 2018b); and (4) providing new datasets for CNNs training (e.g. Li and Deng 2018; Wu et al. 2018).

One of the most recent and illustrative studies employing deep learning for visual sentiment analysis is the one conducted by Hu and Flaxman (2018). The authors proposed a multimodal approach based on using linguistic contents associated with Tumblr posts as self-reported emotions that had to be predicted from image features. For this purpose, they built 3 models: 1 for visual features solely, based on a pretrained network with 22 layers able to recognize images depending on their colours and arrangement of shapes; 1 for textual features, harnessing the power of natural language processing; and 1 based on a neural architecture that concatenated textual and visual information. The multimodal network achieved the best performance. Also, Yang et al. (2018), addressing the issue of multi-labelling in assigning emotions to images, developed a deep learning algorithm able to take into account the hierarchical relation between the emotion labels. They used four well-known datasets: *Flickr and Instagram (FI)*, *IAPSa*, *ArtPhoto*, and *Abstract Paintings*. Their model outperformed the existent methods in terms of both affective image retrieval and classification tasks. Another noteworthy study is the one of Song et al. (2018), which tackled a different challenging issue in image analysis – selecting most representative regions of images for sentiment classification. They presented a novel approach consisted of integrating visual attention into a CNN for this purpose. Visual attention is originated in the human visual system, which uses eye movement to bring into focus the most salient area of a picture/scene (Song et al. 2018). The CNN architecture used to extract the image representations was VGGNet, which was pretrained on about 1.2 million images – this was the first component of their proposed model. The second component was a multilayer neural network aimed to establish the attention distribution (on what region of an image the visual attention focuses on) for sentiment prediction. The third component was designed to produce the saliency map and consisted of a multiscale, fully convolutional network. The end-to-end training process was explained in detail in Song et al. (2018). The testing was made on two datasets: the *ArtPhoto* dataset and a Twitter dataset consisted of 1,269 images. The proposed architecture achieved better results compared to state-of-the-art methods.

4 Conclusions

As more and more of our social currency is becoming visual content stored digitally, image analysis is moving from a niche skill for technical applications to an increasingly valuable skill for social science research. Images are static snippets of real happenings, carrying information that cannot be extracted from other digital traces. They display all at once objects, faces, poses, contexts, colours, and many other elements. Thus, they can go beyond language and metadata in providing psychological meanings.

At the same time, language and other online behaviours are valuable carriers of psychological content, too. In this line of thought, social media images are precious vehicles for self-presentation, since they are shared within large social networks and bring along other valuable footprints such as textual data. Textual analysis can either complement or replicate the results of image analysis or ease the way towards image analysis by providing labels to images.

The state-of-the-art approach of using social media images to infer psychological constructs fits naturally in the evolution of psychology as a science. It comes as a continuation of the never-ending search for better ways to measure and understand human thinking, feeling, and behaviours. By definition, psychology involves the assessment/measurement of variables that cannot be reached directly, which is a challenging task. Two conventional approaches to this issue can be traced in the history of psychology: self-report and implicit measures. The latter approach came as an alternative to the former. However, both are debatable, providing not only advantages but also disadvantages. The digital traces approach, which includes the automatic interpretation of social media images, is the newest approach to measure psychological constructs. This innovative approach is not necessarily entirely disconnected from the traditional approach, as, sometimes, researchers use self-report measures for supervised learning.

Nevertheless, harnessing the power of machine learning to exploit social media images might lead to methodological improvements for various reasons. Most importantly, images shared online contain not only overt information but also latent meanings that can be uncovered only by algorithms. Also, they are produced freely by users worldwide in sync with their real lives, are triggered and recorded in a natural setting, and remain unaltered over the years.

A thorough analysis of the research concerning the link between social media images and various psychological constructs sustains the idea that image processing/computer vision unleashes a golden opportunity for social scientists to advance research in their field. The papers reviewed in the current chapter demonstrated new possibilities not only to study challenging and exciting topics such as personality in context but also to solve some of the most ardent problems of our society, such as depression screening and prevention.

Thus, our literature review showed that a theoretical, topical, and methodological pathway between traditional social science and computational social science has already been established also through computer vision. However, this three-way path needs to be strengthened with more research, as the number of studies conducted so far at the interplay between psychology and image processing/computer vision is relatively small. According to our review, only nine papers addressed depression and mental health. Personality was the focus of 19 studies. Most papers, namely, 44 of 90, cantered on emotions and sentiment analysis.

The connection between social media images and psychological meanings is possible through scripts and algorithms designed to perform three main tasks: (1) extracting basic information from pixel values; (2) detecting faces and objects; and (3) learning the required functionality using the extracted visual features, which involves the understanding of the visual contents at a high level of abstraction.

Understanding the basics of images and their features is the foundation for advanced computer vision analysis. Besides, some pixel-level transformations are required either to preprocess images or to enhance the performance of algorithms for data analysis representing the images differently.

The Viola-Jones algorithm remains one of the most powerful computer vision algorithms, although it was proposed nearly 18 years ago.

The third type of processing required for grasping psychological meanings from images has become possible only recently, with the development of deep learning. The CNNs are the present and the future of computer vision research since they improve classification performance and open doors to new and exciting prospects.

Most studies reviewed in this chapter, regardless of the topic they approached, included the extraction of colour and texture features as a way to obtain the input variables. However, the review presented in the current paper included examples of using other pixel-level features, too. Also, a trend in applying the Viola-Jones algorithm for face detection can be noticed. Deep learning methods have been employed mostly in research concerning emotions and sentiment analysis. Only a few studies incorporated such powerful methods in the prediction of personality and mental health from images, which suggests that much is yet to be done in such areas. Also, the work concerning emotions and sentiment analysis can be a valuable source of inspiration in terms of computer vision methods.

Nevertheless, as one can notice, making the most of social media images to take social science research to a new level requires proficiency in data science, including programming skills in languages such as Python. Acquiring such knowledge is not usually part of the current social science curriculum, which might appear as a problem. Two approaches to this challenge are possible for social scientists. The easiest way is to work in multidisciplinary teams that include computer scientists/data scientists. The alternative solution is for the social scientists to learn by themselves those skills, which is harder but more rewarding as it also broadens one's view when designing research and confers independence in study implementation. The current chapter encourages more the second approach, by introducing the reader to the basics of image processing and convolutional neural networks and showing that such knowledge can be accessible even for researchers without any background in computer science. Today, there is a vast amount of learning resources that kindle the transition of traditional social scientists specialized only in their field to computational social scientists who also have strong knowledge in computer vision and other data science methods.

References

A. Abdellaoui, H.Y. Chen, G. Willemsen, E.A. Ehli, G.E. Davies, K.J. Verweij, … J.T. Cacioppo, Associations between loneliness and personality are mostly driven by a genetic association with neuroticism. J. Pers. **87**(2), 386–397 (2018). https://doi.org/10.1111/jopy.12397

S. Abdullah, E.L. Murnane, J.M. Costa, T. Choudhury, Collective smile: measuring societal happiness from geolocated images, in *Proceedings of the 18th ACM Conference on Computer Supported Cooperative Work & Social Computing* (ACM, 2015), pp. 361–374. https://doi.org/10.1145/2675133.2675186

M. Amencherla, L.R. Varshney, Color-based visual sentiment for social communication, in *The 15th Canadian Workshop on Information Theory (CWIT)* (IEEE, 2017), pp. 1–5. https://doi.org/10.1109/CWIT.2017.7994829

S. Bakhshi, E. Gilbert, Red, purple and pink: the colors of diffusion on Pinterest. PLoS One **10**(2), e0117148 (2015). https://doi.org/10.1371/journal.pone.0117148

M.H. Bharati, J.J. Liu, J.F. MacGregor, Image texture analysis: methods and comparisons. Chemom. Intell. Lab. Syst. **72**(1), 57–71 (2004). https://doi.org/10.1016/j.chemolab.2004.02.005

S.K. Bhatti, A. Muneer, M.I. Lali, M. Gull, S.M.U. Din, Personality analysis of the USA public using Twitter profile pictures, in *2017 International Conference on Information and Communication Technologies (ICICT)* (IEEE, 2017), pp. 165–172. https://doi.org/10.1109/ICICT.2017.8320184

J.I. Biel, O. Aran, D. Gatica-Perez, You are known by how you vlog: personality impressions and nonverbal behavior in YouTube, in *Fifth International AAAI Conference on Weblogs and Social Media* (2011)

J.I. Biel, L. Teijeiro-Mosquera, D. Gatica-Perez, Facetube: predicting personality from facial expressions of emotion in online conversational video, in *Proceedings of the 14th ACM International Conference on Multimodal Interaction* (ACM, 2012), pp. 53–56

D.E. Bloom, E.T. Cafiero, E. Jané-Llopis, S. Abrahams-Gessel, L.R. Bloom, S. Fathima, … C. Weinstein, *The Global Economic Burden of Noncommunicable Diseases* (World Economic Forum, Geneva, 2011)

G. Bradski, The OpenCV library. Dr. Dobb's J. Softw. Tools **120**, 122–125 (2000)

F. Celli, E. Bruni, B. Lepri, Automatic personality and interaction style recognition from Facebook profile pictures, in *Proceedings of the 22nd ACM International Conference on Multimedia* (ACM, 2014), pp. 1101–1104. https://doi.org/10.1145/2647868.2654977

T. Chamorro-Premuzic, A. Furnham, Personality predicts academic performance: evidence from two longitudinal university samples. J. Res. Pers. **37**(4), 319–338 (2003). https://doi.org/10.1016/S0092-6566(02)00578-0

S.K. Chinnamgari, Achieving computer vision with deep learning, in *R Machine Learning Projects* (Packt Publishing, Birmingham, 2019)

Y.G. Cinar, S. Zoghbi, M.F. Moens, Inferring user interests on social media from text and images, in *2015 IEEE International Conference on Data Mining Workshop (ICDMW)* (IEEE, 2015), pp. 1342–1347. https://doi.org/10.1109/ICDMW.2015.208

P.W. Corrigan, B.G. Druss, D.A. Perlick, The impact of mental illness stigma on seeking and participating in mental health care. Psychol. Sci. Public Interest **15**(2), 37–70 (2014). https://doi.org/10.1177/1529100614531398

S. Datta, *Learning OpenCV 3 Application Development* (Packt Publishing, Birmingham, 2016)

J. Deeb-Swihart, C. Polack, E. Gilbert, I. Essa, Selfie-presentation in everyday life: a large-scale characterization of selfie contexts on instagram, in *Eleventh International AAAI Conference on Web and Social Media* (2017)

P. Dewan, A. Suri, V. Bharadhwaj, A. Mithal, P. Kumaraguru, Towards understanding crisis events on online social networks through pictures, in *Proceedings of the 2017 IEEE/ACM International Conference on Advances in Social Networks Analysis and Mining 2017* (ACM, 2017), pp. 439–446 https://doi.org/10.1145/3110025.3110062

S. Dey, *Hands-on Image Processing with Python* (Packt Publishing, Birmingham, 2018)

A. Dhall, J. Hoey, First impressions-predicting user personality from Twitter profile images, in *International Workshop on Human Behavior Understanding* (Springer, Cham, 2016), pp. 148–158. https://doi.org/10.1007/978-3-319-46843-3_10

A. Dhall, R. Goecke, T. Gedeon, Automatic group happiness intensity analysis. IEEE Trans. Affect. Comput. **6**(1), 13–26 (2015). https://doi.org/10.1109/TAFFC.2015.2397456

P.D. Ekstrom, C.M. Federico, Personality and political preferences over time: evidence from a multiwave longitudinal study. J. Pers. **87**(2), 398–412 (2018). https://doi.org/10.1111/jopy.12398

G. Farnadi, J. Tang, M. De Cock, M.F. Moens, User profiling through deep multimodal fusion, in *Proceedings of the Eleventh ACM International Conference on Web Search and Data Mining* (ACM, 2018), pp. 171–179. https://doi.org/10.1145/3159652.3159691

W. Fengjiao, M. Aono, Visual sentiment prediction by merging hand-craft and CNN features, in *2018 5th International Conference on Advanced Informatics: Concept Theory and Applications (ICAICTA)* (IEEE, 2018), pp. 66–71 https://doi.org/10.1109/ICAICTA.2018.8541312

B. Gawronski, J. De Houwer, Implicit measures in social and personality psychology, in *Handbook of Research Methods in Social and Personality Psychology*, ed. by H. T. Reis, C. M. Judd, 2nd edn., (Cambridge University Press, New York, NY, 2014), pp. 283–310

R.D. Goffin, A.C. Boyd, Faking and personality assessment in personnel selection: Advancing models of faking. Can. Psychol. **50**(3), 151–160 (2009). https://doi.org/10.1037/a0015946

C.E. Goodall, An overview of implicit measures of attitudes: Methods, mechanisms, strengths, and limitations. Commun. Methods Meas. **5**(3), 203–222 (2011) https://doi.org/10.1080/19312458.2011.596992

S.D. Gosling, O.P. John, K.H. Craik, R.W. Robins, Do people know how they behave? Self-reported act frequencies compared with on-line codings by observers. J. Pers. Soc. Psychol. **74**(5), 1337–1349 (1998). https://doi.org/10.1037/0022-3514.74.5.1337

E.K. Graham, J.P. Rutsohn, N.A. Turiano, R. Bendayan, P.J. Batterham, D. Gerstorf, ... E.D. Bastarache, Personality predicts mortality risk: an integrative data analysis of 15 international longitudinal studies. J. Res. Pers. **70**, 174–186 (2017). https://doi.org/10.1016/j.jrp.2017.07.005

S.C. Guntuku, W. Lin, J. Carpenter, W.K. Ng, L.H. Ungar, D. Preoţiuc-Pietro, Studying personality through the content of posted and liked images on Twitter, in *Proceedings of the 2017 ACM on Web Science Conference* (ACM, 2017), pp. 223–227. https://doi.org/10.1145/3091478.3091522

S.C. Guntuku, J.T. Zhou, S. Roy, W. Lin, I.W. Tsang, Who likes what and, why? 'Insights into modeling users' personality based on image 'likes'. IEEE Trans. Affect. Comput. **9**(1), 130–143 (2018). https://doi.org/10.1109/TAFFC.2016.2581168

H.A. Han, S. Czellar, M.A. Olson, R.H. Fazio, Malleability of attitudes or malleability of the IAT? J. Exp. Soc. Psychol. **46**(2), 286–298 (2010) https://doi.org/10.1016/j.jesp.2009.11.011

S.E. Hampson, G.W. Edmonds, M. Barckley, L.R. Goldberg, J.P. Dubanoski, T.A. Hillier, A Big Five approach to self-regulation: personality traits and health trajectories in the Hawaii longitudinal study of personality and health. Psychol. Health Med. **21**(2), 152–162 (2016). https://doi.org/10.1080/13548506.2015.1061676

S. Haseena, S. Bharathi, I. Padmapriya, R. Lekhaa, Deep learning based approach for gender classification, in *2018 Second International Conference on Electronics, Communication and Aerospace Technology (ICECA)* (IEEE, 2018), pp. 1396–1399. https://doi.org/10.1109/ICECA.2018.8474919

A. Hu, S. Flaxman, Multimodal sentiment analysis to explore the structure of emotions, in *Proceedings of the 24th ACM SIGKDD International Conference on Knowledge Discovery & Data Mining* (ACM, 2018), pp. 350–358. https://doi.org/10.1145/3219819.3219853

F. Huang, X. Zhang, Z. Zhao, J. Xu, Z. Li, Image–text sentiment analysis via deep multimodal attentive fusion. Knowl.-Based Syst. **167**, 26–37 (2019). https://doi.org/10.1016/j.knosys.2019.01.019

W.M.K.S. Ilmini, T.G.I. Fernando, Computational personality traits assessment: a review, in *2017 IEEE International Conference on Industrial and Information Systems (ICIIS)* (IEEE, 2017), pp. 1–6. https://doi.org/10.1109/ICIINFS.2017.8300416

J. Islam, Y. Zhang, Visual sentiment analysis for social images using transfer learning approach, in *2016 IEEE International Conferences on Big Data and Cloud Computing (BDCloud), Social Computing and Networking (SocialCom), Sustainable Computing and Communications (SustainCom) (BDCloud-SocialCom-SustainCom)* (IEEE, 2016), pp. 124–130. https://doi.org/10.1109/BDCloud-SocialCom-SustainCom.2016.29

K.S. Jennings, J.H. Cheung, T.W. Britt, K.N. Goguen, S.M. Jeffirs, A.L. Peasley, A.C. Lee, How are perceived stigma, self-stigma, and self-reliance related to treatment-seeking? A three-path model. Psychiatr. Rehabil. J. **38**(2), 109–116 (2015). https://doi.org/10.1037/prj0000138

Kang, K., Yoon, C., & Kim, E. Y., Identifying depressive users in Twitter using multimodal analysis, in *2016 International Conference on Big Data and Smart Computing (BigComp)* (2016), pp. 231–238. doi:https://doi.org/10.1109/BIGCOMP.2016.7425918

R. Karim, M. Sewak, P. Pujari, *Practical Convolutional Neural Networks. Implement Advanced Deep Learning Models Using Python* (Packt Publishing, Birmingham, 2018)

S. Kasper, Anxiety disorders: under-diagnosed and insufficiently treated. Int. J. Psychiatry Clin. Pract. **10**(sup1), 3–9 (2006). https://doi.org/10.1080/13651500600552297

R.C. Kessler, E.J. Bromet, The epidemiology of depression across cultures. Annu. Rev. Public Health **34**, 119–138 (2013). https://doi.org/10.1146/annurev-publhealth-031912-114409

Y. Kim, J.H. Kim, Using computer vision techniques on Instagram to link users' personalities and genders to the features of their photos: an exploratory study. Inf. Process. Manag. **54**(6), 1101–1114 (2018). https://doi.org/10.1016/j.ipm.2018.07.005

J.H. Kim, M.S. Kim, Y. Nam, An analysis of self-construals, motivations, Facebook use, and user satisfaction. Int. J. Hum.–Comput. Interact. **26**(11–12), 1077–1099 (2010). https://doi.org/10.1080/10447318.2010.516726

S.Y. Kim, R. Stewart, K.Y. Bae, S.W. Kim, I.S. Shin, Y.J. Hong, … J.M. Kim, Influences of the Big Five personality traits on the treatment response and longitudinal course of depression in patients with acute coronary syndrome: a randomised controlled trial. J. Affect. Disord. **203**, 38–45 (2016). https://doi.org/10.1016/j.jad.2016.05.071

M. Klassen, M.A. Russell, *Mining the Social Web*, 3rd edn. (O'Reilly Media, Inc, Sebastopol, 2019)

V. Kovalevsky, Image segmentation and connected components, in *Modern Algorithms for Image Processing* (Apress, Berkeley, 2019). https://doi.org/10.1007/978-1-4842-4237-7_9

K. Kroenke, R.L. Spitzer, J.B. Williams, P.O. Monahan, B. Löwe, Anxiety disorders in primary care: prevalence, impairment, comorbidity, and detection. Ann. Intern. Med. **146**(5), 317–325 (2007). https://doi.org/10.7326/0003-4819-146-5-200703060-00004

Y. LeCun, L. Bottou, Y. Bengio, P. Haffner, Gradient-based learning applied to document recognition. Proc. IEEE **86**(11), 2278–2324 (1998)

S. Li, W. Deng, Blended emotion in-the-wild: Multi-label facial expression recognition using crowdsourced annotations and deep locality feature learning. Int. J. Comput. Vis. (2018). https://doi.org/10.1007/s11263-018-1131-1

Z. Li, Y. Fan, B. Jiang, T. Lei, W. Liu, A survey on sentiment analysis and opinion mining for social multimedia. Multimed. Tools Appl. **78**(6), 1–29 (2018). https://doi.org/10.1007/s11042-018-6445-z

H. Lin, J. Jia, Q. Guo, Y. Xue, Q. Li, J. Huang, ... L. Feng, User-level psychological stress detection from social media using deep neural network, in *Proceedings of the 22nd ACM International Conference on Multimedia* (ACM, 2014), pp. 507–516

D. Lin, L. Li, D. Cao, Y. Lv, X. Ke, Multi-modality weakly labeled sentiment learning based on explicit emotion signal for Chinese microblog. Neurocomputing **272**, 258–269 (2018). https://doi.org/10.1016/j.neucom.2017.06.078

Y. Liu, D. Zhang, G. Lu, W.Y. Ma, A survey of content-based image retrieval with high-level semantics. Pattern Recogn. **40**(1), 262–282 (2007). https://doi.org/10.1016/j.patcog.2006.04.045

T. Liu, F. Jiang, Y. Liu, M. Zhang, S. Ma, Do photos help express our feelings: incorporating multimodal features into microblog sentiment analysis, in *Chinese National Conference on Social Media Processing* (Springer, Singapore, 2015), pp. 63–73

L. Liu, D. Preoţiuc-Pietro, Z.R. Samani, M.E. Moghaddam, L. Ungar, Analyzing personality through social media profile picture choice, in *Tenth International AAAI Conference on Web and Social Media* (2016)

R. Malhoski, A. Rock, Hue, saturation, value (HSV), in *Mapping with ArcGIS Pro* (Packt Publishing, Birmingham, 2018)

L. Manikonda, M. De Choudhury, Modeling and understanding visual attributes of mental health disclosures in social media, in *Proceedings of the 2017 CHI Conference on Human Factors in Computing Systems* (ACM, 2017), pp. 170–181. https://doi.org/10.1145/3025453.3025932

N. Mavridis, W. Kazmi, P. Toulis, Friends with faces: how social networks can enhance face recognition and vice versa, in *Computational Social Network Analysis*, ed. by A. Abraham, A.E. Hassanien, V. Snášel (Springer, London, 2010), pp. 453–482

R.R. McCrae, P.T. Costa, The five-factor theory of personality, in *Handbook of Personality: Theory and Research*, 3rd edn., ed. by O.P. John, R.W. Robbins, L.A. Pervin (Guilford, New York, 2008), pp. 159–181

G. Muhammad, M.F. Alhamid, User emotion recognition from a larger pool of social network data using active learning. Multimed. Tools Appl. **76**(8), 10881–10892 (2017). https://doi.org/10.1007/s11042-016-3912-2

C.J. Murray, T. Vos, R. Lozano, M. Naghavi, A.D. Flaxman, C. Michaud, ... V. Aboyans, Disability-adjusted life years (DALYs) for 291 diseases and injuries in 21 regions, 1990–2010: a systematic analysis for the Global Burden of Disease Study 2010. Lancet **380**(9859), 2197–2223 (2012). https://doi.org/10.1016/S0140-6736(12)61689-4

J. Nie, L. Huang, Z. Li, C. Wei, B. Hong, W. Zhu, Thinking like psychologist: learning to predict personality by using features from portrait and social media, in *2016 4th International Conference on Cloud Computing and Intelligence Systems (CCIS)* (IEEE, 2016), pp. 21–26 https://doi.org/10.1109/CCIS.2016.7790218

R. Pang, A. Baretto, H. Kautz, J. Luo, Monitoring adolescent alcohol use via multimodal analysis in social multimedia, in *2015 IEEE International Conference on Big Data (Big Data)* (IEEE, 2015), pp. 1509–1518. https://doi.org/10.1109/BigData.2015.7363914

C.P. Papageorgiou, M. Oren, T. Poggio, A general framework for object detection. In Sixth International Conference on Computer Vision (IEEE, 1998), pp. 555–562. https://doi.org/10.1109/ICCV.1998.710772

D.L. Paulhus, S. Vazire, Handbook of research methods in personality psychology, in *The Self-Report Method*, ed. by R. W. Robins, R. C. Fraley, R. F. Krueger, (The Guilford Press, New York, NY, 2007), pp. 224–239

X. Peng, J. Luo, C. Glenn, L.K. Chi, J. Zhan, Sleep-deprived fatigue pattern analysis using large-scale selfies from social media, in *2017 IEEE International Conference on Big Data (Big Data)* (IEEE, 2017), pp. 2076–2084. https://doi.org/10.1109/BigData.2017.8258154

D. Preoţiuc-Pietro, J. Carpenter, S. Giorgi, L. Ungar, Studying the Dark Triad of personality through Twitter behavior, in *Proceedings of the 25th ACM International on Conference on Information and Knowledge Management* (ACM, 2016), pp. 761–770. https://doi.org/10.1145/2983323.2983822

L. Rainie, J. Brenner, K. Purcell, Photos and videos as social currency online. Pew Internet & American Life Project (2012). Retrieved from https://www.pewinternet.org/2012/09/13/photos-and-videos-as-social-currency-online/

A.G. Reece, C.M. Danforth, Instagram photos reveal predictive markers of depression. EPJ Data Sci. **6**(1), 15 (2017). https://doi.org/10.1140/epjds/s13688-017-0110-z

R.W. Robins, J.S. Beer, Positive illusions about the self: Short-term benefits and long-term costs. J. Pers. Soc. Psychol. **80**(2), 340–352 (2001). https://doi.org/10.1037/0022-3514.80.2.340

M.G. Rothstein, R.D. Goffin, The use of personality measures in personnel selection: What does current research support? Hum. Resour. Manag. Rev. **16**(2), 155–180 (2006). https://doi.org/10.1016/j.hrmr.2006.03.004

A. Roy, A. Paul, H. Pirsiavash, S. Pan, Automated detection of substance use-related social media posts based on image and text analysis, in *2017 IEEE 29th International Conference on Tools with Artificial Intelligence (ICTAI)* (IEEE, 2017), pp. 772–779. https://doi.org/10.1109/ICTAI.2017.00122

J.C. Russ, F.B. Neal, Image enhancement in the spatial domain, in *The Image Processing Handbook*, 7th edn. (Taylor & Francis Group, Boca Raton, 2016), pp. 243–319

Z.R. Samani, S.C. Guntuku, M.E. Moghaddam, D. Preoţiuc-Pietro, L.H. Ungar, Cross-platform and cross-interaction study of user personality based on images on Twitter and Flickr. PLoS One **13**(7), e0198660 (2018). https://doi.org/10.1371/journal.pone.0198660

F.A. Sava, L.P. MaricuToiu, S. Rusu, I. Macsinga, D. Vîrgă, C.M. Cheng, B.K. Payne, An inkblot for the implicit assessment of personality: The semantic misattribution procedure. Eur. J. Personal. **26**(6), 613–628 (2012) https://doi.org/10.1002/per.1861

C. Segalin, D.S. Cheng, M. Cristani, Social profiling through image understanding: personality inference using convolutional neural networks. Comput. Vis. Image Underst. **156**, 34–50 (2016). https://doi.org/10.1016/j.cviu.2016.10.013

C. Segalin, F. Celli, L. Polonio, M. Kosinski, D. Stillwell, N. Sebe, ... B. Lepri, What your Facebook profile picture reveals about your personality, in *Proceedings of the 25th ACM International Conference on Multimedia* (ACM, 2017a), pp. 460–468. https://doi.org/10.1145/3123266.3123331

C. Segalin, A. Perina, M. Cristani, A. Vinciarelli, The pictures we like are our image: continuous mapping of favorite pictures into self-assessed and attributed personality traits. IEEE Trans. Affect. Comput. **8**(2), 268–285 (2017b). https://doi.org/10.1109/TAFFC.2016.2516994

D.V. Sheehan, Depression: underdiagnosed, undertreated, underappreciated. P&T Digest, **13**(6 Suppl Depression), 6–8 (2004)

G. Shen, J. Jia, L. Nie, F. Feng, C. Zhang, T. Hu, ... W. Zhu, Depression detection via harvesting social media: a multimodal dictionary learning solution, in *IJCAI* (2017), pp. 3838–3844

H. Singh, *Practical Machine Learning and Image Processing for Facial Recognition, Object Detection, and Pattern Recognition Using Python* (Apress, New York, 2019)

V.K. Singh, S. Hegde, A. Atrey, Towards measuring fine-grained diversity using social media photographs, in *Eleventh International AAAI Conference on Web and Social Media* (2017)

K. Song, T. Yao, Q. Ling, T. Mei, Boosting image sentiment analysis with visual attention. Neurocomputing **312**, 218–228 (2018). https://doi.org/10.1016/j.neucom.2018.05.104

C.J. Soto, Is happiness good for your personality? Concurrent and prospective relations of the big five with subjective well-being. J. Pers. **83**(1), 45–55 (2015). https://doi.org/10.1111/jopy.12081

G. Spacagna, I. Vasilev, D. Slater, V. Zocca, P. Roelants, *Python Deep Learning*, 2nd edn. (Packt Publishing, Birmingham, 2019)

J. Tan, M. Xu, L. Shang, X. Jia, Sentiment analysis for images on microblogging by integrating textual information with multiple kernel learning, in *Pacific Rim International Conference on Artificial Intelligence* (Springer, Cham, 2016), pp. 496–506. https://doi.org/10.1007/978-3-319-42911-3_41

L. Teijeiro-Mosquera, J.I. Biel, J.L. Alba-Castro, D. Gatica-Perez, What your face vlogs about: expressions of emotion and big-five traits impressions in YouTube. IEEE Trans. Affect. Comput. **6**(2), 193–205 (2015). https://doi.org/10.1109/TAFFC.2014.2370044

K. Tieu, P. Viola, Boosting image retrieval. Int. J. Comput. Vis. **56**(17) (2004). https://doi.org/10.1023/B:VISI.0000004830.93820.78

R. Torfason, E. Agustsson, R. Rothe, R. Timofte, From face images and attributes to attributes, in *Asian Conference on Computer Vision* (Springer, Cham, 2017), pp. 313–329. https://doi.org/10.3929/ethz-a-010811115

R. Tourangeau, The science of self-report: Implications for research and practice, in *Remembering What Happened: Memory Errors and Survey Reports*, ed. by A. A. Stone, J. S. Turkkan, C. A. Bachrach, J. B. Jobe, H. S. Kurtzman, V. S. Cain, (Lawrence Erlbaum Associates Publishers, Mahwah, NJ, 2000), pp. 29–47

J. Viinikainen, K. Kokko, Personality traits and unemployment: evidence from longitudinal data. J. Econ. Psychol. **33**(6), 1204–1222 (2012). https://doi.org/10.1016/j.joep.2012.09.001

A.F. Villán, *Mastering OpenCV 4 with Python* (Packt Publishing, Birmingham, 2019)

P. Viola, M. Jones, Rapid object detection using a boosted cascade of simple features. CVPR **1**(1), 511–518 (2001)

P.S. Wang, S. Aguilar-Gaxiola, J. Alonso, M.C. Angermeyer, G. Borges, E.J. Bromet, ... J.M. Haro, Use of mental health services for anxiety, mood, and substance disorders in 17 countries in the WHO world mental health surveys. Lancet **370**(9590), 841–850 (2007). https://doi.org/10.1016/S0140-6736(07)61414-7

X. Wei, D. Stillwell, How smart does your profile image look?: estimating intelligence from social network profile images, in *Proceedings of the Tenth ACM International Conference on Web Search and Data Mining* (ACM, 2017), pp. 33–40. https://doi.org/10.1145/3018661.3018663

H.A. Whiteford, L. Degenhardt, J. Rehm, A.J. Baxter, A.J. Ferrari, H.E. Erskine, ... R. Burstein, Global burden of disease attributable to mental and substance use disorders: findings from the global burden of disease study 2010. Lancet, **382**(9904), 1575–1586 (2013). https://doi.org/10.1016/S0140-6736(13)61611-6

D. Won, Z.C. Steinert-Threlkeld, J. Joo, Protest activity detection and perceived violence estimation from social media images, in *Proceedings of the 25th ACM International Conference on Multimedia* (ACM, 2017), pp. 786–794. https://doi.org/10.1145/3123266.3123282

World Bank Group, & World Health Organization, Out of the shadows: making mental health a global development priority (2016). Retrieved from http://pubdocs.worldbank.org/en/391171465393131073/0602-SummaryReport-GMH-event-June-3-2016.pdf

World Health Organization, Depression (2018). Retrieved from https://www.who.int/news-room/fact-sheets/detail/depression

A.G. Wright, Current directions in personality science and the potential for advances through computing. IEEE Trans. Affect. Comput. **5**(3), 292–296 (2014). https://doi.org/10.1109/TAFFC.2014.2332331

L. Wu, M. Qi, H. Zhang, M. Jian, B. Yang, D. Zhang, Establishing a large scale dataset for image emotion analysis using Chinese emotion ontology, in *Chinese Conference on Pattern Recognition and Computer Vision (PRCV)*, ed. by J.-H. Lai et al. (Springer, Cham, 2018), pp. 359–370. https://doi.org/10.1007/978-3-030-03341-5_30

X. Xiong, M. Filippone, A. Vinciarelli, Looking good with Flickr faves: Gaussian processes for finding difference makers in personality impressions, in *Proceedings of the 24th ACM*

International Conference on Multimedia (ACM, 2016), pp. 412–415. https://doi.org/10.1145/2964284.2967253

H. Xiong, H. Liu, B. Zhong, Y. Fu, *Structured and Sparse Annotations for Image Emotion Distribution Learning* (Association for the Advancement of Artificial Intelligence, 2019)

Y. Yang, J. Jia, B. Wu, J. Tang, Social role-aware emotion contagion in image social networks, in *Thirtieth AAAI Conference on Artificial Intelligence* (2016), pp. 65–71

J. Yang, D. She, M. Sun, Joint image emotion classification and distribution learning via deep convolutional neural network, in *IJCAI* (2017), pp. 3266–3272

J. Yang, D. She, Y. Lai, M.H. Yang, Retrieving and classifying affective images via deep metric learning. Association for the Advancement of Artificial Intelligence (2018)

M. Yazdani, L. Manovich, Predicting social trends from non-photographic images on Twitter, in *2015 IEEE International Conference on Big Data (Big Data)* (IEE, 2015), pp. 1653–1660. https://doi.org/10.1109/BigData.2015.7363935

A.H. Yazdavar, M.S. Mahdavinejad, G. Bajaj, W. Romine, A. Monadjemi, K. Thirunarayan, … J. Pathak, Fusing visual, textual and connectivity clues for studying mental health (2019) *arXiv*

Q. You, S. Bhatia, T. Sun, J. Luo, The eyes of the beholder: gender prediction using images posted in online social networks, in *2014 IEEE International Conference on Data Mining Workshop* (IEEE, 2014), pp. 1026–1030. https://doi.org/10.1109/ICDMW.2014.93

Q. You, D. García-García, M. Paluri, J. Luo, J. Joo, Cultural diffusion and trends in Facebook photographs, in *Eleventh International AAAI Conference on Web and Social Media* (2017)

Y. Zhang, L. Shang, X. Jia, Sentiment analysis on microblogging by integrating text and image features, in *Pacific-Asia Conference on Knowledge Discovery and Data Mining* (Springer, Cham, 2015), pp. 52–63. https://doi.org/10.1007/978-3-319-18032-8_5

S. Zhao, H. Yao, Y. Gao, G. Ding, T.S. Chua, Predicting personalized image emotion perceptions in social networks. IEEE Trans. Affect. Comput. **9**(4), 526–540 (2018a). https://doi.org/10.1109/TAFFC.2016.2628787

S. Zhao, X. Zhao, G. Ding, K. Keutzer, EmotionGAN: unsupervised domain adaptation for learning discrete probability distributions of image emotions, in *2018 ACM Multimedia Conference on Multimedia Conference* (ACM, 2018b), pp. 1319–1327. https://doi.org/10.1145/3240508.3240591

Validating Simulation Models: The Case of Opinion Dynamics

Klaus G. Troitzsch

1 Introduction

Social science is about complex, nested systems consisting on the lowest level of human beings who, unlike the elements at the bottom of other complex systems, have properties and capabilities which, to the best of our knowledge, are unique. Among them are the use of symbolic languages of very high versatility and high ambiguity, a much longer memory than any other living things, let alone non-living things, the ability to communicate about the counterfactual (for instance, angels and unicorns), to lie and hide or counterfeit their internal states in their communicative acts, and the ability to derive regularities from what they observe and, moreover, to learn from others about regularities of processes they never observed themselves (Troitzsch 2012). These properties and capabilities lead to what has been called "second-order emergence": "people are routinely capable of detecting, reasoning about and acting on the macro-level properties (the emergent features) of the societies of which they form part" (Gilbert 1995, p. 151). And as new features emerge on the macro and intermediate levels of societies, societies evolve and are rarely in an equilibrium (Anzola et al. 2018), hence it is worthwhile to analyse societies from a non-equilibrium perspective—which makes it difficult to do so with the symbol systems (Ostrom 1988) of natural language and, even, of mathematics. Mathematics are helpful for analysing systems far from equilibrium but often do not provide closed solutions such that numerical treatment is necessary, but this is already very similar to simulation, and if mathematical methods are to be used on multilevel problems, they can usually only be applied to systems consisting

K. G. Troitzsch (✉)
Institut für Wirtschafts- und Verwaltungsinformatik (retired), Universität Koblenz-Landau, Mainz, Germany
e-mail: klaus.g.troitzsch@bluewin.ch

T. Rudas, G. Péli (eds.), *Pathways Between Social Science and Computational Social Science*, Computational Social Sciences,
https://doi.org/10.1007/978-3-030-54936-7_6

of homogeneous elements—and human beings in most interesting settings are far from all alike.

This is where a third symbol system comes into the play which, derived from mathematics, can deal with very heterogeneous settings: the programming languages used for designing algorithms that can be executed by machines. Hence computational social science becomes the way to solving some of the problems of social science which are difficult to treat in the symbol systems of natural language and systems of ordinary, partial and stochastic differential equations—and this dates back to the 1960s when first simulation models were written by psychologists, sociologists and political scientists, as it were computational social scientists *avant le mot*. As Abelson (1964, p. 159) put it, "One advantage of the simulation approach over the mathematical models approach is that no restrictive simplicity considerations are imposed on the functional forms—linearity is not necessarily preferred. On the other hand, one apparent disadvantage is that so large a number of degrees of freedom are available to the model-builder that the model does not clearly stand or fall in any empirical test." Beside the latter argument of the too many degrees of freedom, one should note that because of the restriction that empirical knowledge of these features of the real-world actors can be obtained with adequate research methods, not all problems which the social sciences have to face can be solved with the help of agent-based models: not all of the (too) many degrees of freedom of agent-based models can be covered with sufficient measurements of important features of human actors or of corporate actors such as firms or municipalities, as the empirical methods to describe this multitude of features, properties and action rules are still defective at least in cases when human actors are not even themselves aware of the rules which govern their behaviour and actions. This makes validation of models of all kind difficult enough.

Anyway, another attempt will be made in this chapter as it tries to link research of more than six decades of empirical social research to agent-based modelling in a field which lends itself to quantitative analysis to formulate hypotheses about human behaviour and actions which are difficult, if at all, to observe: the forming of attitudes as the result of observation and communication, where the human actors are often not aware themselves how their attitudes and opinions came about. Taking into account that attitudes were found measurable as early as in the 1930s (Abelson 1964, p. 142 praises Thurstone (1928) for having declared that "attitudes can be measured"), the history of attitude research can be traced even 90 years back.

The next Sect. 2 discusses the problem of validating models (of all kinds) against observation, before Sect. 3 gives an overview of approaches to model opinion and attitude dynamics, mostly from a hardly empirical perspective, and leads to an extended model which has empirical applications as it uses terms which can be (and have been) observed in many empirical studies. Some of these studies are mentioned and analysed in depth in Sect. 4, while validation and calibration issues are discussed in Sect. 5. The final Sect. 6 discusses the results of the simulation and of the empirical analysis and opens a perspective on the validation—and a necessary further extension—of the extended model presented in Sect. 3.3. An appendix gives additional information mainly about the empirical analysis.

2 Types of Validity: Validation Against Empirical Data and Against Stylised Facts

2.1 Types of Validity

Zeigler (1985) distinguished three kinds of validity (and hence ways to validate models), namely, replicative, predictive and structural validity. Assessing replicative validity is possible wherever a model describes a part of reality which has been observed and documented with measurements of as many properties of the target system and its elements as possible, and validity is assured to the degree in which the model output is equal or similar to the documented measurements. At first sight, this seems simple and easy to achieve, but usually the model— and particularly the simulation model formulated in a programming language—is a full model of its underlying theory in the sense of the "non-statement view" Balzer et al. (1987), i.e., does not need to make a difference between observable and theoretical variables in the sense of, for instance, Nagel (1961, p. 146) who distinguished between "the smells of cooking" (physically real) and " the pains a man feels when he turns an ankle (not physically real) or in the sense of the "non-statement view" which distinguishes terms which are "theoretical with respect to a certain theory" and others which are not. The target system whose measured properties are compared to the attributes of the models, however, cannot be described with all terms in the model, as the latter contains terms which have never been measured—as for instance the rules according to which, anticipating the examples in Sect. 4, a person changes her attitudes towards a political issue or a political party after having discussed about this issue or party with other persons, as most interviewees will not be in a position to give a detailed account of why they changed their opinions as a result of such discussions. Anyway, according to Zeigler's definition of replicative validity, it seems to be sufficient to have whatever can be observed in the target system similar enough to the outcome of the simulation model.

This replicative validity, however, is not enough, as the discussion of spurious correlation shows when regression models replicate the regressand with some high precision where it is clear that the regressor is not the cause of the regressand (as in the classical example of the storks and the babies (Matthews 2000)). And even an extension of such applications (in which spurious correlations occur) to structural equations models (SEM) (Holland 1988) does not lead to structural validity in the sense of Zeigler, as these typically do not make the dynamics of the underlying processes explicit but restrict themselves to introducing intervening variables (such as, in the storks-and-babies case, the percentage of farmers in the active population, the agricultural area where storks can find food, the percentage of people with or without a compulsory insurance against the risks of old age and the readiness of children to care for their ancestors). But the emergence of institutions such as pension funds and behaviours of children going abroad instead of stepping into their parents' shoes typically remains undiscussed in approaches like SEM. And even time series analysis of the ARIMA and other sophisticated approaches, although taking the dynamical changes of the variables—for instance demographic ones—in question into account only rarely discuss the mechanisms behind demographic processes, and these processes obviously violate the "stable unit treatment value assumption (SUTVA)" (Sobel 2008, p. 117, 127). Only sophisticated microsimulation models bridge this gap, but these are very similar to agent-based models (Troitzsch 2020).

The step from replicative to predictive validity is not very far: A predictive model describes the future state of the target system it represents, and if one waits long enough to be able to compare the prediction to the state which the target system has meanwhile reached and finds that predicted and reached state are similar enough, then predictive validity is achieved. This can be entirely misleading, as the example of the precise prediction of solar eclipses by Babylonian priests easily shows—the underlying model was not structurally valid.

Zeigler's structural validity is not properly described when he writes that it is achieved when a model "not only reproduces the observed real system behaviour, but truly reflects the way in which the real system operates to produce this behaviour" (Zeigler 1985, p. 5). To make this operational one must be able to compare "the way in which the real system operates" to the way in which the model operates—the latter being no problem at all, the former being impossible in certain cases, and particularly in the social sciences which deal with human beings who often enough do not know themselves the way in which they operate. Hausman's analogy Hausman (1994, p. 219) of the mechanic who examines the engine of a used car to provide relevant and useful information about the future performance of this used car just illustrates Zeigler's postulate to find out "the way in which the real system operates" or, in Hedström's terms, "focus[es] on the mechanisms that generate change in social entities" (Hedström 2005, p. 24). Hence, predictive power is not sufficient for validation, or as (Hausman 1994, p. 220) puts it: *"Even if all one cares about is predictive success in some limited domain, one should still be concerned about the realism of the assumptions of an hypothesis and the truth of its irrelevant or unimportant predictions"* (his italics)—and the latter are typically suppressed in stylised facts (see below).

But perhaps agent-based modelling opens a way to make a step forward at explaining what happens in persons (individual or corporate) when they make their decisions, change their minds or start an activity—by describing such events in a simulation models with sufficient replicative and predictive validity and afterwards designing new experiments on real persons to find out to which degree the assumptions about the "the way in which the real system operates" are in line with what the new experiments or new surveys, perhaps inspired by the models discussed in this chapter, found out.

A majority of agent-based models up to now do not have a solid empirical base, "are too abstract and too far from reality" (Waldherr and Wijermans 2013, p. 3.8) (see also Troitzsch (2017, p. 413)—to name only a few critical voices from inside the ABM community) and try only to replicate so-called "stylised facts" (Heine et al. 2005; Kaldor 1968; Meyer 2019; Solow 1970)—the latter are mental models made from repeated occasional and sometimes anecdotical observations of the macro behaviour of some system, (one of the classical stylised facts being " a continued increase in the amount of capital per worker, whatever statistical measure of 'capital' is chosen in this connection." (Kaldor 1968, p. 178)). Meyer (2019, p. 391), quoting Grimm et al. (2005) argues that "stylized facts can be used as a starting point to identify the basic assumptions and parameters in the model that are responsible for the reproduction of stylized facts" and comes to the conclusion that "ideally, at the end of such a process, a consensus emerges, at least with regard to some stylized facts. Before such a consensus is reached, the transparency of derivation, *the amount and consistency of empirical results* and the independence from specific theories or streams of literature may serve as supporting indicators of stylized fact quality" (my italics). If modellers restrict themselves to take such stylised facts for granted and implement them into a running simulation model they will expect that such a model is to a certain degree valid in all three senses of Zeigler's, but this is not necessarily the case, as the examples in Sect. 3, compared to the findings in Sect. 4, will show where the validation is not against a stylised fact but against the empirical data of a specific case. These examples were taken from a field of social science (somewhere between sociology and political science) which raised a lot of interest in the past six decades—even outside scientific journals (Deffuant et al. 2003; von Randow 2003), namely, polarisation effects and the rise of extremism in populations—and its empirical counterparts reflected in many surveys carried out worldwide and in many countries whose results are nearly immediately published to a wide audience at least in the times of ongoing elections.

3 Models of Opinion and Attitude Dynamics

3.1 One-Dimensional Models

One-dimensional models of opinion dynamics go back at least to Downs (1957) who, building on Hotelling's spatial theory of competition (Hotelling 1929) (who, in turn, applied this theory to two cider merchants but also, in an aside, to two political parties). Whereas Hotelling was mainly interested in the changing positions of sellers and parties, Downs also discussed "movements of men along the political scale" when he tried to find out the micro-foundation of the "political cycle typical of revolution" but did not discuss in detail how "voters can somehow be moved to the center of the scale to eliminate their polar split" (Downs 1957, p. 120). Voter movements on the ideological landscape seem to have first been analysed by Abelson and Bernstein (1963, p. 108: rules A18–A22); see also Davis et al. (1970) who discussed the effect of bimodal distributions but restricted it to a one-dimensional landscape (without using this word); see also Enelow and Hinich (1984) and Rabinowitz and Macdonald (1989).

Hegselmann and Krause (2002, 2006) were the first to give these earlier models a strict mathematical treatment and to convert them to simulation models (but see

Abelson 1964; Abelson and Bernstein 1963 with simulation models whose code is no longer available). However, they mainly "consider a group of agents (or experts or individuals of some kind) among whom some process of opinion formation takes place" (Hegselmann and Krause 2002, p. 2), "e.g., a group of experts asked by the UN to merge their different assessments on, say, the magnitude of the world population in the year 2020, into one single judgement",[1] but not so much a group of the size of "an entire society in which the individuals ruled by various networks of social influences develop a wide spectrum of opinions" [p. 1]. Their models are restricted to a one-dimensional opinion space (with the exception of their Appendix C [p. 30]), and they are deterministic, except that they start from random initial distributions.

Lorenz (2017) seems to have been the first to compare output from such a one-dimensional model of opinion dynamics to empirical data, building on earlier research (Fowler and Laver 2008; Kollman et al. 1992, 1998; Laver 2005; Laver and Schilperoord 2007) on party competition, and tuned his NetLogo model (Lorenz 2012) to produce output quite similar to distributions of left-right ideology landscapes[2] Lorenz (2017, p. 256) of Sweden and France from 2002 to 2012. In his model he uses the algorithm first introduced by Deffuant et al. (2000), restricted it to the mixing parameter $\mu = 0.5$ and added a random element which, with a user-set probability, had an agent change its opinion to a random value with no influence, the latter idea first introduced by Pineda et al. (2009)—they argue that "quite generally, this rule is equivalent to allowing each agent to return to a specific, basal, opinion preferred by him, provided that the basal opinions are randomly distributed amongst the agents". But this quotation does not explain why an agent should "return" to a random opinion which is likely to be different from "a specific, basal, opinion preferred by" this agent (which, as a software object has neither sex nor gender). Hence, a plausible alternative would have been a random effect added to the opinion which came out of the meeting between the agent and its partner. Anyway, this resulted In Lorenz' paper in a visually acceptable similarity between the empirical and the simulated frequency distribution over the one-dimensional left-right ideology (Pineda and his colleagues did not care for empirical validation).

3.2 Multiple Attitude Dynamics

More often than not, one finds a multidimensional space in which opinions can be represented, one of the earliest hint at such multidimensionality being Stokes (1963, p. 370) who argued that "the axiom of unidimensionality is difficult to reconcile

[1]This is particularly true for Krause (2005).

[2]The term "landscape" seems to have first been used by Kollman et al. (1998) where a first landscape with hills and valleys above a plane spanned by two issue dimensions appears on p. 148–149, much resembling the diagrams in Troitzsch (1987, p. 179–185).

with the evidence from multiparty systems as well. The support for the parties of a multiparty system is often more easily explained by the presence of several dimensions of political conflict than it is by the distribution of the electorate along any single dimension" (see also Davis et al. 1970, p. 429). Downs himself doubted that a linear scale from 0 for full government control over the economy to 100 for a completely free market would be appropriate, as he found "this apparatus ... unrealistic for the following two reasons: (1) actually each party is leftish on some issues and rightish on others, and (2) the parties designated as right wing extremists in the real world are for fascist control of the economy rather than free markets" (Downs 1957, p. 116), but he "ignore[d] these limitations temporarily" and never came back to his doubts (instead he gave different political issues different weights and positions, but still on a unidimensional scale, both for two-party and multiparty systems). And Lorenz' examples (France and Sweden) were taken from multiparty systems. Moreover, many surveys ask for opinions, values and attitudes with batteries of questions from which, via multidimensional scaling, factor analysis and other statistical methods quantitative scales can be derived, and often these batteries yield more than one scale (see the results in Tables 1 and 2 below in Sect. 4 where the first two eigenvalues of the correlation matrices are less than 50 per cent—41.0 and 38.4 per cent, respectively—of the total variance). This is sufficient reason to extend the models in Deffuant et al. (2000), Pineda et al. (2009) and Lorenz (2012) to two dimensions and experiment with some random effects (as without these only non-empirical results can be generated with only one or very few distinct opinions shared by all agents).

Jager and Amblard (2005a,b) seem to have been the first (but see Troitzsch 1987) who have analysed "multiple attitude dynamics" in the recent past, when they argue "up till now there has hardly been any attention for the dynamics of multiple attitude dynamics" among the "increasing number of scientists [who] study attitude or opinion dynamics using multi-agent models" (Jager and Amblard 2005a, p. 16). They were mainly interested in the dynamics in two attitude dimensions when agents communicated about both in their narrow spatial neighbourhood, and they were not able to test their assumptions about individual behaviour against individual data. They argue "Experimenting with the dynamics of attitude or opinion dynamics is not possible using laboratory studies. Field data on the contrary are too complex to identify the causalities of observed dynamical processes" Jager and Amblard (2005a, p. 1). For the laboratory studies, this is certainly true, as it seems to be impossible to construct an experimental design which keeps a considerable number of test persons over some considerable time free from any influence from outside the lab situation and have them communicate only with a small number of the other test persons. Moreover such an experimental setting, if it were feasible, would be far from representative for a population such as the one taken into consideration in their paper (the population of France and the Netherlands having just rejected the European Constitution). And field data in representative studies usually do not contain information about the "neighbours" of the interviewees (perhaps with the exception of household panels such as the German Socioeconomic Panel, but even there it is difficult "to identify the causalities of observed dynamical processes", and this does not only apply because only the attitudes of the few other household members—and not those of friends, colleagues and neighbours—are available for analysis, but also as nothing is actually known about the communication about these attitudes within a household). For this reason no attempt was made in Sect. 4.1 to use the information about other household members' attitudes, but after all, the individual histories of panel members' attitudes could be traced in both panel studies used in Sect. 4 and compared to the results in Jager and Amblard (2005a) whose diagrams "relation A and B" in their Figs. 1 and 3 show the paths of all agents through their common attitude space, just as in Figures 1 (diagram at the far right), 3 and 6. Comparing all these diagrams shows that the replicative validity of these two-dimensional models with respect to the agents' movements through their common attitude space is rather good, as the distances of these movements are more or less the same in the models and the panels.

3.3 A Two-Dimensional Model Along the Lines of Hegselmann-Krause and Deffuant-Weisbuch

If one analyses the algorithms of opinion change in one dimension adopted by Hegselmann and Krause (2002) and Deffuant et al. (2000) (for instance, in the versions given in Lorenz (2012)) one finds

$$x_i(t+1) = x_i(t) + \alpha \frac{1}{|X_e|} \sum_{j \in X_e} \mu(x_i(t) - x_j(t)) + (1-\alpha)\tau \tag{1}$$

where

- x_i is the position of the agent i currently changing its opinion,
- α is the fraction of the influence other agents' opinions have in comparison to other sources of influence (this is 1 in Deffuant et al. (2000) and in the DW version of Lorenz (2012)),
- X_e is the set of agents who influence i's opinion (which has only one element in Deffuant et al. (2000) and in the DW version of Lorenz (2012)),
- μ is the size of the influence which is usually set to 0.5 (definitely so in Lorenz (2012) and Hegselmann and Krause (2002))
- x_j is the position of any agent j within X_e (including x_i, but as $x_i(t) - x_i(t) = 0$, this does not really matter),
- τ is the position of an alternative source of influence on i (τ only exists in Hegselmann and Krause (2006) as the "true value"; in the current context it could represent the position of the party preferred by agent i).

Lorenz (2017) introduced a random effect which, with a certain probability, overwrites the opinion which had been changed immediately before with a new opinion which is randomly distributed over the whole opinion landscape, just as if

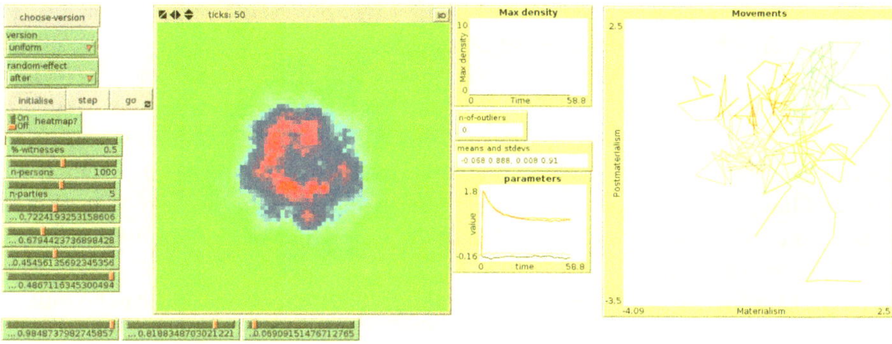

Fig. 1 Interface of the model

the agent had been replaced by an entirely new agent (and this is obviously also the reason to call this function `entry-exit`). As this element of his model can be considered to be not entirely plausible, an alternative is integrated into the model presented in this paper—it adds this random movement simultaneously with the adaptation to the peers, which is perhaps even better in line with equation 1, as the influence of the $x_j \in X_e$ and τ is only blurred a little by the stochastic influence and not entirely replaced. The sensitivity analysis will have to show which of these two random causes has which effects, and the validation will have to show which of them best replicates empirical data both on the micro and the macro level, but without any of these random effects, the agent population will move to one or only very few points of the attitude landscape as used to be the case with (Deffuant et al. 2000; Hegselmann and Krause 2002, 2006).

For reasons already discussed above, the model presented in this paper, equation 1 is replaced by equation 2 with an additional random term. In this equation the symbols have mostly the same meaning as above, but the opinions are no longer scalars but vectors—and for allowing a display of the results on paper, these vectors have just two elements, such that the opinion space is now two-dimensional (which is much more like a "landscape" than a one-dimensional space). The more or less technically motivated restriction to two dimensions might be justified when one considers that usually the first two eigenvalues of batteries of questions about values, attitudes, opinions and concerns cover at least two thirds of the overall variance (as Tables 1 and 2 below in Sect. 4 show).

The parameter τ or τ_i was taken from Hegselmann and Krause (2006) and their "assumptions that *there is a true value T* in our opinion space" which was later on called τ in Douven and Riegler (2010) (the role of α is used as in the latter, not as in the former). In a two-dimensional landscape, $\boldsymbol{\tau}$ can be interpreted as the position of a party, and $\boldsymbol{\tau}_i$ would be the position of i's preferred party (see Downs 1957). The current model allows for runs with and without parties:

$$\mathbf{x_i}(t+1) = \mathbf{x_i}(t) + \alpha \frac{1}{|X_e|} \sum_{j \in X_e} \mu(\mathbf{x_i}(t) - \mathbf{x_j}(t)) + (1-\alpha)\nu(\mathbf{x_i} - \boldsymbol{\tau}_i) + \boldsymbol{\epsilon} \qquad (2)$$

with the same symbols as in equation 1 and $\boldsymbol{\epsilon}$ a random vector whose orientation is uniformly distributed between 0 and 360 degrees and whose length may be either uniformly distributed between 0 and ϵ_U or normally distributed with mean 0 and standard deviation ϵ_σ. X_e includes all agents with similar opinions, i.e., within a circle with radius r around $\mathbf{x_i}(t)$. Unlike Lorenz (2012) the initial distribution of the agents over the opinion plane can either be normally distributed with its mean in the centre of the plane or standard deviation of a size that no agent lies beyond the limits of the plane (this restriction is only due to the limitations of the tool used, namely, NetLogo (Wilensky 1999).[3])

With the alternatives offered by the model, eight different versions can be run:

- with (five) parties or without parties,
- with agents' initial attitudes distributed following either a standard normal distribution ($\boldsymbol{\mu}_\mathbf{x} = \mathbf{O}$ and $\boldsymbol{\Sigma} = \boldsymbol{I}$) or a uniform distribution (uncorrelated between -3 and $+3$ in both directions),
- with random attitude changes either simultaneously with or with a certain probability ϵ_S separately instead of the adaptation to the attitudes of similar agents.

For sensitivity analysis and calibration, the parameters of equation 2 are varied as follows (in terms of a Monte Carlo approach):

α $\in [0.5, 1.0]$ if there are any other sources of information like $\boldsymbol{\tau}$, otherwise $\alpha = 1$,

X_e containing all agents x_j within a distance of $\rho \in [0.05, 1.0]$ in attitude space (called "peers" in the following),

μ the strength of the attracting force of the peers,

ν the strength of the attracting force of the nearest party if there is any,

ϵ_U the upper limit of the uniform distribution of $\boldsymbol{\epsilon}$ in the version controlled by the uniform distribution, the lower limit being 0,

ϵ_σ the standard deviation of the normal distribution of $\boldsymbol{\epsilon}$ in the version controlled by the normal distribution, the mean being 0,[4]

ϵ_S the probability of a random movement instead of a systematic movement triggered by peers and party—the movement itself will then be a random move from the current position as if μ and ν were 0[5]

and in this phase some parameters of the final distribution of the agents in their attitude space—those which can easily be compared to empirical data—are chosen,

[3]To be in line with the model in Sect. 4, the landscape has its centre at (0, 0) and it has no bounds; for representing the agents on NetLogo's patches, their coordinates are multiplied with a factor such that (nearly) all of them occupy patches within the view. This makes it possible to represent outliers in the model without violating NetLogo's restrictions.

[4]ϵ_U and ϵ_σ control the length of the vector $\boldsymbol{\epsilon}$, whereas the direction of this vector is always uniformly distributed between 0° and 360°.

[5]This is different from the approach in Lorenz (2012) where the new position is entirely random over the whole landscape.

namely, the means, standard deviations and the correlation of the agents' positions are measured, together with more visual information about this two-dimensional distribution. These have to be compared to empirical cases in the next section to find out which of the eight versions and, for each of these versions, which parameter combinations yield a final distribution which is most similar to a distribution in an empirical survey as well a realistic model of the size of the individual movements of the agents on the attitude landscape.

4 Opinion and Attitude Dynamics: Empirical Findings

There are few datasets available reporting opinions or attitudes for the same interviewees over a longer period of time. The following two subsections use the German Socio-Economic Panel (GSOEP) (Schupp 2009; Socio-Economic Panel (SOEP) 2017) and the 2017 German National Election Campaign Panel (Roßteutscher et al. 2018).

For both studies a number of items were selected, a factor analysis was calculated for both groups of items (principal component extraction of two factors, varimax rotation), and the two-dimensional frequency density functions above the plane spanned by these two factors were estimated with an algorithm described in Cobb (1978), Herlitzius (1990), and Troitzsch (2018). The form of these frequency density functions is always

$$\begin{aligned} f(x,y;\boldsymbol{\theta}) = \exp\{ \; & \theta_{00} & + \theta_{10}x & + \theta_{20}x^2 + \cdots + \theta_{n0}x^n \\ & + \theta_{01}y & + \theta_{11}xy & + \cdots \quad + \theta_{n-1,1}x^{n-1}y \\ & + \cdots & & \\ & + \cdots & + \theta_{ij}x^iy^j & + \cdots \\ & + \theta_{0n}y^n\} & & \end{aligned} \tag{3}$$

with n even, $i+j \leq n$, $i, j \geq 0$, $\theta_{n0}, \theta_{0n} < 0$, and $\int_{-\infty}^{\infty}\int_{-\infty}^{\infty} f(x,y;\boldsymbol{\theta})dydx = 1$

4.1 Reported Concerns in the German Socio-Economic Panel from 1984 till 2016

There are five questions about attitudes which were present in all 33 waves so far; see Table 1. A factor analysis with these five variables produced two factors with loadings and eigenvalues also shown in Table 1 (see also Table 4 in the Appendix). The factor analysis was run over all 33 waves yielding zero means and unit variances for both, but the lower right diagram in Fig. 2 shows that between waves there are severe differences between means and slight differences between standard deviations, once more making clear that these two parameters can be

Table 1 Variables about concerns available for all 33 waves of the GSOEP

How concerned are you about the following issues? Very concerned (3)/Somewhat concerned (2)/Not concerned at all (1) [a]

		Factor loadings	
Concern	Communality	Materialism	Postmaterialism
Worried about economic development	0.485	0.605	0.344
Worried about finances	0.727	0.850	0.066
Worried about environment	0.730	0.004	0.855
Worried about peace	0.683	0.169	0.809
Worried about job security	0.659	0.811	-0.022
Eigenvalues	(65.690%)	1.776	1.508

[a]In the original questionnaire the codes were 1 for very concerned and 3 for not concerned at all

misleading for heavily skewed and/or multimodal distributions (not in 1 year was the distribution similar to a normal distribution—which seems to be not only due to the poor quality of the underlying variables). These two factors resemble the concepts of materialism and postmaterialism first discussed by Inglehart (1977), but they describe concerns derived from these two attitudes. The two-dimensional frequency density functions above the plane spanned by these two factors can be seen in Fig. 2 (see also Fig. 10 in the Appendix).

Over most of the period so far covered by the GSOEP, people are very concerned about problems connected with environment and peace, as the highest maximum of the frequency density function (FDF for short) is always—until 2003—in the upper left corner of the coordinate system, meaning that these interviewees were mainly "very concerned" about environment and peace, but not so much about material issues. Often there is a smaller maximum of the FDF in the upper right corner containing interviewees "very concerned" about all four issues, but in all 33 years the FDF extends into the lower left corner where interviewees are found who are not concerned at all about all five issues. From 2004 till 2006, the maximum of the FDF moves to the upper right corner, signalising that concerns about economic development and own economic situation are nor becoming stronger; 2007 the main maximum of the FDF moves back to the upper left quadrant. From 2011 to 2014 the concern about peace and environment becomes more or less medium, whereas the concern about material issues is still low; in 2015 and 2016 both begin to grow.

Figure 2 shows the history of the distribution of the concerns of the panel members over 33 years. Whereas the details—and also the minima, maxima, saddle points and also the means—of the frequency distribution function change considerably, the standard deviation remains surprisingly constant. This is an effect which is difficult to achieve in the model of Sect. 3, at least in the version using the normal distribution for initialisation and opinion changes.

Figure 3 shows a few diagrams showing individual movements of interviewees in attitude space which make clear that the movements are rather long and each of them covers a large part of the attitude space. To make the extent of these movements comparable between different applications, an ellipse spanned by eigenvalues of

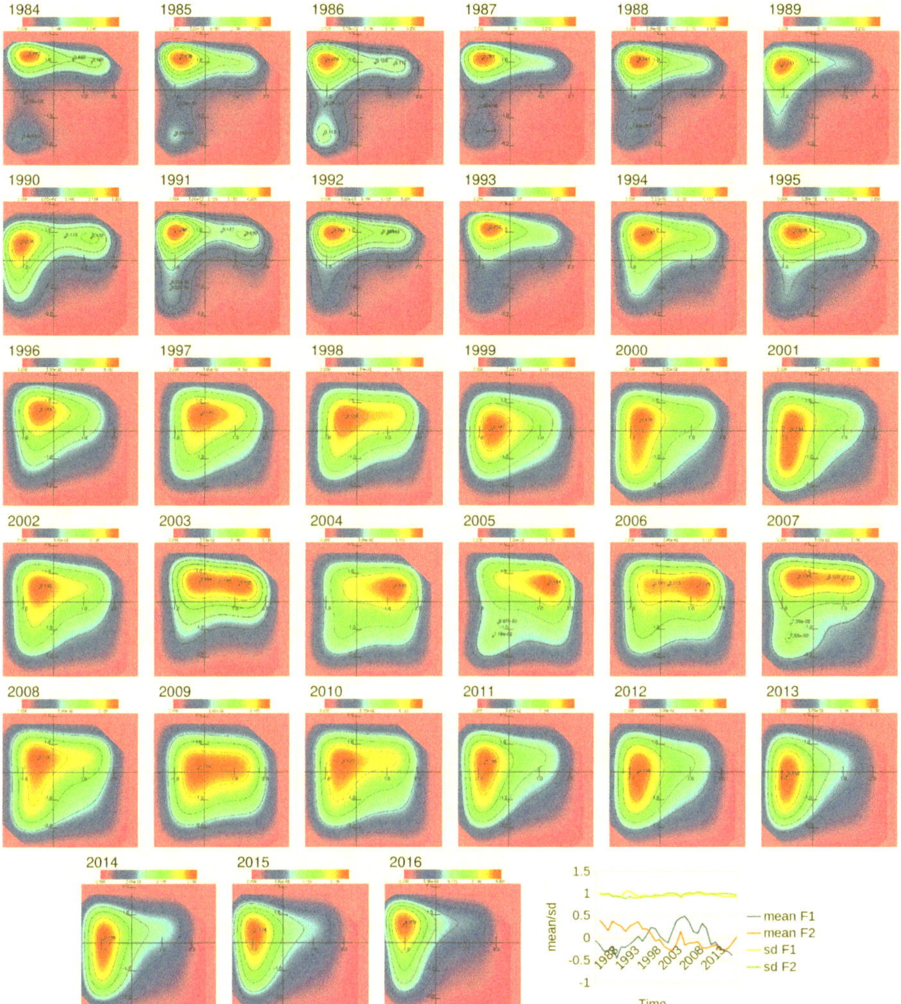

Fig. 2 Frequency density functions above the plane spanned by two factors derived from questions about worries for economic development, finances, environment and peace for the 33 waves of the German Socio-Economic Panel (horizontal, F1 = materialism; vertical, F2 = postmaterialism), the last diagram shows the changes of mean and standard deviation over time

the covariance matrix of all positions occupied by an interviewee over time was built, its area calculated and averaged over all interviewees with more than one participation.[6] The average area of the interviewees' movement ellipses is 0.303598

[6]This is measured as follows; first calculate the covariance matrix of the coordinates of the (up to 33) subsequent attitude points, and then draw an ellipse with the semimajor and the semiminor axis

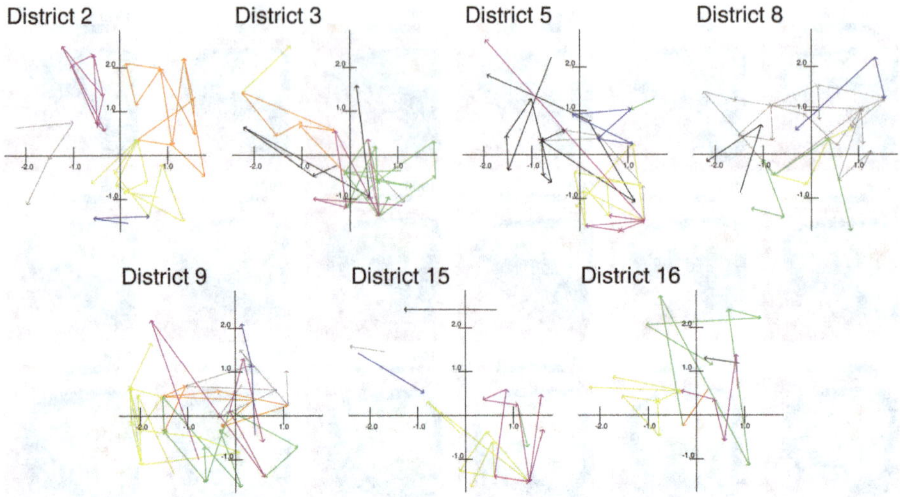

Fig. 3 Attitude changes of individual interviewees above the plane spanned by two factors derived from questions about worries for economic development, finances, environment and peace for the 33 waves of the German Socio-Economic Panel. (The diagrams show the movements of up to ten interviewees from the federal states of Hamburg (2), Niedersachsen (3), Nordrhein-Westfalen (5), Baden-Württemberg (8), Bayern (9), Sachsen (15) and Thüringen (16); colours are only used to distinguish the interviewees)

(and its median is 0.449369) square units of the diagrams in Fig. 3, which means that the movements happen within a square with an edge length of about 0.551 or a rectangle of 1.0 × 0.3 or a circle with radius of 0.3107.

4.2 *Party Scalometers in the German Election Panel 2016–2018*

Scalometer questions were first applied by Jan Stapel of the Netherlands Institute of Public Opinion in the early 1950s. The answer set originally consisted of values from −5 to −1 and from 1 to 5 (without 0) but was adopted later on as an 11-point scale from −5 to 5 including the 0 (Alwin 1997); (Pappi 1998, p. 257). In Germany it has been used since the early 1970s and is still being used in the German Longitudinal Election Study (GLES) (Schmitt-Beck 2011) (more recently a scale from 1 to 11 instead from −5 to 5 was used, but this does not make a difference).

---

equal to the eigenvalues of the covariance matrix and the orientation according to the correlation coefficient. If the nine points in attitude space were bivariate normally distributed, about two thirds of the points were within this ellipse whose areas is $\pi\sigma_x\sigma_y$. The reported 0.303598 is the mean of the individual area measures over all interviewees.

Table 2 Party scalometers for the nine waves of the German Election Panel 2016–2018

		Factor loadings	
	Communalities	CDU/CSU affinity	SPD and Grüne affinity
CDU	0.809	0.866	0.242
CSU	0.785	0.878	−0.118
SPD	0.667	0.284	0.765
FDP	0.613	0.754	0.211
Bündnis 90/Die Grünen	0.749	0.157	0.851
Die Linke	0.484	−0.363	0.593
AfD	0.347	−0.087	−0.583
Eigenvalues	63.615 %	2.334	2.119

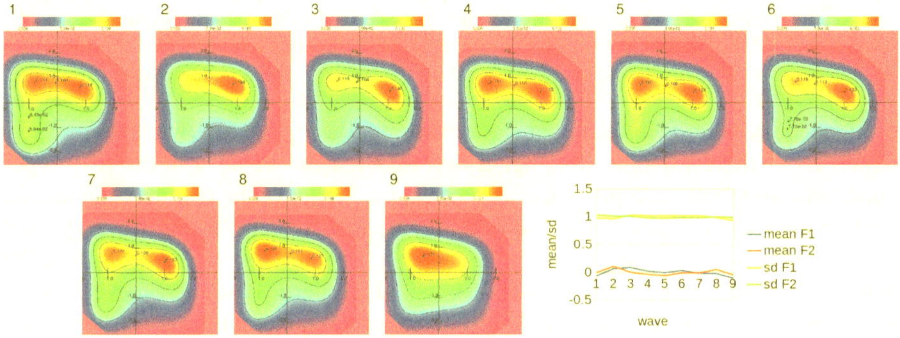

Fig. 4 Frequency density functions above the plane spanned by two factors derived from scalometer questions for the nine waves of the GLES Election Panel (horizontal, F1 = CDU affinity; vertical, F2 = SPD affinity), the last diagram shows the changes of mean and standard deviation over time

For the purpose of this subsection, the scalometers for the German parties CDU, CSU, SPD, FDP, Bündnis 90/Die Grünen, Die Linke and AfD were used for a factor analysis yielding two factors (see Table 2) covering 63.6 % of the total variance of the seven party scalometers. Again, a factor analysis was run over all nine waves, and again the lower right diagram in Fig. 4 shows how the means and variances changed over time—much less than in the GSOEP case, but again the distributions are far from normal.[7]

Figure 4 shows the changes between the nine waves from autumn 2016 till spring 2018: In the wave from mid-October to mid-December 2016, there are two equally high maxima in the upper right and upper left quadrants of the attitude landscape; in the two waves reaching from mid-December 2016 till mid-April 2017, the only

[7]This is a finding which differs from findings in 1980 (Troitzsch 1987) and also from more recent findings of the Politbarometer series (Forschungsgruppe Wahlen Mannheim 2018); see "Politbarometer: Selected Results from Scalometers from 1994 to 2016" in the Appendix for more details.

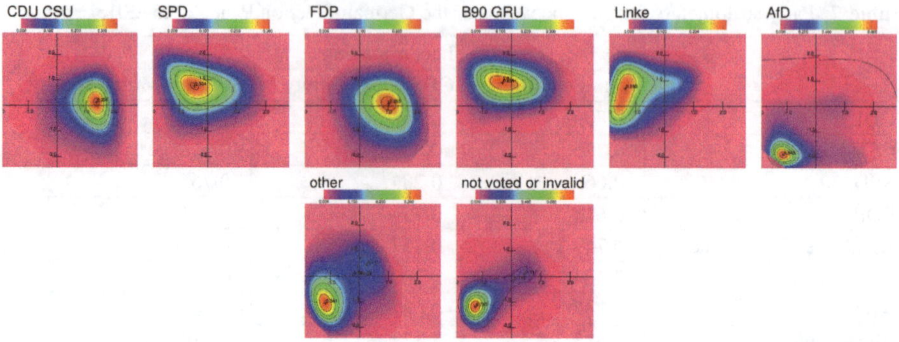

Fig. 5 Frequency density functions for voters of parties (most recent party preference articulated) and non-voters above the plane spanned by two factors derived from scalometer questions for the seventh wave of the GLES Election Panel (horizontal, CDU affinity; vertical, SPD affinity)

or higher maximum is in the upper right quadrant; during the following four waves, which cover the campaign time, again two maxima can be seen in the two upper quadrants, and in the two post-election waves, the higher of the two maxima moved to the upper left quadrant of the attitude landscape. In some of the waves (1 and 6), there is also a third maximum in the lower left corner, but in all waves the FDF extends in this direction. Figure 5 shows for which parties the interviewees voted who were near the maxima of the FDF: The upper left quadrant is mainly the region where voters of the SPD and the Green Party can be found, whereas the voters of CDU and CSU populate a region with its centre at (0, 1)—slightly "below" the overall FDF maximum in the upper right quadrant. The FDF maximum, however, is the region where voters of the right-wing AfD and of other, smaller parties (which did not enter parliament) and non-voters can be found.

Figure 6 shows the movements of some real voters for the election districts where the parties enjoyed their highest percentages (voters of different parties are marked in different colours, mainly to distinguish them from each other). All seven diagrams show that voters cover long distances in the attitude landscape, even between waves which are only 2 months apart. Part of this effect may be due to the relatively coarse-grained measurement on the seven scalometers, but even then it is surprising that the area covered by the nine points in the attitude space of an interviewee is an ellipse of 0.043865 square units in the units of the diagrams.[8] At first sight, 0.04 square units seem to be a small amount, but this would still be a square of 0.2×0.2 or a rectangle of 0.8×0.05 or a circle with radius 0.12. The distribution of the individual ellipse areas is, of course, extremely skewed with only some 20 per cent above the mean (the median being 0.012688, which still would be a circle with radius 0.06355). The difference in the sizes between the GSOEP and GLES panels can easily be

[8]See Footnote 6 for an explanation how the area of this ellipse is calculated.

Fig. 6 Attitude changes of individual interviewees in seven election districts above the plane spanned by two factors derived from scalometer questions for the seventh wave of the GLES Election Panel (The election districts with highest percentages for the parties were selected: 032-Cloppenburg/Vechta (CDU), 240-Kulmbach (CSU), 024-Aurich/Emden (SPD), 281-Freiburg (Grüne), 158-Sächsische Schweiz/Osterzgebirge (AfD), 106-Düsseldorf I (FDP) and 084-Berlin-Treptow-Köpenick (Linke). Colours represent the parties for which the interviewees had voted: black, CDU/CSU; orange, SPD; yellow, FDP; green, Grüne; magenta, Linke; blue, AfD; brown, other parties)

explained with the fact that the former covers 33 years whereas the latter covers only 18 months.

4.3 A First Conclusion from the Empirical Evidence

The two cases showed that the variances of the attitude scales remain remarkably constant even over long times. This may partly be due to the statistical procedures to derive just two orthogonal scales from three-point scales or scalometers which have a restricted length. Means of the attitude scales are much more variable as particularly the 33-years panel has shown, but the changes of the means over the 18 months around the federal election of 2017 are anyway much greater than the changes of the respective standard deviations. This describes a first requirement regarding the parameterisation and calibration of the two-dimensional opinion dynamics model: only runs with more or less stable variances and standard deviations qualify as replicatively valid.

A second requirement results from the analysis of individual attitude changes which turned out to be considerably wide. The results from both the 33 years and the 18 months (average ellipse areas of 0.30 for 33 years and 0.04 for 1.5 years) are more or less compatible, as the movements over 18 months are considerably shorter than the movements over 33 years—although there is no proportionality, but this could not be expected as the measured attitudes are by no means the same, but the GSOEP is one of the longest panels available; hence structural validity can only be stated if the simulation model yields comparable results for different run lengths.

The background for these two requirements seems to be the fact that the scales in the two panels are limited and non-continuous. In the five-dimensional space with only $243 = 3^5$ allowable positions, a movement has at least the length of the distance between the two nearest among these 243 positions. In the seven-dimensional space with 11^7 (about 19.5 million) allowable positions, the smallest possible movements are certainly shorter but still of a certain minimum length—whereas the model with its (nearly) continuous opinion space allows for arbitrarily short movement distances.

These empirical findings make an extension of the model presented in Sect. 3.2 necessary as most of the first several thousand runs of those eight versions failed to reproduce the more or less constant variances and covariances of the two empirical panels; those runs which had more or less constant covariance matrices started from the same parameters as the others which lead to the conclusion that the continuous model is strongly path dependent. All this will be discussed, and the additional extensions of the model will be presented, in the next section.

5 Calibrating and Validating the Model Against the Empirical Cases

Several thousand runs of the eight versions of the model outlined in Sect. 3.2 revealed that one of the central empirical findings—namely, that the covariance matrix of the positions of the agents remains more or less constant—is violated in about 50 to 75 per cent of the Monte Carlo runs with varying parameters α, ρ, μ, ν and ϵ_σ, ϵ_U and ϵ_S, respectively. With sets of constant values of these parameters, very different outcomes were possible, which is a clear hint at the path dependence of the model: depending on the stochastic initialisation and the first few steps of this partly stochastic model, the population of the agents evolves in different directions. As already mentioned at the end of the previous section, it turned out to extend the model in a way that it takes account of the circumstances of the empirical measurement of opinions or attitudes. Apart from continuous magnitude scaling (Lodge 1978; Lodge and Tursky 1981) (a technique which seems to have fallen into oblivion since the mid-1980s), there is no technique which yields interviewees' opinions on a continuous scale, and scales with more than the eleven entries of the scalometer questions were not very successful as, for example, in a numerical scale from 0 to 100 (thermometer scale), mostly multiples of 10 and 25 were used such that in practice not much more than 11 scale points were actually used (Thomas and Bremer 2012).

This is why 2 additional versions were added to the model, with agents having 5 opinions on a 3-point scale and with agents having 7 opinions on an 11-point scale, just as seen in the 2 panels used in Sect. 4.

5.1 The GLES Version of the Model

The GLES version of the model is initialised with a subsample of the GLES panel (1000 cases from all nine waves randomly selected with the seven scalometers) and the positions of the agents thus endowed on the two-dimensional plane of NetLogo with start values calculated with the help of the factor score coefficients which were the result of the factor analysis described in Sect. 4.2. Positions are updated as follows: in every time step each agent

- selects one of seven scalometers for a possible change,
- with probability ϵ_S selects another agent from its neighbours within a radius ρ and copies this agent's scalometer value in the selected position into its own list of scalometer values,
- otherwise, i.e., with probability $(1 - \epsilon_S)$, changes the value of this scalometer by a random integer $\in \{-\chi..\chi\}$ where χ is a new parameter of the model in a way replacing ϵ_U or ϵ_σ of the original versions of Sect. 3.2; if the result of this change would lead to a number < 1 or > 11, it is replaced to either 1 or 11.

It turns out that runs with $\rho \in [0.1, 0.9)$ and $\epsilon_S \in [0.775, 0.860)$ are stable with respect to the standard deviations of the factor scores (with $\texttt{varstab} = \sqrt{(\sigma_x - 1)^2 + (\sigma_y - 1)^2} < 0.1$) and have similar distributions over the $x - y$-plane as the scalometer factors in Sect. 4.2. The parameter χ does not seem to play a major role. Figure 7 (left diagram) shows the results of 1000 runs with varying $\rho \in [0.1, 1.6)$ and $\epsilon_S \in [0.1, 1.0)$. The results in the area mentioned above have more or less stable variances and have a kurtosis comparable to that of a normal distribution, whereas the runs (also marked blue and violet) in the region with $\rho \in [0.5, 0.8)$ and $\epsilon_S \in [0.1, 0.6)$ are extremely leptokurtic.

5.2 The GSOEP Version of the Model

The GSOEP version of the model is initialised with a distribution calculated from the GSOEP data. The five 3-point scales allow for 243 different positions (from 1-1-1-1-1 to 3-3-3-3-3), and their empirical frequencies are used to initialise the agents.[9] Position are updated as follows: in every time step each agent

[9] For instance, as the cumulative frequency for 3-1-2-3-1 is 0.87574041 and the cumulative frequency for 3-1-2-3-2 is 0.8759526, an agent which receives a value between these two numbers

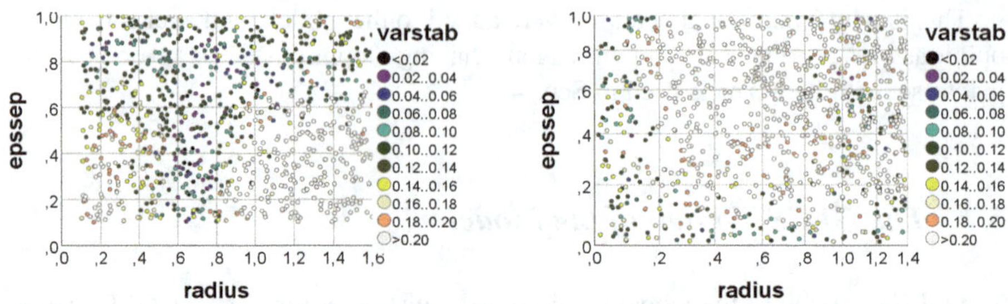

Fig. 7 Outcomes of runs of the empirically founded versions of the two-dimensional opinion dynamics model, left, GLES version; right, GSOEP version; the latter has a logarithmic scale for ρ

- selects one of its five answers for a possible change,
- with probability ϵ_S selects another agent from its neighbours within a radius ρ and copies this agent's answer in the selected position into its own set of opinions,
- otherwise, i.e., with probability $(1 - \epsilon_S)$, changes the value of this answer by a random integer $\in [-\chi..\chi]$ where χ is a new parameter of the model in a way replacing ϵ_U or ϵ_σ of the original versions of Sect. 3.2; if the result of this change would lead to a number < 1 or > 3, it is replaced to either 1 or 3.

It turns out that there are two regions in the $\rho - \epsilon$ space where the variance is more or less constant, but only one of these regions—namely, for $\rho < 0.2$ or $\epsilon < 0.2$—contains runs with a kurtosis comparable to the kurtosis of a normal distribution, whereas the other—namely, for $\rho \approx 1.1$ and $\epsilon \approx 0.5$—contains extremely leptokurtic distributions (see Fig. 7, right diagram).

5.3 *The Original Versions of the Model*

5.3.1 Initialisation and Other Stochastic Effects with a Normal Distribution

As already mentioned, practically all combinations of the parameters α, ρ, μ, ν, ϵ_S (the latter two if applicable) and ϵ_U or ϵ_σ led both to stable more or less covariance matrices and to distributions with decreasing or increasing variances, even for identical sets of parameters. It turned out for the version with an initially normal distribution and normal stochastic changes that for certain combinations of ρ, μ, α, ν and ϵ_S (the latter three if applicable) on one hand and ϵ_σ on the other hand the percentage of variance-stable and mesokurtic runs was maximised—in these cases the stochastic element defined by ϵ_σ compensated for the systematic influence

from the pseudorandom number generator with a uniform distribution between 0 and 1 will receive 3-1-2-3-1 as its initial set of opinions.

Table 3 Parameter combinations which match the empirical findings for the continuous model initialised and run according to a normal distribution (regressions coefficient with ϵ_σ as dependent parameter)

Distribution	Normal				Uniform			
n of parties	No parties		Five parties		No parties		Five parties	
Random effect	Simulta-neously	Separa-tely	Simulta-neously	Separa-tely	Simulta-neously	Separa-tely	Simulta-neously	Separa-tely
Constant	−0.055	0.144	0.012	0.163	0.101	.	−0.195	.
ρ	0.224	0.028	0.153	0.004	0.262	.	0.346	.
$\alpha\mu$	0.181	0.050	0.140	−0.034	2.038	.	1.910	.
$(1-\alpha)\nu$	n.a.	n.a.	0.769	−0.034	n.a.	n.a.	2.424	.
ϵ_S	n.a.	−0.180	n.a.	−0.162	n.a.	.	.	
R^2	0.729	0.278	0.749	0.315	0.796	.	0.853	.
P[a]	18.9	14.0	22.8	30.4	5.7	0.0	9.9	0.0

[a]Percentage of runs with more or less mesokurtic and variance stable distributions (`vartab`< 0.2)

of the agents with similar opinions which usually led to even more similarity among all agents. These combinations were found out by restricting a linear regression of ϵ_σ on ρ, $\alpha\mu$, $(1-\alpha)\nu$[10] and ϵ_S to those runs which turned out to be more or less variance stable and mesokurtic. The regression coefficients for the six versions can be found in Table 3.

Figure 8 (first row, left diagram A) shows this for the coefficient combinations of the first column of Table 3, where the runs represented below the coloured dots (these always represent runs with more or less mesokurtic and variance stable distributions) produced leptokurtic distributions, whereas the runs represented above the coloured dots produced platykurtic distributions. The right diagram of the first row of Fig. 8 (B) shows this for the version of the respective column of Table 3, and again the runs represented below the chain of coloured dots produced leptokurtic distributions, hence for both cases for ϵ_σ less than the right-hand side of an equation constructed from the parameters of the respective column of Table 3, a leptokurtic distribution can be expected—the influence of the peers leads to one or more narrow clusters—whereas for ϵ_σ greater than that a platykurtic distribution can be expected—the influence of the peers is overcompensated by the stochastic opinion change.

The second row of Fig. 8 (diagrams C and D) shows more or less the same as the first row, but it becomes clear that the versions with uniformly distributed initialisation and relocation produce far fewer mesokurtic distributions.

The case with alternative systematic and stochastic opinion changes— an agent changes its opinion following peers with similar opinions with a certain probability and otherwise changes its opinion purely at random—is different from the four version discussed so far, as the respective columns (2, 4 and 6) of Table 3 show:

[10]According to equation 2, α and μ and ν, respectively, cannot influence any result separately.

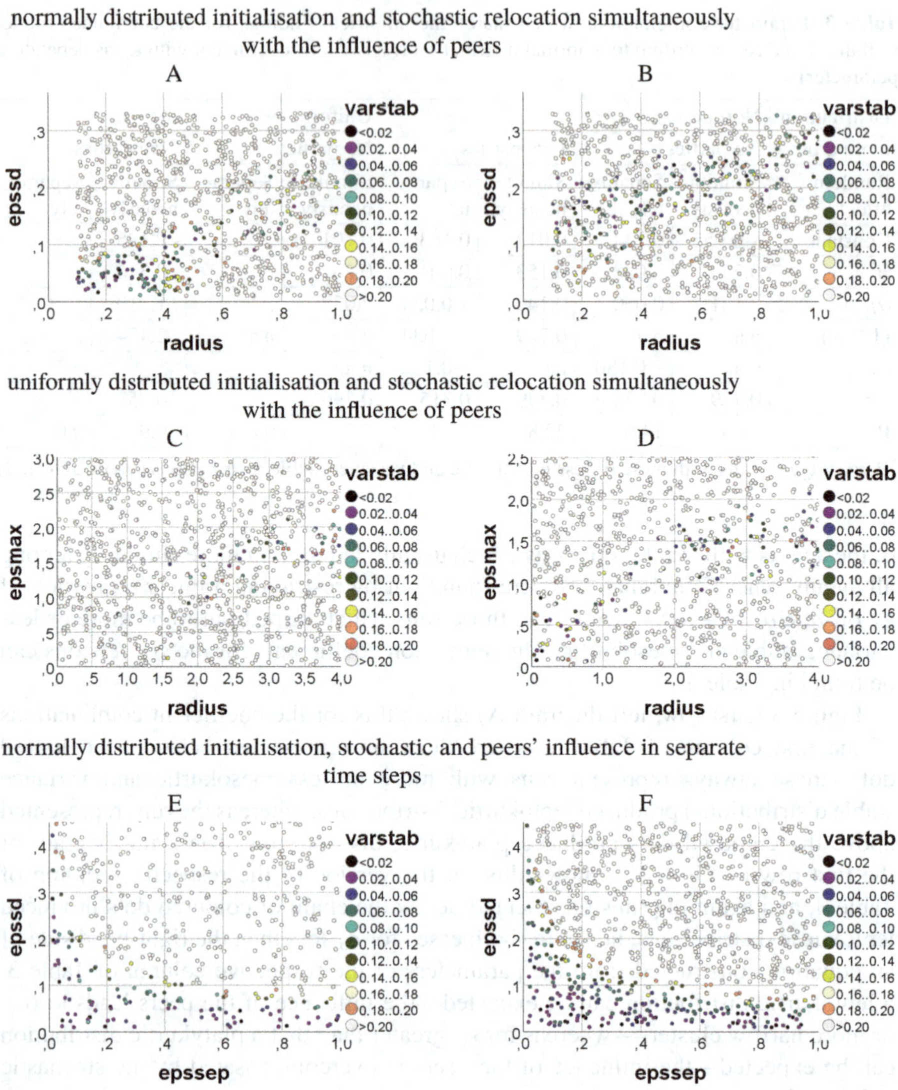

Fig. 8 Outcomes of runs of the continuous version of the two-dimensional opinion dynamics model. (**a**) no parties (**b**) five parties (**c**) no parties (**d**) five parties (**e**) no parties (**f**) five parties

the only significant regression coefficient in both equations is the one −0.180 with ϵ_S for the ϵ_σ in runs with more or less mesokurtic and variance-stable runs (the ones marked in colour in the bottom left diagram (E) of Fig. 8). The bottom row of Fig. 8 shows the $\epsilon_S - \epsilon_\sigma$ combinations for which most runs are mesokurtic and variance stable (as the relation between ϵ_S and ρ (with a standardised regression

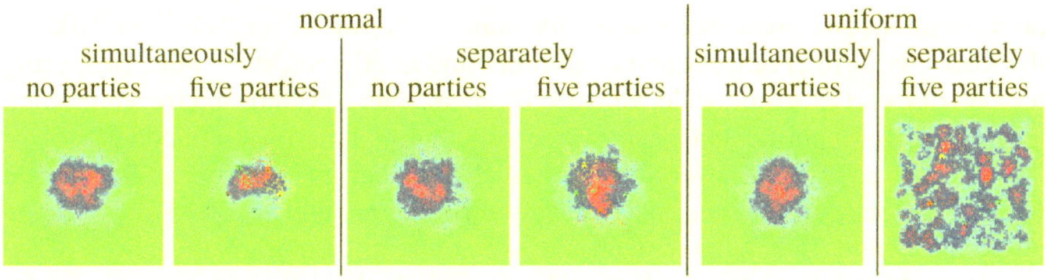

Fig. 9 Some final distributions (heatmaps) of runs of the continuous version of the two-dimensional opinion dynamics model

coefficient 0f 0.008) is so weak that any value of ρ is compatible with the postulate of mesokurtic and variance stable distributions).

5.3.2 Initialisation and Other Stochastic Effects with a Uniform Distribution

Less than ten per cent of the 2,000 runs represented in Fig. 8 (last row) yielded mesokurtic and variance stable distributions for the uniform version without and with parties, and the stochastic influence applied simultaneously with the influence of the peers was consulted; the two last diagrams in Fig. 8 show under which parameter constellations they could be observed (the coefficients of the regression equation can be found in the respective column of Table 3). The version in which the stochastic influence is applied alternatively to the systematic influence of the peers does not yield any result which is mesokurtic and variance stable. Diagrams for the case of uniformly distributed initialisation, stochastic and peers' influence in separate time steps are not even shown, as columns 6 and 8 of Table 3 do not show any mesokurtic and variance stable distributions: In all 2,000 runs, the standard deviations increase from about 1 to at least about 1.7, and the distribution is extremely platykurtic; an example of the heatmap for one of these 2.000 runs is shown as the last diagram in Fig. 9—which looks slightly degenerate and bears no similarity to any of the empirical distributions of Sect. 4.

6 Conclusion

The comparison between different versions of a two-dimensional model of opinion dynamics with empirical panels showed that model validation was exposed to several problems some of which are due to empirical measurement problems—but one could also argue that more or less constant covariance matrices (or just more or less constant standard deviations) are only due to the restriction that interviewees have to reveal their attitudes on a fixed scale with three or eleven (or perhaps 100) points without any guarantee that the meaning of "1" or "3" or "11"[11] or "100" is the same for all interviewees and at all times and in all contexts. Unlike measuring temperatures on a Celsius scale, where 100° and 0° mean boiling and freezing water, a sympathy for a politician (for instance, for Hillary Clinton) expressed as "−5" or "1" can mean "I would never vote for her", "Lock her up"[12] or even "She should be tried and hanged for treason"[13]; widely differing meanings of "+5" or "11" are also conceivable. Hence the outcomes of the non-restricted original versions of the model might be even more realistic than the seemingly validated outcomes of the GLES and GSOEP versions with their restriction to the 3-point and 11-point scales.

The original one-dimensional versions of the opinion dynamics models were, all of them, restricted to attitude scales ranging from 0.0 to 1.0 (although one can always think of an extremist who can become even more extremist, which is definitely excluded, for instance, in Lorenz 2012), and the update algorithms made sure that attitudes outside this scale were impossible but on the other hand led to highly leptokurtic distributions. In most cases these probability density functions had values >0 only for one or few arguments—which will empirically be found in negotiations about a numerical value such as the VAT rate in parliamentary debate or a salary increase in collective bargaining negotiations, but not in election campaigns where attitude scales are used to contribute to the prediction of election results or in comparative value studies such as (EVS 2011; World Values Survey Association 2015).

The two-dimensional extensions discussed in this paper (and also the extension in Lorenz (2017) with the stochastic influence added to the influence of other agents with more or less similar attitudes) revealed that it is possible with some of these versions and with some parameter constellations to replicate empirical findings. The versions adapted to the two empirical panels showed relatively wide areas within parameter space with outcomes similar to the empirical distributions. In the case of the GSOEP version, it was a little surprising to see that a rather high probability of random shocks exerted on the agents produced the best correspondence between model results and empirical data on the level of the distributions, whereas for the GLES version, the contrary was the case—probably the reason is that the 3-point scale allows for or even necessitates for a higher random compensation than the 11-point scales. For the continuous-scale versions, the results are in favour of the assumption of normally distributed random influences, while they cannot support a decision whether the simultaneous or the separate application of the random influence is the better model. The movement of the agents over their

[11] Note that in the questionnaires the interviewees are shown a list of values ranging from −5 to +5, whereas in the dataset this is recoded to a scale from 1 to 11.

[12] Among many others Jeff Sessions and dozens of students according to https://www.theguardian.com/us-news/2018/jul/24/jeff-sessions-lock-her-up-chant-trump-clinton, accessed 2019-01-05 16:29.

[13] Ted Nugent, according to https://www.washingtonexaminer.com/ted-nugent-i-stand-by-saying-hillary-clinton-should-be-hanged-for-treason, accessed 2019-01-05 16:26.

attitude space cannot easily compared between model and data, mostly because the transformation of the 3-point or 11-point scales to continuous scales—necessary to show approximate movements on a two dimensional space—blurs all effects. But both Figs. 3 and 6 show movements covering quite a great part of the attitude space, and Fig. 1 shows that this is also true for the model, but a quantitative comparison between model and data in this respect does not seem possible.

The latter observation—comparing movements in attitude space between model and panel—is a first step at asserting structural validity, as it shows that both model agents and real persons make relatively great leaps in their attitude spaces. This is an effect that has never been shown in the traditional versions of opinion dynamic models without stochastic influences. On the other hand, it is exactly these stochastic influences in the current model versions which only stand in for all the unknown and unmodelled causes of opinion changes in human minds—unfortunately without representing them materially.

It goes without saying that all opinion dynamics models of this kind have their weaknesses, mainly because the communications between human actors is much more complex than the communication between the agents in all of these models: Humans do not discuss their attitudes in terms of points on 3- or 11-point scales (not even when they learn from survey results published every other week over the TV that the average evaluation of a certain politician is 2.6 on a scale from -5 to $+5$). Instead, they exchange arguments trying to convince others that—for instance, in the case of the items analysed from the GSOEP panel—it is important to care for peace more than is currently the case, for these or other reasons. Moreover it is not only the individual others who influence people but also (and perhaps even more powerfully) the media. All this is not modelled in all these attempts at modelling opinion dynamics, so perhaps Abelson and Bernstein (1963) were structurally nearer to reality in their model of attitude changes with respect to drinking water fluoridation. In their model—see the chapter on formal design methods in this volume—agents exchange assertions of different kinds among each other which lead to attitude change in the addressees of these assertions, and they listen to assertions which come from public channels. Still this kind of communication is poor enough, but it is anyway superior to the early mathematical methods of opinion change—and this kind of communication (and in the future perhaps more sophisticated communication) can only be modelled in agent-based models, perhaps supported by a replication of the empirical research that was the background of Abelson's and Bernstein's model, a replication that would have to be also inspired by the opinion dynamics models discussed in this chapter in order to validate them against newly collected data.

Appendix: Results of Data Transformations

GSOEP Variables About Concerns

To make clear what the dimensions in Fig. 2 mean, Fig. 10 shows separate PDF diagrams for the nine combinations of answers to the "worried about finances" and "worried about environment" questions, from which it is clear that the upper right quadrant contains those that are worried about both materialism and postmaterialism concerns.

To check whether the results of the factor analysis covering all 33 waves are compatible with separate factor analyses, the history of the eigenvalues (both

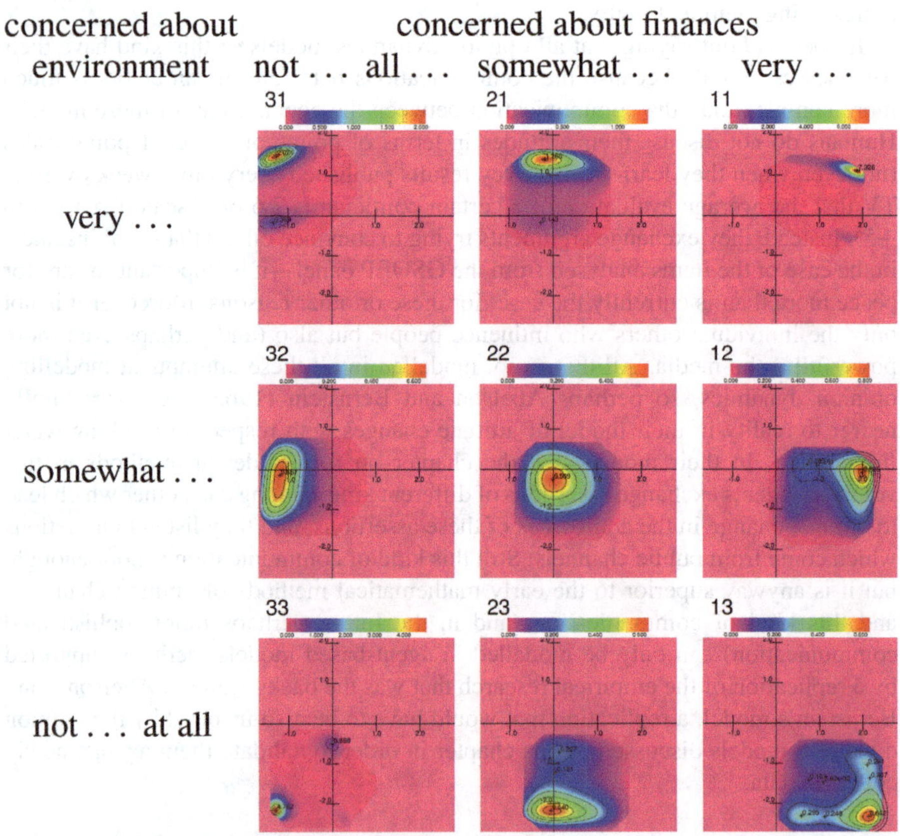

Fig. 10 Frequency density functions above the plane spanned by two factors derived from questions about worries for economic development, finances, environment and peace for the 33 waves of the German Socio-Economic Panel (horizontal, materialism; vertical, postmaterialism, the codes 11. . . 33 refer to the original codes in the questionnaire, not the ones used for factor analysis)

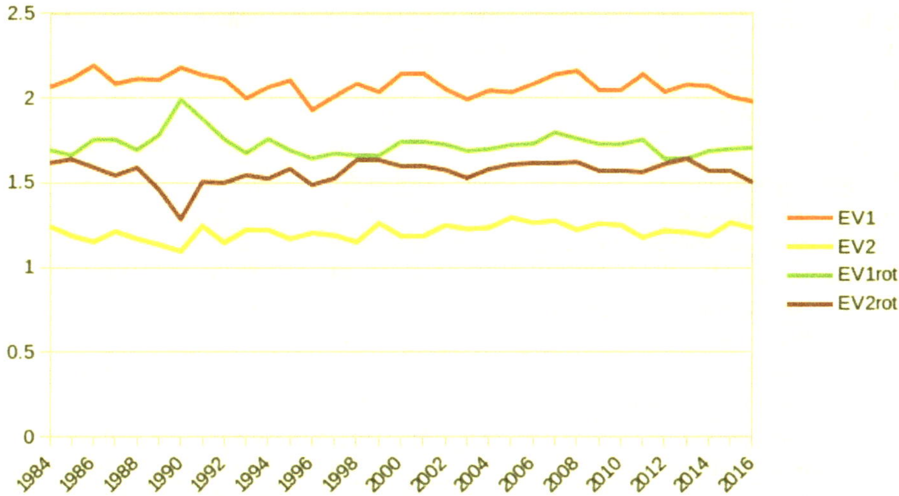

Fig. 11 Eigenvalues of separate factor analyses for the 33 GSOEP waves of the German Socio-Economic Panel (EV1rot, materialism; EV2rot, postmaterialism; EV1 and EV2 are the eigenvalues of the unrotated solutions of the separate factor analyses)

unrotated and Varimax rotated) are given in Fig. 11: eigenvalues are nearly constant, and the eigenvalues of the rotated solution are almost equal, such that the normalised diagrams in the other figures are not misleading.

To document more details, Table 4 contains the correlation coefficients between all variables about concerns and the factor scores calculated from the four variables available in all waves, together with information when the additional questions were asked.

German Longitudinal Election Study (GLES): Scalometers and Party Preferences of the Campaign Panel 2017

The Campaign Panel 2017 is available as ZA6804 (Roßteutscher et al. 2018) and contains the answers of some 20,000 interviewees over nine panel waves which were in the field for about 1 months each, centred around mid-November of 2016, mid-January, mid-March, mid-May, mid-July, mid-September, mid-November of 2017 and mid-January and mid-March of 2018. Each row of the file contains all information about one interviewee. To calculate factor scores on all scalometers over all nine waves, this file was split vertically to cover each wave separately, and then the nine wave files were combined. The correlations between scalometers and factor scores per wave can be found in Table 5. As not all interviewees answered questions about their planned or actual vote in all waves, the most recent party preference known was reconstructed for Fig. 5. To check the consistency of all these

Table 4 Correlation coefficients between all variables about concerns and the factor scores calculated from the four variables available in all waves

How concerned are you about the following issues? Very concerned (3)/Somewhat concerned (2)/Not concerned at all (1)

Concern	When asked	Materialism	Postmaterialism
Worried about adapting to change	1990–1991	0.390	0.101
Worried about childcare	1990–1998	0.320	0.120
Worried about consequences of EU enlargement	2004–2008	0.276	0.222
Worried about dwelling	1993–1998	0.440	0.113
Worried about economic development	All waves	0.605	0.344
Worried about environment	All waves	0.004	0.855
Worried about finances	All waves	0.850	0.066
Worried about introduction of Euro	1999–2003, 2011	0.287	0.158
Worried about job security	All waves	0.811	−0.022
Worried about own health	1999–2016	0.347	0.197
Worried about peace	All waves	0.169	0.809
Worried about rights of use and assets	1990–1991	0.277	0.109
Worried about global terrorism	2008–2013	0.195	0.526
Worried about consequences from climate change	2009–2016	0.055	0.702
Worried about crime in Germany	1994–2016	0.244	0.376
Worried about stability of financial markets	2009–2014	0.305	0.352

Table 5 Correlations (loadings) between factors and scalometers in the waves of the German Longitudinal Election Study

Wave	Factor	CDU	CSU	SPD	FDP	Grüne	Linke	AfD
1	1	0.910	0.899	0.412	0.641	0.189	−0.149	−0.101
1	2	0.295	−0.100	0.714	0.316	0.950	0.495	−0.484
2	1	0.917	0.911	0.371	0.663	0.204	−0.171	−0.099
2	2	0.290	−0.103	0.738	0.305	0.951	0.513	−0.495
3	1	0.926	0.917	0.254	0.659	0.135	−0.242	−0.081
3	2	0.179	−0.164	0.733	0.217	0.944	0.478	−0.466
4	1	0.923	0.916	0.310	0.697	0.155	−0.256	−0.097
4	2	0.223	−0.149	0.739	0.229	0.948	0.482	−0.447
5	1	0.924	0.916	0.302	0.688	0.152	−0.244	−0.109
5	2	0.220	−0.147	0.734	0.220	0.948	0.484	−0.443
6	1	0.924	0.919	0.270	0.690	0.142	−0.272	−0.123
6	2	0.204	−0.159	0.737	0.187	0.947	0.487	−0.460
7	1	0.923	0.918	0.266	0.700	0.168	−0.254	−0.128
7	2	0.230	−0.142	0.747	0.179	0.947	0.489	−0.474
8	1	0.919	0.911	0.243	0.690	0.204	−0.245	−0.151
8	2	0.265	−0.123	0.736	0.201	0.946	0.515	−0.514
9	1	0.913	0.923	0.314	0.519	0.149	−0.249	−0.089
9	2	0.237	−0.140	0.712	0.114	0.948	0.507	−0.510

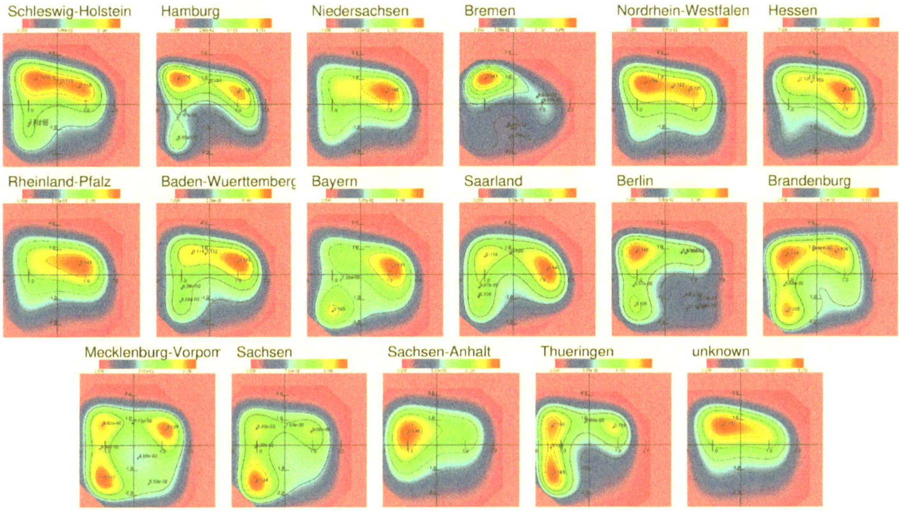

Fig. 12 Frequency density functions for voters in German federal states above the plane spanned by two factors derived from scalometer questions of the GLES Election Panel (horizontal, CDU affinity; vertical, SPD affinity)

calculations, frequency density functions were also calculated for the interviewees of the federal states of Germany (see Fig. 12). Table 5 shows that the correlations between factors and scalometers do not change much between November 2016 and March 2018.

To check whether the results of the factor analysis covering all nine waves are compatible with separate factor analyses, the history of the eigenvalues (both unrotated and Varimax rotated) are given in Fig. 13: eigenvalues are nearly constant, and the eigenvalues of the rotated solution are almost equal, such that the normalised diagrams in the other figures are not misleading.

Politbarometer: Selected Results from Scalometers from 1994 to 2016

As mentioned in Sect. 4.2, the frequency density functions generated from the Politbarometer series of surveys look different from the ones generated from the GLES panel. This can easily be seen from the selected diagrams in Fig. 14.

In most (all with the exception of August 1994) of the 18 monthly surveys documented in Fig. 14, the frequency density function has only one maximum in the upper right quadrant, whereas the respective diagrams generated from the GLES panel usually showed two or sometimes even three maxima. The first diagram in Fig. 4 and the last diagram of Fig. 14 both cover the same field time (December 2016) and look entirely different. This is not the place to try and find out what the reason for this surprising difference is—one reason might be that the Politbarometer

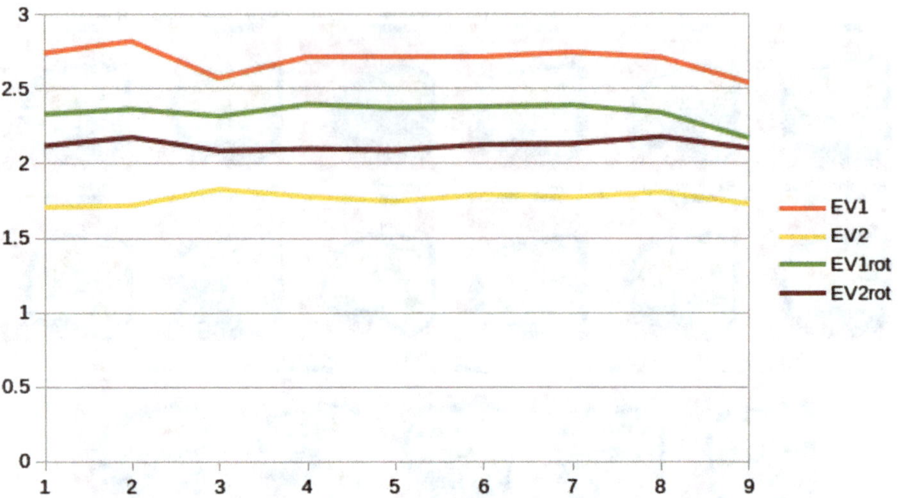

Fig. 13 Eigenvalues of separate factor analyses for the nine GSOEP waves of the German Socio-Economic Panel (EV1rot, materialism; EV2rot, postmaterialism; EV1 and EV2 are the eigenvalues of the unrotated solutions of the separate factor analyses)

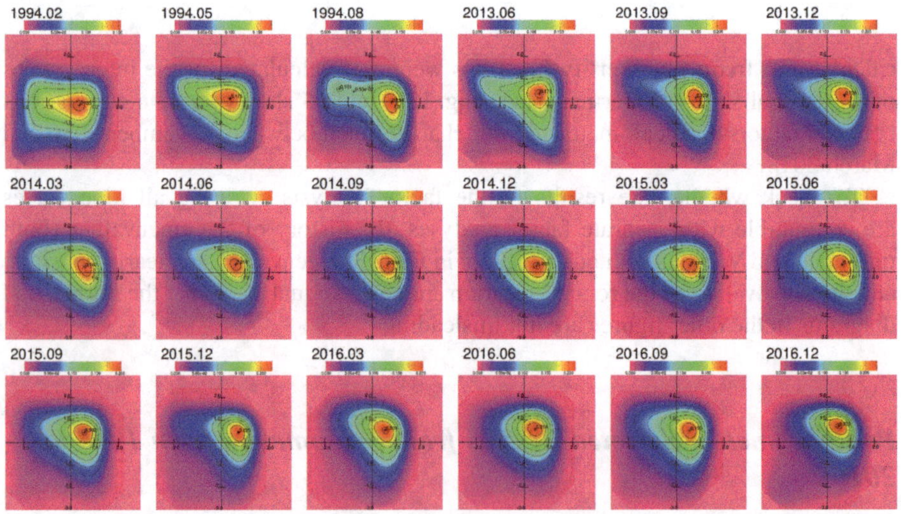

Fig. 14 Frequency density functions for voters in Germany above the plane spanned by two factors derived from scalometer questions of the Politbaromer survey series (horizontal: CDU affinity, vertical: SPD affinity)

surveys are done by telephone interviewing whereas GLES uses online surveys, so perhaps there is a self-selection effect which is not annihilated by re-weighting and which leads to an over-representation of people with opinions represented in the two quadrants in the left half of the attitude space.

References

R.P. Abelson, Mathematical models of the distribution of attitudes under controversy, in *Contributions to Mathematical Psychology*, ed. by N. Frederiksen, L.L. Thurstone, H. Gulliksen (Holt, Rinehart and Winston, Inc., New York, 1964), pp. 141–160

R.P. Abelson, A Bernstein, A computer simulation model of community referendum controversies. Public Opin. Q. **27**(1), 93–122 (1963)

D.F. Alwin, Feeling thermometers versus 7-point scales: which are better? Sociol. Methods Res. **25**(3), 318–340 (1997)

D. Anzola, P. Barbrook-Johnson, M. Salgado, N. Gilbert, Sociology and non-equilibrium social science, in *Non-Equilibrium Social Science and Policy, Introduction and Essays on New and Changing Paradigms in Socio-Economic Thinking*, ed. by J. Johnson, A. Nowak, P. Ormerod, B. Rosewell, Y.-C. Zhang. Understanding complex systems (Springer, Cham, 2018), pp. 59–69

W. Balzer, C.U. Moulines, J.D. Sneed, *An Architectonic for Science. The Structuralist Program*, volume 186 of *Synthese Library* (Reidel, Dordrecht, 1987)

L. Cobb, Stochastic catastrophe models and multimodal distributions. Behav. Sci. **23**, 360–374 (1978)

O.A. Davis, M.J. Hinich, P.C. Ordeshook, An expository development of a mathematical model of the electoral process. Am. Polit. Sci. Rev. **64**(2), 426–448 (1970)

G. Deffuant, F. Amblard, G. Weisbuch, T. Faure, Simple is beautiful—and necessary. J. Artif. Soc. Soc. Simul. (2003). http://www.soc.surrey.ac.uk/JASSS/6/1/6.html

G. Deffuant, D. Neau, F. Amblard, G. Weisbuch, Mixing beliefs among interacting agents. Adv. Complex Syst. **03**(01n04), 87–98 (2000)

I. Douven, A. Riegler, Extending the hegselmann-krause model I. Log. J. IGPL **18**(2), 323–335 (2010)

A. Downs, *An Economic Theory of Democracy* (Addison-Wesley, Boston, 1957)

J.M. Enelow, M.J. Hinich, *The Spatial Theory of Voting: An Introduction* (Cambridge University Press, New York, 1984)

EVS, European values study 2008: Integrated dataset (EVS 2008). GESIS Data Archive. ZA4800 Data file version 3.0.0. (2011)

Forschungsgruppe Wahlen Mannheim, Politbarometer 1977-2017 (Partielle Kumulation). GESIS Datenarchiv, Köln. ZA2391 Datenfile Version 9.0.0. (2018)

J.H. Fowler, M. Laver, A tournament of party decision rules. J. Confl. Resolut. **52**(1), 68–92 (2008)

N. Gilbert, Emergence in social simulation, in *Artificial Societies: The Computer Simulation of Social Life*, ed. by N. Gilbert, R. Conte (UCL Press, London, 1995), pp. 144–156

V. Grimm, E. Revilla, U. Berger, F. Jeltsch, W.M. Mooij, S.F. Railsback, H.-H. Thulke, J. Weiner, T. Wiegand, D.L. DeAngelis, Pattern-oriented modeling of agent-based complex systems: lessons from ecology. Science **310**(5750), 987–991 (2005)

D.M. Hausman, Why look under the hood, chapter 11, in *The Philosophy of Economics: An Anthology. Second Edition*, ed. by D.M. Hausman (Cambridge University Press, Cambridge, 1994), pp. 217–221

P. Hedström, *Dissecting the Social. On the Principles of Analytic Sociology* (Cambridge University Press, Cambridge, 2005)

R. Hegselmann, U. Krause, Opinion dynamics and bounded confidence: models, analysis and simulation. J. Artif. Soc. Soc. Simul. **5**(3), 1–33 (2002)

R. Hegselmann, U. Krause, Truth and cognitive division of labor: first steps towards a computer aided social epistemology. J. Artif. Soc. Soc. Simul. **9**(3), 10 (2006)

B.-O. Heine, M. Meyer, O. Strangfeld, Stylised facts and the contribution of simulation to the economic analysis of budgeting. J. Artif. Soc. Soc. Simul. 8(4) (2005). http://jasss.soc.surrey.ac.uk/8/4/4.html

L. Herlitzius, Schätzung nicht-normaler Wahrscheinlichkeitsdichtefunktionen, in *Computer Aided Sociological Research. Proceedings of the Workshop "Computer Aided Sociological Research"*

(CASOR'89), Holzhau/DDR, Oct 2nd–6th, 1989, ed. by J. Gladitz, K.G. Troitzsch (Akademie-Verlag, Berlin, 1990), pp. 379–396

P.W. Holland, Causal inference, path analysis and recursive structural equations models. Technical Report TR 88-81, RR 88-14 (Educational Testing Service, Princeton, 1988)

H. Hotelling, Stability in competition. Econ. J. **39**(53), 41–57 (1929)

R. Inglehart, *The Silent Revolution. Changing Values and Political Styles Among Western Publics* (Princeton University Press, Princeton, 1977)

W. Jager, F. Amblard, Multiple attitude dynamics in large populations, in *Paper presented at the Agent 2005 Conference on: Generative Social Processes, Models, and Mechanisms, Argonne National Laboratory* The University of Chicago, 13–15 Oct 2005. (2005a)

W. Jager, F. Amblard, Uniformity, bipolarization and pluriformity captured as generic stylized behavior with an agent-based simulation model of attitude change. Comput. Math. Organ. Theory **10**(4), 295–303 (2005b)

N. Kaldor, Capital accumulation and economic growth, in *The Theroy of Capital*, ed. by C.F. und Douglas, A.L. Hague (Macmillan, London, 1961/1968, Reprint), pp. 177–222

K. Kollman, J.H. Miller, S.E. Page, Adaptive parties in spatial elections. Am. Polit. Sci. Rev. **86**(4), 929–937 (1992)

K. Kollman, J.H. Miller, S.E. Page, Political parties and electoral landscapes. Br. J. Polit. Sci. **28**(1), 139–158 (1998)

U. Krause, Time-variant consensus formation in higher dimensions, in *Proceedings of the Eighth International Conference on Difference Equations and Applications*, ed. by S. Elaydi, G. Ladas, B. Aulbach, O. Dosly (Chapman and Hall/CRC, Taylor and Francis, Boca Raton, 2005)

M. Laver, Policy and the dynamics of political competition. Am. Polit. Sci. Rev. **99**(2), 263–281 (2005)

M. Laver, M. Schilperoord, Spatial models of political competition with endogenous political parties. Philos. Trans. R. Soc. B Biol. Sci. **362**(1485), 1711–1721 (2007)

M. Lodge, *Magnitude Scaling. Quantitative Measurement of Opinions*, volume 07–025 of *Sage University Paper Series on Quantitative Applications in the Social Sciences* (Sage, Beverly Hills/London, 1978)

M. Lodge, B. Tursky, On the magnitude scaling of political opinion in survey research. Am. J. Polit. Sci. **25**(2), 376–419 (1981)

J. Lorenz, Continuous opinion dynamics under bounded confidence. NetLogo User Community Models (2012)

J. Lorenz, Modeling the evolution of ideological landscapes through opinion dynamics, in *Advances in Social Simulation 2015*, volume 526 of *Advances in Intelligent Systems and Computing*, ed. by W. Jager, R. Verbrugge, A. Flache, G. de Roo, L. Hoogduin, C. Hemelrijk (Springer International Publishing Switzerland, Cham, 2017), pp. 255–266

R. Matthews, Storks deliver babies (p = 0.008). Teach. Stat. **22**(2), 36–38 (2000)

M. Meyer, How to use and derive stylized facts for validating simulation models, in *Computer Simulation Validation—Fundamental Concepts, Methodological Frameworks, and Philosophical Perspectives*, ed. by C. Beisbart, N.J. Saam (Springer, Cham, 2019), pp. 383–403

E. Nagel, *The Structure of Science. Problems in the Logic of Scientific Explanation* (Hartcourt Brace World, New York/Burligame, 1961). Zitiert nach der bei Routledge Kegan Paul 1979/1982 in London erschienenen Ausgabe

T. Ostrom, Computer simulation: the third symbol system. J. Exp. Soc. Psychol. **24**, 381–392 (1988)

F.U. Pappi, Political behavior: reasoning voters and multi-party systems,, chapter 9, in *Introduction to Political Science*, ed. by R.E. Goodin, H.-D. Klingemann (Oxford University Press, Oxford, 1998), pp 254–274

M. Pineda, R. Toral, E. Hernández-García, Noisy continuous-opinion dynamics. J. Stat. Mech: Theory Exp. **2009**(08), P08001 (2009)

G. Rabinowitz, S.E. Macdonald, A directional theory of issue voting. Am. Polit. Sci. Rev. **83**(1), 93–121 (1989)

S. Roßteutscher, R. Schmitt-Beck, H. Schoen, B. Weßels, C. Wolf, M. Preißinger, A. Kratz, A. Wuttke, L. Gärtner, Wahlkampf-Panel 2017 (GLES). GESIS Datenarchiv, Köln. ZA6804 Datenfile Version 6.0.0 (2018). https://doi.org/10.4232/1.13150

R. Schmitt-Beck, Political participation—national election study, in *Building on Progress: Expanding the Research Infrastructure for the Social, Economic and Behavioral Sciences*, ed. by German Data Forum (Budrich UniPress, Opladen/Farmington Hills, 2011), pp. 1123–1137

J. Schupp, 25 Jahre Sozio-oekonomisches Panel—Ein Infrastrukturprojekt der empirischen Sozial- und Wirtschaftsforschung in Deutschland. Zeitschrift für Soziologie **38**(5), 350–357 (2009)

M. Sobel, Causation and causal inference: defining, identifying, and estimating causal effects, chapter 8, in *Handbook of Probability. Theory and Applications*, ed. by T. Rudas (Sage, London, 2008), pp. 113–129

Socio-Economic Panel (SOEP), Data for years 1984-2016, version 33, SOEP (2017)

R.M. Solow, *Growth Theory: an Exposition* (Oxford University Press, New York, 1970)

D.E. Stokes, Spatial models of party identification. Am. Polit. Sci. Rev. **57**, 368–377 (1963)

R.K. Thomas, J. Bremer, I got a feeling: comparison of feeling thermometers with verbally labeled scales in attitude measurement, in *Presented at the 67th Annual Conference of AAPOR*, Orlando, May 16–May 20, 2012

L.L. Thurstone, Attitudes can be measured. Am. J. Sociol. **33**(4), 529–554 (1928)

K.G. Troitzsch, *Bürgerperzeptionen und Legitimierung. Anwendung eines formalen Modells des Legitimations-/Legitimierungsprozesses auf Wählereinstellungen und Wählerverhalten im Kontext der Bundestagswahl 1980* (Lang, Frankfurt, 1987)

K.G. Troitzsch, Simulating communication and interpretation as a means of interaction in human social systems. Simul. Trans. Soc. Model. Simul. Int. **99**(1), 7–17 (2012)

K.G. Troitzsch, Using empirical data for designing, calibrating and validating simulation models, in *Advances in Social Simulation 2015*, volume 526 of *Advances in Intelligent Systems and Computing*, ed. by W. Jager, R. Verbrugge, A. Flache, G. de Roo, L. Hoogduin, C. Hemelrijk (Springer International Publishing Switzerland, Cham, 2017), pp. 413–428

K.G. Troitzsch, Can lawlike rules emerge without the intervention of legislators? Front. Sociol. **3**, 2 (2018)

K.G. Troitzsch, Mikrosimulationsmodelle und agentenbasierte Simulation, in *Mikrosimulationen. Methodische Grundlagen und ausgewählte Anwendungsfelder*, ed. by M. Hannappel, J. Kopp (Springer VS Verlag für Sozialwissenschaften, 2020), pp. 85–107

G. von Randow, When the centre becomes radical. J. Artif. Soc. Soc. Simul. **6**(1). Originally *Frankfurter Allgemeine Sonntagszeitung*, 10 Nov 2002, No. 45, p. 63, and *Courier International*, 12 Dec 2002, No. 632 (2003)

A. Waldherr, N. Wijermans, Communicating social simulation models to sceptical minds. J. Artif. Soc. Soc. Simul. **16**, 13 (2013)

U. Wilensky, NetLogo. (1999). http://ccl.northwestern.edu/netlogo

World Values Survey Association, World Values Survey 1981–2014 Longitudinal Aggregate v.20150418. Aggregate File Producer: JDSystems (Madrid, Spain, 2015)

B.P. Zeigler, *Theory of Modelling and Simulation*. Krieger, Malabar. Reprint, first published in 1976 (Wiley, New York, 1985)

Part III
New Look on Old Issues: Research Domains Revisited by Computational Social Science

A Spatio-Temporal Approach to Latent Variables: Modelling Gender (im)balance in the Big Data Era

Franca Crippa, Gaia Bertarelli, and Fulvia Mecatti

1 Introduction

The 2015 deadline of the United Nations (UN) Millennium Development Goals (MDGs), still unmet, has resulted in an extraordinary growth in the availability of gender-related datasets and statistical knowledge. An engendered statistical reasoning appears motivated by the increasing demand for gender-sensitive statistical information, gender balance being globally recognized as a crucial objective for economic growth and human development for society as a whole, for both women and men. The 2000 UN Millennium Declaration and the eight MDGs have engaged the world in a strict agenda to systematically monitor and report country progresses, on the basis of a shared system of measurable parameters and statistical indicators, which has provided, in the process, voluminous data of controlled quality at the country level and comparable in time and space. Moreover, gender equality is included in the 2030 UN Agenda of Sustainable Development Goals (SDGs), both as a particular goal (5), namely, 'To achieve gender equality and empower all women and girls', and as fundamental to delivering on the promises of sustainability, peace and human progress. The passage from the 2015 MDGs to the 2030 SDGs appears as a key turning point in the availability of

F. Crippa (✉)
Department of Psychology, University of Milano-Bicocca, Milan, Italy
e-mail: franca.crippa@unimib.it

G. Bertarelli
Department of Economics and Management, University of Pisa, Pisa, Italy
e-mail: gaia.bertarelli@ec.unipi.it

F. Mecatti
Department of Sociology & Social Research, University of Milano-Bicocca, Milan, Italy
e-mail: fulvia.mecatti@unimib.it

T. Rudas, G. Péli (eds.), *Pathways Between Social Science and Computational Social Science*, Computational Social Sciences,
https://doi.org/10.1007/978-3-030-54936-7_7

gender-sensitive data, with an unprecedented outbreak of good quality and easily accessible gender-sensitive data. This phenomenon represents a huge step in the development of gender statistics and a peculiar challenge ever since its original interpretation as a mere data disaggregation between men and women. In this chapter, a critical review of the current data richness, accessibility and usability will be given. In the present scenario, the World Bank's Gender Data Portal covers a wide spectrum of dimensions, and it gathers data from multiple sources, both internal and from several supranational agencies. It represents the first example of gender-related big data that is standardised and comparable. Therefore, the Gender Portal offers relevant analytical opportunities linked to three key points, calling for methodological innovations, which will be discussed further on in the chapter: (1) the potential of computational methods for massive amounts of data to develop improved and effective multivariate gender statistics, able to go beyond simple and composite indicators, which are so far mainly used in this field; (2) the need for effective methodologies for selecting relevant data streams and significant variables; and (3) the emerging possibility for highlighting the 'two-speed road to gender equity' (European Commission 2015) usually revealed by the gender condition when looking at the world path towards fully realized democracies, namely, from denied basic rights to equal opportunities between men and women. From the perspective of the gender gap as a latent construct, these key points imply taking into account a large variety of information in assessing both the measure and the causes of the gap. Multivariate latent Markov models (MLMMs) meet these requirements, as they allow including covariates, on any scale, both in the measurement and in the latent part of the model, unlike more customary statistical analysis that imposes several restrictions, as in the case of structural equation models (SEMs). MLMM results yield new insight, allowing the comparison of areas in time and in space, even when the gender gap measures do not hold a unique order, therefore tracing paths and patterns. It would therefore be possible to study in depth the causation and determinants of the 'transition' from a situation of maximum disparity to either the most equitable one reached by a nation at present or some way in between.

2 The Gender Data Revolution: Setting a New Frontier in Engendered Statistics

Originally, gender statistics were conceived for, when not confined to, the collection of women characteristics only. Under the impulse of the three initial Conferences on Women, starting from Mexico City in 1975, then in Copenhagen in 1980 and finally in Nairobi in 1985, the beginning of an international systematic work on gender statistics started as the declination by gender of measures that were previously collected for the general population. It soon became clear, though, that the primary requirement was to enshrine all aspects affected by gender issues, a perspective that required both conceptual definitions and collection methods to capture topics and aspects where gender inequalities hardly come to the surface, due to the likely embedment of stereotypes and bias (Corner 2003; Hedman et al. 1996). The gender statistics production process has therefore embraced a process that goes under the umbrella term 'engendering data', where the aim is to develop methods and techniques for showing gender differentials in data, from more clear-cut gender issues to indirectly perceivable differences. This perspective encompasses women's and men's stance in addressing hindrance to their civil participation in society, even when in dramatically different directions and to different extents. The whole historical gender statistics process develops throughout this path, moving from the fundamental, first acknowledgement of statistics on women in 1975, in the UN International Year of Women, which initially focused on the female standpoint of available data, collected using a previous gender-insensitive standard (Corner 2003). Major boosts to the development and affirmation of gender statistics came from the 1995 Beijing World Conference and from the 2000 Millennium Declaration. With reference to the former, its impulse is symbolized by Strategic Objective H3 of the Platform for Action (http://www.un.org/womenwatch/daw/beijing/platform/institu.htm). In particular, at point 206.c, national statistical services are asked to 'involve (...) research organizations in developing and testing appropriate indicators and research methodologies to strengthen gender analysis, as well as in monitoring and evaluating the implementation of the goals of the Platform for Action'. Two supranational actions are currently crucial for addressing gender equality, by means of policies that need highly gender-sensitive data: the UN's The Millennium Declaration and the mandatory EU Gender Action Plan (European Council 2015) 2016–2020. The former establishes international standards for producing and disseminating gender-sensitive statistics. The eight MDGs United Nations Member States (2015) are to be monitored according to approximately 60 statistical indicators (United Nations Member States 2015). The latter includes mainstreaming as an effective strategy that promotes the incorporation of a gender equality perspective into all policies at all levels and all stages. Both actions rely on the establishment of a cooperation between data producers and policymakers (Corner 2003). The 2000 UN Millennium Declaration goals on gender equality were not fully achieved by the preset 2015 deadline. As a consequence, the 2030 Agenda has once again raised the concern for gender statistics as an indispensable information system for setting civil goals and monitoring paths towards them. The 17 SDGs (United Nations Member States 2019) constitute in truth a plan where these goals are interwoven, a necessary approach to the complexity of reality, where issues more very often are reciprocally associated. SDG5, pertaining to gender equality, is described as follows: 'Achieve gender equality and empower all woman and girls', and it is divided into 9 targets and 15 indicators. At any rate, roughly a quarter of all SDG indicators (53 out 230) address gender equality. There are indicators in SDG5 and also indicators across the framework that explicitly refer to sex, gender, women and girls and/or are specifically or generally targeted at women and girls.

The attempt to grasp the multifaceted nature of the gender gap in full, coupled with the huge regular production of data induced by the UN MDGs system, has given rise, for the first time ever, to a massive number of measures of gender disparities. The potential of this vast collection of engendered measures lies primarily in the chance of describing the multifaceted nature of the gender gap, calling the variety of sources for harmonization capable of a full integration. This huge information availability can be ascribed to the so called Big Data Revolution (Kitchin 2014), an ongoing process roughly summarizable as the explosion in the volume of data, and in their further 'V's, such as the variety and velocity, that is, in the speed with which these data are produced, in the number of producers and in the range of issues covered. At the core of the integration need for gender gap measures lies the feasibility of harmonizing traditional and innovative informative sources in a way that is apt to enhance the quality and timeliness of the data production paths. With the urgency being to reach an in-depth quantitative summary, easily readable and accessible to social scientists, this instance cannot be faced from a traditional perspective. Some challenges await the construction of this summary, which is meant to foster a more gender-responsive system for societal equity and for sustainable development (Lopes and Bailur 2018). One requirement consists of handling the widest possible number of aspects at the roots of this phenomenon, with reference not only to a given place and to a specific time point but also to its change in time and space. Another question consists of accounting for spatio-temporal dynamic change measurements for explicative and intervention variables, both in terms of a relational system and of assessing the dynamic measure itself. The latter statistical aspect represents an innovative methodological contribution to the gender bias issue. As a matter of fact, the engendering data process aforementioned has generated, as a major outcome, a wide and structured on-line search and availability of gender gap measures, originally produced by several distinct supranational agencies. As mentioned in the previous section, at present, possibly the richest source is represented by the Gender Data Portal, an open access tool on-line since 2017.

> The Gender Data Portal World Bank (2019) is the World Bank Group's comprehensive source for the latest sex-disaggregated data and gender statistics covering demography, education, health, access to economic opportunities, public life and decision-making, and agency. (Gender Statistics, The World Bank)

Data and resources come from the World Bank Group itself as well as other agencies, such as the United Nations Economic Commission for Europe (UNECE), Food and Agriculture Organization (FAO), International Labour Organization (ILO), Organisation for Economic Co-operation and Development (OECD), World Health Organization's Department of Gender, Women and Health Network (WHO's GWH), United Nations Children's Fund (UNICEF) and others. The database is updated four times a year, namely, in April, July, September and December, and it provides access to several types of gender indicators and indexes and to their time series. It provides topic dashboards for a single country and cross-country comparisons, including data visualization and analysis, and it allows searching for resources from the World Bank and other agencies.

3 The Rise of Computational Approaches from Recent Statistical Advances

The unprecedented availability of engendered big datasets, of controlled quality at the country level and comparable in time and space, appears as a great opportunity for adopting an advanced computational approach to the measure of gender bias. As anticipated in the Introduction, we rate particularly promising MLMMs for which, to our knowledge, there are no applications in the specialized literature.

MLMMs were introduced in the 1970s for the analysis of longitudinal data related to a latent response variable, i.e. a variable of interest that can be measured only indirectly. In the last few years, researchers (Bertarelli 2015; Bertarelli et al. 2018a,b, 2019) started to develop advancements to extend the application of MLMMs to some latent traits in the social sciences (see Bartolucci et al. 2012 for a general review on MLMMs).

The approach brings added value, on the one hand, when the latent response variable is supposed to have both a distribution in space and a time dynamic of its own. On the other hand, there are some advantages in choosing latent Markov modelling over more popular approaches in the social sciences. Indeed, a methodological relevant contribution of MLMMs consists of overtaking a well-known restriction of SEMs (Hausman 1983) (see the Introduction). Both SEMs and MLMMs consist of a two-part system: (1) a measurement model; (2) a structural model for the SEMs and a latent model for the MLMMs. In SEMs, at any rate, covariates can be introduced in the structural equations only, because of their contribution to the causal interpretation of the phenomenon under study. In MLMMs, instead, they apply to either submodel component, depending on the main objective of the study. This analytical concept expresses the assumption of covariates affecting the measure itself, as it is the case of gender bias and not only its hypothesized explanation, very often with no feedback or retroaction. This aspect potentially represents a groundbreaking advance with respect to up-to-date methodologies such as the descriptive construction of composite indicators and structural equation modelling.

3.1 The Multivariate Latent Markov Model for Spatio-Temporal Studies at a Glance

The basic idea is to look at the true measure of the variable under study at the spatial level of interest as a latent response indirectly measured, via related dimensions and longitudinal covariates, at successive time points, i.e., to consider it an underlying latent process. Regarding this latent process, both the distribution in space and its evolution in time are modelled by means of a (finite-state discrete first-order) Markov chain, a popular probability distribution for a process. As aforesaid, MLMMs consist of a two-part system, (1) the measurement model, which concerns the measurable part of the response variable, via available indirect measures, conditioned on the underlying latent process, and (2) the latent model, which concerns the probability distribution of the latent process itself, that is, the mentioned discrete Markov chain.

The methodological approach is hierarchical Bayes. The measurement component is introduced at the top of the hierarchy. It models the probability distribution of any number of measurable (response) variables observed for each area unit included in a larger region of study, at every occasion of a certain time period. This distribution is affected by (conditioned on) the underlying latent process. The latent model is introduced at the lower level of the hierarchy, and it states the probability distribution of the underlying latent process.

As discussed below, a major advantage of MLMMs as an innovative computational social sciences (CSS) tool is the fact that covariates can be introduced either in its measurement or in its latent component or in both. As a matter of fact, MLMMs allow one to introduce several explanatory variables, with no restrictions on the measurement scale. When covariates are included in the measurement part, the latent component would account for the unobserved heterogeneity. When the covariates are included in the latent model, we essentially assume that the observed response variables do measure the individual characteristic of interest embedded in the latent attribute. The latter, in addition to not being directly observable, may also evolve over time so that the primary research interest lies on modelling the effect of covariates on the latent process. Moreover, in this way, the latent model would detect the variability of the latent response that is unobserved in the measurement model, as well as catch all the residual heterogeneity. The next section illustrates a gender statistics application with covariates included in the latent model.

The underlying latent process is assumed to follow a (first-order) Markov chain characterized by three parameters: the number K of latent states, the $(K \times 1)$ vector of initial probabilities and the $(K \times K)$ matrix of transition probabilities. Each of these parameters plays a crucial role in the modelling and represents a powerful tool on its own for interpreting the model results. A main asset of a MLMM is its rather rich output, which includes the spatial distribution of the latent response, its evolution in time and a prediction engine. Moreover, this output is easily readable and ready to use, despite the methodological and computational complexity of the modelling approach. In more detail, the spatial distribution of the latent response is provided as a classification of all area units into K clusters of increasing intensity with respect to the latent response variable. It is visualized in a map of the region under investigation, where homogeneous areas are represented by the same colour and between-area variability is highlighted by different colours. The model provides a map with the aforesaid latent clusterization for every time point in the observational period, which shows the evolution in time of the response variable. Finally, the transition probabilities will predict, for any area classified in a given cluster, the chance to change a class in the future, moving either forward or backward across the cluster ranking.

Essential details of the model specification are given below. Interested readers may find complete details and an extensive discussion in Bartolucci et al. (2009).

Let J be the number of measurable (response) variables $Y_j^{(t)}$ observed for each area unit at every occasion of the $T-$length time span, where $j = 1 \ldots J$ and $t = 1 \ldots T$. Let $\{U^{(t)},\ t = 1 \ldots T\}$ denote the latent process with K latent states. Finally, let $\mathbf{X}^{(t)}$ be the set of longitudinal covariates available at time t. The measurement model states the probability distribution of $Y_j^{(t)}|U^{(t)}$, while the latent model states the distribution of $U^{(t)}$ with covariates that affect both its initial and transition probabilities:

$$\pi_k = P(U^{(1)} = k|\mathbf{X}^{(1)} = \mathbf{x})$$

where $\mathbf{x}$ denotes a vector of values of $\mathbf{X}^{(t)}$ and k is a realization (a latent state) of the latent process $U^{(t)}$, $k = 1 \ldots K$. The vector of K initial probabilities informs us about the overall average probability, for a given area, to be classified into each one of the K increasing levels of the latent response variable at the beginning $t = 1$ of the observational period.

Let us consider now the probability for a given area to belong to cluster k at time t, given it belongs to $\bar{k}$ at time $(t-1)$. These probabilities, for all given areas and all time points, define the *transition* probabilities of the latent process as given by

$$\pi_{k\bar{k}}^{(t)} = P(U^{(t)} = k|U^{(t-1)} = \bar{k},\ \mathbf{X}^{(t)} = \mathbf{x})$$

$t = 2 \ldots T$; $\bar{k} \neq k = 1 \ldots K$. A multinomial parsimonious logit parameterization is adopted for both kinds of probabilities above (see Agresti 2002, p. 267, for a detailed illustration of multinomial logistic models and related parameterizations).

For the CSS application, a Bayesian approach consistent with the hierarchy above is advised. Under this approach, an MLMM can be fitted by a data augmentation Markov chain Monte Carlo (MCMC), based on a Gibbs sampler for the measurement part of the model and on a Metropolis-Hastings sampler for the latent component of the model, where covariates are accommodated. Thus, the model fitting needs to run a computational statistics algorithm, which requires a minimum of 30,000 Monte Carlo runs on top of a burn-in period of at least 20,000 runs.

After choosing a suitable range of values for K, usually between 1 and 5, all combinations of K values per available covariate, either the whole set or sub-groups of them, are then fitted. This procedure leads to a collection of possible models from which the best model can be selected.

Model selection, as is usually the case for complex statistical models with several parameters, is defined in terms of an appropriate information criterion, i.e., a mechanism that uses data to assign each candidate model a certain score, usually based on (some version of) the maximum likelihood. We suggest selecting the best model based on the Bayesian information criterion (BIC) and then validating this choice by computing the more familiar Akaike information criterion (AIC), which should be in accordance.

4 Towards a Computational Approach to the Gender Gap Issue in the Network Age

Several questions remain unsolved in approaching gender studies on a common ground. The debate raised on alternative indexes capable of reaching the same synthesis and worldwide powerfulness of the human development index (HDI) (Hedman et al. 1996; Permanyer 2008, 2010, 2013; Sagar and Najam 1998) has not yet found a completely satisfactory answer. This illustrates clearly why descriptive measures of gender disparities are still vital. In some areas, the very availability of data is an issue, as in the case of time use, where its complete lack involves 135 countries out of 217, a situation denounced by the World Bank Group as a serious obstacle to knowledge (Gender Portal, World Bank 2018). In addition, only 39 countries can rely on two or more data points from time-use surveys (*ibidem*). Indeed, this data non-existence cuts some basic knowledge out, for instance, in the prototypicality of gender roles.

Notwithstanding these grey areas, gender investigation in social studies has witnessed a surge of massive amounts of data, in the very same manner that has interested some areas of other disciplines, for instance, biology or physics. The latter, though, has pondered how to turn this data eruption into scientific knowledge computationally, while other sciences, among which are social and behavioural ones, have maintained a rather sceptical attitude towards the so-called data-driven approach. Inherently, the risk for big data is to become the domain of various agencies or restricted research groups (Lazer et al. 2009). As a consequence, the restitution to people of their personal data output could be negated, as well as the verification and reproduction of the results (*ibidem*). Data-driven computational methods can actually increase knowledge provided that the results are disseminated and the materials and methods, i.e., the database used and the estimation procedures, are given public access.

4.1 When Current Gender Gap Indexes Do Not Support Disambiguation of Societal Trends

The application to the gender gap has the main purpose of showing the added value offered, with respect to traditional gender statistics, by using innovative (CSS) tools such as the MLMM proposed in Sect. 4.1 when coupled with the use of available engendered big datasets, as provided by the Gender Data Portal illustrated previously, a groundbreaking advance in data production that leads the analysis towards nontraditional computational approaches. A starting point of this application has been the familiar controversial output and consequent interpretation of available composite indicators of the (same) national gender gap. Though based on different choices of measurable dimensions of the latent primary outcome, they usually offer different quantitative results, which ultimately lead to different statistical pieces of evidence, rankings and gender perspectives of the world. For this purpose, we considered the two popular official gender gap measures aforementioned: the global gender gap index (GGGI) (Hausmann et al. 2007) by the World Economic Forum and the gender inequality index (GII) (Gaye et al. 2010). The GGGI was first introduced by the World Economic Forum in 2006 as a framework for capturing the magnitude of gender-based disparities and tracking their progress. There are three basic concepts underlying the GGGI. First, the index focuses on measuring gaps rather than levels. Second, it captures gaps in the outcome variables rather than gaps in the input ones. Third, it ranks countries according to the gender gap rather than to women's empowerment. It measures the gender gap under four aspects, namely, economic participation and opportunity, educational attainment, health and political empowerment, via 14 observable variables, each measured as the ratio of females to males (for a complete list, see, for instance, Mecatti et al. (2012), Table 1). Instead, the GII is an inequality index introduced by the United Nations Human Development (UNDP) Reports in 2010. It measures gender inequality from three important aspects of human development: reproductive health (measured by maternal mortality ratio and adolescent birth rates); empowerment (measured by the proportion of parliamentary seats occupied by females and the proportion of adult females and males aged 25 years and older with at least some secondary education); and economic status (expressed as labour market participation and measured by the labour force participation rate of female and male populations aged 15 years and older). The GGGI and GII, even if measures of the same latent notion, differ both in the point indicators composition and in the aggregation method; thus, they produce different results (see, for instance, Mecatti et al. (2012) for a comparative review). In fact, rankings based on these indicators neither coincide nor are constant over time. Instead of privileging one index or the other, on the basis of some a priori criterion, both the GGGI and GII enter the model as observable response variables, each contributing to the measurement component with its own perspective and indicator choice.

The application concerns longitudinal-spatial data for countries in the Schengen Area over a 7-year time period, namely, from 2010 to 2016. We took into account all Scandinavian countries and Iceland, even if the latter and Norway are not in the European Union (EU). The same remark holds for Switzerland, which is included too. The great availability of data from the aforementioned Gender Data Portal enables the adjustment for discrepancies between the two response variables, GGGI and GII, by means of a set of covariates in two steps. In the first step, we considered 32 mixed-nature (i.e., measured either on a qualitative or quantitative scale) covariates that the authors regarded as sensitive to gender discrimination as provided by the Gender Data Portal. We considered, in the first place, few demographics and general covariates, including total population, life expectancy at birth, literacy rate for adults, human development and Internet usage, to control for different dimensions of the very diverse countries included in the analysis. Towards the same purpose, we then took into consideration two sets of covariates, consisting of a basket of economic and social security measures. Economic covariates regarded mainly the overall population and comprised Gini, poverty and inequality indexes, unemployment, self-employment, home ownership and female family workers. Social security measures mainly focused on women, as they concerned prenatal care and maternity leave.

The third set of covariates has been of interest to legislation against gender discrimination. This emphasis on legislative intervention aims to highlight the efforts, together with their outcomes, that several countries have undergone to accelerate, when not beginning, their process towards gender equity, as an expression of democracy and civil participation, thanks to the propelling inspirational values of the integration of Europe. Towards this point, to fully comprehend the implications of our analysis, let us remember how Sweden and all Scandinavian countries, likewise the Iceland, have come a long way. As a matter of fact, the levels of gender equality were reached in these countries far sooner than elsewhere. In addition, their role model has become so deeply rooted in society that common knowledge does not call for a reflexion in their legislation, whereas most of the Schengen Area countries still require it. We introduced a set of dummy variables as indicators of the presence or absence of the legislative issue, ranging from forms of personal violence or abuse, such as marital rape, domestic violence or sexual harassment, to nondiscrimination in the work place, such as equal remuneration (*coeteris paribus*) and hiring, to paid maternity leave, to ownership equality between spouses and, finally, to seats held by women in parliaments. The complete list of all variables, extracted from the Gender Data Portal and introduced as covariates in our analysis, is provided in Table 5 in the Appendix.

We set the range from 3 to 5 for the number K of possible latent states and then fit all combinations of K values per available covariate, either the whole collection or sub-groups of them. According to both the BIC and AIC criteria explained above, the selected best fit is given by a 4-cluster model with 16 covariates, the latter both significantly explicative of the resulting latent clusterization and, as it will be soon illustrated, also significantly predictive of (future) transition probabilities, either forwards or backwards across clusters. As mentioned in Sect. 3.1, the output of the MLMM is informatively rich. The array of seven maps in Fig. 1 (see the Appendix) shows both the (spatial) classification of the Schengen Area countries according to the four increasing levels of national gender balance attained, from the worst situation at level 4 to the best at level 1, and its evolution in time over the observational period 2010–2016, allowing for a cross-country discussion.

Countries improved in the direction of the most gender-balanced cluster 1 in 2013, while towards cluster 4 in 2010, similarly for the shift from 2013 to 2016. Decreases in the latent gap are possible, though only when starting from a good parity situation. The leading situation of some Northern European countries that were in cluster 1 in 2010 remains for the whole observational period of 7 years. This is the case of Norway, Sweden, Finland and Switzerland. At the opposite extreme, Romania, Bulgaria, Hungary and Malta adhere to cluster 4. In between, definite shifts to upper clusters or, sometimes, to lower ones in the overall observational period concern mainly the central clusters 2 and 3, with somehow limited improvements and some fluctuations that will be discussed later on.

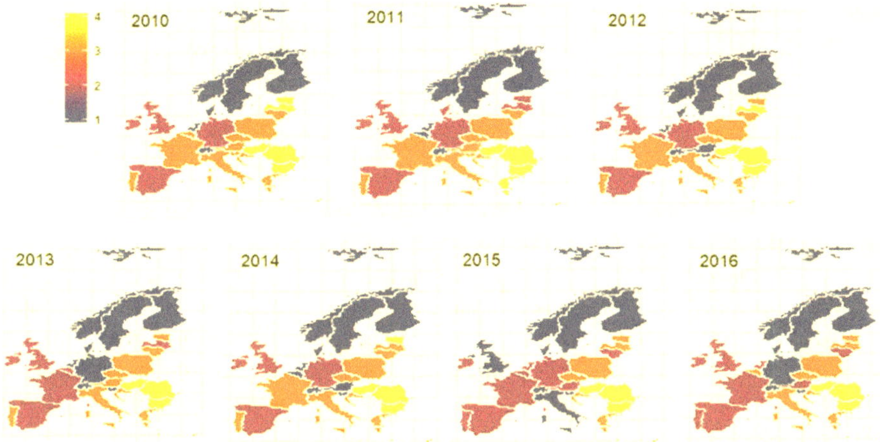

Fig. 1 Evolution of the latent clusterization over the 7-year observational period for Schengen Area countries, classified from the most gender-biased cluster 4 to the most gender-equal cluster 1

Table 1 Estimated parameters of the measurement model of the observed response variables GGGI and GII. SE = standard error

	GGGI		1− GII	
	Intercept	SE	Intercept	SE
Cluster 1	0.786	0.039	0.936	0.015
Cluster 2	0.754	0.018	0.864	0.039
Cluster 3	0.699	0.019	0.865	0.024
Cluster 4	0.689	0.022	0.742	0.055

Table 1 shows (estimated) coefficients of the measurement component of the selected model, namely, mean and standard errors of the two observable response variables (notice that the complement 1−GII is considered for the sake of comparability).

As mentioned above, with respect to gender balance as represented by the two measures GGGI and GII, clusters 1 and 4 represent the two extreme situations, the fourth one being the least balanced and the first one being the most balanced. Clusters 2 and 3 lie in between them, respecting the order relation for the GGGI, with an almost identical value with respect to the GII. In these clusters, both indexes move towards the same situation of improvement in gender parity, with respect to all dimensions. It should be noted how the relative distance between the two response variables remains higher for the GII rather than for the GGGI in the two extreme clusters, not in the central ones. The worst situation, represented by cluster 4, is .097 lower than cluster 1 with respect to the GGGI and by .194 with respect to the GII. Even though the two responses tend to provide the same order, for the GGGI and GII, the latter indicates an almost identical behaviour to that in the central clusters 2 and 3. Empowerment and strength in the labour market, reflected in the GII, seem therefore to act as a steering force especially in the two extreme situations, allowing to move away from the worst and to upgrade.

Table 2 Transition probabilities

	Cluster 1	Cluster 2	Cluster 3	Cluster 4
Cluster 1	1	0.000	0.000	0.000
Cluster 2	0.000	0.966	0.034	0.000
Cluster 3	0.000	0.016	0.984	0.001
Cluster 4	0.000	0.017	0.011	0.972

Table 3 Latent model estimated (standardized) coefficients of covariates affecting the transition probabilities (excluding legislation). Cluster 4 = benchmark, SE = standard error

	Cluster 4 to 1			Cluster 4 to 2			Cluster 4 to 3		
Covariate	Coeff	*SE*	*p*-value	Coeff	*SE*	*p*-value	Coeff	*SE*	*p*-value
HDI female	3.677	0.006	<.00001	13.910	9.434	.140213	41.326	17.762	.019987
Gini index	−8.988	1.213	<.00001	−0.248	0.001	<.00001	6.587	2.082	.001558
Gini index (adj edu)	−2.051	0.184	<.00001	2.322	2.293	.311538	31.859	21.364	.135909
Education	4.500	1.405	.0014	−8.087	6.8054	.234715	−31.759	22.277	.154157
Unemploy ment	0.577	13.059	.85041	19.044	18.223	.296023	22.727	18.1511	.21057
HDI	18.562	3.878	<.00001	9.655	4.996	.053298	0.068	1.004	.946582
Pop total	−0.194	0.003	<.00001	−4.024	0.299	<.00001	−14.533	3.072	<.00001
Internet users	0.316	0.000	<.00001	0.610	0.001	<.00001	−33.137	6.341	<.00001
Life expect. at birth	3.670	1.106	.00091	1.064	0.432	.01378	5.075	3.910	.194322
Labour force	1.181	0.549	.3054	6.193	4.219	.142159	5.5543	6.985	.426614

As illustrated in Sect. 3.1, in addition to the description of the countries' paths towards gender balance during the seven time points of the observational interval 2010–2016, MLMMs also provide a predictive engine allowing for a discussion with respect to a hypothetical future point in time. With regard to our best model, we now illustrate two main tools of this prediction machinery. In the first place, the matrix of transition probabilities (see Table 2) for each country included in a given cluster (i.e., by pointing a matrix row) predicts the chance to improve or, in principle, even to worsen, at a future time point, by moving to another cluster (i.e., forth and back across the matrix columns). The values on the main diagonal are very close to 1, with a predominant permanence in the same cluster. Otherwise, the predicted retroceding situation is from cluster 2 to 3; the improving ones are even rarer.

Still in line with the prediction step, Tables 3 and 4 show the estimated coefficients for the covariates that affect the transition probabilities, along with their standard error and significance. In both tables, the estimated coefficients are comparable (standardized) and have cluster 4 as the baseline for the sake of parsimonious parametrization. Therefore, they suggest the direction and intensity with which covariates affect the probability of a country to be headed from the benchmark cluster to another in the near future.

Table 4 Latent model estimated (standardized) coefficients of law mandate covariates affecting the transition probabilities. Cluster 4 = benchmark, SE = standard error

	Cluster 4 to 1			Cluster 4 to 2			Cluster 4 to 3		
Covariate	Coeff	*SE*	*p*-value	Coeff	*SE*	*p*-value	Coeff	*SE*	*p*-value
Prenatal care	1.787	0.097	<.00001	4.352	1.146	.000151	4.479	1.013	<.00001
Gender discrimination in hiring	−1.398	0.003	<.00001	−0.843	0.008	<.00001	−14.399	0.267	<.00001
Child or early marriage	−4.227	0.264	<.00001	−5.220	1.002	<.00001	−0.968	0.764	.205155
Domestic violence	−5.546	2.002	.005606	−2.553	1.001	.012419	−3.170	1.459	.029856
Paid maternity leave	3.850	1.003	.000124	23.506	4.568	<.00001	−6.506	3.128	.037617
Sexual harassment	−6.073	2.379	.010772	−5.430	2.001	.006728	23.407	8.001	.003439

In Table 3, we excluded legislative intervention for simplicity's sake, while in Table 4, we take into account the set of variables regarding explicit legislative interventions against gender discrimination.

Regarding non-legislative intervention, the human development index for both sexes and education in the predictive output of the model act only in favour of the transition from cluster 4 to cluster 1. The female version of the HDI supports also the shift to cluster 3. The Gini inequality index acts strongly on the upgrade to cluster 1 and 2, as greater gains in gender equality are reasonably easier in contexts that promote equality as a general social value. Instead, the penalization for the total population dimension becomes stronger the closer the shift. This result might suggest an attrition in reaching the general population for the lowest improvement.

Legislation is generally in favour of predictive transitions when supporting maternity, even if mandatory maternity leave becomes penalizing for the transition from cluster 4 to 3. In the case of non-discrimination rules in the workplace, estimates are not too reassuring, as they do not seem to act effectively in the desired direction.

As a general remark, the Schengen Area shows some resistance in abandoning the lowest positions in the cluster ranking. To sketch an overall pattern, we refer in the first place to the rankings in the middle (2013) and at the end of the observational period (2016), compared with the first observational year (2010). As mentioned above, countries that belong to the worst cluster in 2010 tend to remain there in 2013 and 2016. This is the case for Romania, Hungary, Bulgaria and Malta. Only two countries starting from cluster 4 were able to improve their gender balance. Estonia moved to cluster 3 in 2013, remaining there 3 years later, while Latvia improved to cluster 2 before regressing to cluster 3 in 2016. Countries ranked in cluster 3 in 2010 showed a slightly more dynamic pattern. On the one hand, once again, some of them held the same position in the three reference time points, namely, Croatia, Cyprus, Greece, Italy, the Czech Republic and Poland. On the other hand, a shift towards better clusters involved the paths of some countries, with different outcomes. Austria, France, Luxembourg and Belgium moved to cluster 2 in 2013, the former three staying there through 2016, with the latter moving back to cluster 3 at the end of the observational period. Portugal and Lithuania, instead, moved from cluster 3 to cluster 2 in 2016. In this scenario, Slovenia represented an exception, being the only country ranked in cluster 3 in 2010 to reach cluster 1 from 2013 on. Ireland, Spain and the United Kingdom, ranking in cluster 2 in 2010, showed practically no dynamics for the whole observational period, whereas Germany reached cluster 1 in 2013 and was still there in 2016. Almost all the countries in the most gender-balanced cluster kept their leading role steadily, with Denmark losing one position in 2011 but permanently regaining the best situation immediately after. The exception to this permanence in the leading situation was the Netherlands, which lost a position in the rankings, joining cluster 2 in the last 2 years at the end of the observational period. The yearly trend allows one to focus on some specific paths that tend to show more fluctuations, as in the case of Estonia, which was able to move from the worst cluster to the second in 2011, then sliding back to the third one later on, with a regression to the worst cluster just for the year of 2015. These yearly shifts in ranking are of particular interest when involving cluster 1. This is the case for Belgium, which moved up from cluster 3 in 2010, reaching cluster 1 in 2 years, 2011 and 2014, before finally returning to its initial situation. Austria started again from cluster 3 and ranked in cluster 1 both in 2012 and 2014, losing one position after that. The United Kingdom, moving from cluster 2, improved in 2015 to cluster 1 and then went back to its stable position. Even Italy, always in cluster 3, reached the best position once in 2015. Reaching the most gender-balanced situation occurs more often via an isolated upward shift than a stable gain. Several initiatives have been carried out, ranging from interventions in terms of the work-life balance to the presence of women in companies' advisory boards. Their impact relies heavily, on the one hand, on the organization and general culture of work. On the other hand, stagnation of the gender gap in employment has characterized the last few years, and countries with lower women participation in the labour market, such as Greece, Malta, Romania and Italy, have not filled the gap (European Commission 2018).

5 Conclusions

The interpretative power of traditional gender statistics, in our case of the GGGI and GII, is improved by the CSS approach, allowing for in-depth insight into the situation and the path of countries over time. MLMMs proved to be flexible and apt to integrate data from different sources and on different measurement scales, leading towards the absorption and full exploitation of every advancement in data collection that can add knowledge to our comprehension of gender disparities. In our application, some major impressions can be drawn. The role model for the gender equality of leading countries is reasserted as a target aimed at but yet not reached on a stable ground. Apart from Germany and Slovenia, at the end of the observational period, no other countries were in cluster 1, while the Netherlands lost one position. The countries in the Schengen Area, though, show differentiated paths, some with little to no shifts, especially when at extreme positions, some with wider movements, either towards improvements or in some cases even towards situations characterized by a diminished gender equity. To this respect, the latent gender gap seems apt, to some extent, to reflect some sort of resistance to interventions in favour of gender parity. This distance between formally stated directions and underlying everyday practice has been especially reckoned in the workplace (United Nations News 2019). In fact, even though the private sector has been undergoing a strenuous effort in favour of promoting equal opportunities for both genders, a major obstacle to the transposition into current, widespread practice has been the mentality that still segregates some positions between women and men. As a matter of fact, horizontal segregation is considered one of the main causes of the gender pay gap (European Commission 2018), and it is even fostered on the basis that gains obtained by women could result in some sort of loss for the companies themselves. Legislative intervention in favour of gender equity is far more plausibly found where a right is reachable by governance action, but it is not yet rooted in society as acquired. In countries with a long-standing historical tension towards the full expression of civil rights, e.g., Sweden, there is no need for some laws that can act as a propeller, as in the case of other countries.

In our opinion, the cluster dynamics that some countries show in the CSS model are a sign of the contradictory forces coming from the governance on the one hand and from some social environments on the other hand. This contradiction has been captured, in our application, as a difficulty for a country either in departing from unfavourable initial positions, or in maintaining acquired better ones in terms of gender equality. In our CSS model, we were able to include a large number of observed variables with no restrictions on them. We relied on a very large availability of gender data, provided by the Gender Data Portal. Despite all this massive information and the capability of the computational method to embrace it, the identification of sources of gender disparities still needs some fine tuning with respect to aspects that influence matters and issues that do not come from some governance but are the outcome of more complex processes, for instance, the balance between working and private life. In this respect, we consider that there is still a lack of adequately gender-sensitive data to express sometimes subtle but decisive opportunity differentials in everyday experiences (Table 5).

Appendix

Table 5 32 variables selected from the Gender Data Portal to be used as model covariates. * indicates the subset of 16 covariates in the selected best model

Model covariate	Variable	Browsing the gender data portal
* Pop. total	Population, total	SP.POP.TOTL
Rural	Rural population	
	female (% of total)	SP.RUR.TOTL.FE.ZS
Urban	Urban population	
	female (% of total)	SP.URB.TOTL.FE.ZS
* Life expect. at birth	Life expectancy at birth	
	total (years)	SP.DYN.LE00.IN
Life expect. at birth	Life expectancy at birth	
female	female (years)	SP.DYN.LE00.FE.IN
* Internet users	Internet users	
	(per 100 people)	IT.NET.USER.P2
* Education	Literacy rate (%)	
	of people aged ≥ 15	SE.ADT.LITR.ZS
Smoking	Smoking prevalence	
	females (% of adults)	SH.PRV.SMOK.FE
Alcohol	Total alcohol consumption	
	per capita, females (15+)	SH.ALC.PCAP.FE.LI
* Labour force	Labour force, total	SL.TLF.TOTL.IN
* Unemployment	Unemployment, total (% of total	
	labour force national estimate)	SL.UEM.TOTL.NE.ZS
Unemployment	Unemployment, female (% of female	
female	labour force national estimate)	SL.UEM.TOTL.FE.NE.ZS
Family workers	Contributing family workers, female	
	(% of female employment)	SL.FAM.WORK.FE.ZS
Self-employment	Self-employed, female	
	(% of female employment)	SL.EMP.SELF.FE.ZS
Three decisions	Women participating in the 3 decisions:	
	health care, major household purchases,	
	visiting family (% of women aged 15-49)	SG.DMK.ALLD.FN.ZS
Nonpregnant	Nonpregnant and non-nursing	
	women can do the same jobs as men	SG.JOB.NOPN.EQ
Women in parliaments	Proportion of seats held by	
	women in national parliaments (%)	SG.GEN.PARL.ZS
Ministerial positions	Proportion of women	
	in ministerial positions (%)	SG.GEN.MNST.ZS

(continued)

Table 5 (continued)

Model covariate	Variable	Browsing the gender data portal
* Gini index	GINI index	
	(World Bank estimate)	SI.POV.GINI
* Gini index	GINI index in educational	
(adj edu)	attainment (World Bank estimate)	SI.POV.EDU.GINI
* HDI	Human development index	UNDP.HDI.XD
* HDI	Human development	
female	index female	UNDP.HDI.XD.FE
Maternity leave	Maternity leave (days paid)	SH.MMR.LEVE
* Prenatal care	% of pregnant women	
	receiving prenatal care	SH.STA.ANVC.ZS
Prenatal care $\geq$ 4	% of pregnant women receiving	
	prenatal care (at least 4 visits)	SH.STA.ANV4.ZS
* Gender discrimination	Law mandates nondiscrimination	
in hiring	based on gender hiring	SG.LAW.NODC.HR
* Child or early	Law prohibits or invalidates	
marriage	child or early marriage	SG.LAW.CHMR
* Domestic violence	Legislation exists on	
	domestic violence	SG.LEG.DVAW
* Paid maternity leave	Law mandates paid or	
	unpaid maternity leave	SG.LAW.LEVE.PU
Sexual harassment	Legislation specifically	
	addresses sexual harassment	SG.LEG.SXHR
Marital rape	Legislation explicitly	
	criminalizes marital rape	SG.LEG.MRRP
Equal remuneration	Law mandates equal remuneration for females	
	and males for work of equal value	SG.LAW.EQRM.WK

References

A. Agresti, *Categorical Data Analysis* (Wiley Publication, 2002)

H. Akaike, Maximum likelihood identification of Gaussian autoregressive moving average models. Biometrika, **60**(2), 255–265 (1973)

UN General Assembly, *Transforming our world: the 2030 agenda for sustainable development* (New York, 2015) https://sustainabledevelopment.un.org/post2015/transformingourworld. Accessed 25 Oct 2019

F. Bartolucci, M. Lupparelli, G.E. Montanari, Latent Markov model for longitudinal binary data: an application to the performance evaluation of nursing homes. Ann. Appl. Stat. **3**(2), 611–636 (2009)

F. Bartolucci, A. Farcomeni, F. Pennoni, *Latent Markov models for longitudinal data* (Chapman and Hall/CRC, 2012)

G. Bertarelli, Latent Markov models for aggregate data: application to disease mapping and small area estimation (2015). https://boa.unimib.it/handle/10281/96252. Accessed 25 Oct 2019

G. Bertarelli, F. Crippa, F. Mecatti, Measuring latent variables in space and/or time: a gender statistics exercise, in C.H. Skiadas, C. Skiadas, *Demography and Health Issues. Population Aging, Mortality and Data Analysis* (Springer, Cham/New York, 2018a), pp. 133–142

G. Bertarelli, M. Pratesi, C. Giusti, F. Bartolucci, M.G. Ranalli, L. Quattrociocchi, Modelling educational poverty by area-level SAE latent Markov models, in *Computational and Methodological Statistics book of Abstract (CMStatistics 2018)*, pp. 113 (2018b). http://cmstatistics.org/CMStatistics2018/docs/BoACFECMStatistics2018.pdf?20181120232255. Accessed 25 Oct 2019

G. Bertarelli, C. Giusti, M. Pratesi, Spatial distribution of multidimensional educational poverty using SAE, in *Data Science & Social Research 2019 – Book of Abstracts*, p. 23. (2019). https://drive.google.com/file/d/1W9MeCB8cHfO0QlbhsXmSBVb-hS0n8Hdq/view. Accessed 25 Oct 2019

L. Corner, From margins to mainstream. From gender statistics to engendering statistical systems. Bangkok, Fondo de Desarrollo de las Naciones Unidas para la Mujer (UNIFEM), en Asia y el Pacifico y los Estados Arabes (2003). https://www.researchgate.net/profile/Lorraine_Corner/publication/265540874_From_Margins_to_Mainstream_From_Gender_Statistics_to_Engendering_Statistical_Systems/links/55fbc82908aeafc8ac41beff/From-Margins-to-Mainstream-From-Gender-Statistics-to-Engendering-Statistical-Systems.pdf. Accessed 25 Oct 2019

European Council, Council Conclusions on the Gender Action Plan 2016–2020 (2015). https://www.consilium.europa.eu/en/press/press-releases/2015/10/26/fac-conclusions-gender-development/. Accessed 25 Oct 2019

European Commission, Two-speed road to gender equality – unequal labour market attachment of mothers (2015). https://ec.europa.eu/social/main.jsp?catId=89&furtherNews=yes&newsId=2362&langId=en. Accessed 25 Oct 2019

European Commission, 2018 Report on equality between women and men in the EU (2018). https://www.google.com/search?client=firefox-b-d&q=why+do+european+countries+have+difficulties+in+gender+equity+. Accessed 25 Oct 2019

A. Gaye, J. Klugman, M. Kovacevic, S. Twigg, E. Zambrano, Measuring key disparities in human development: the gender inequality index. *Human Development Research Paper*, 46 (UNDP-HDRO, New York, 2010)

J.A. Hausman, Specification and estimation of simultaneous equation models. Handb. Econ. **1**, 391–448 (1983)

R. Hausmann, L.D. TySon, S. Zahidi, The global gender gap index 2007, in *The global gender gap report* (2007), pp. 3–27. https://members.weforum.org/pdf/gendergap/index2007.pdf. Accessed 25 Oct 2019

B. Hedman, S. Centralbyraan, F. Perucci, P. Sundstroem, *Engendering Statistics: A Tool for Change*. Sweden Stockholm Statistics (Oakland Publications, Sweden, 1996)

D. Lazer, A. Pentland, L. Adamic, S. Aral, A.L. Barabasi, D. Brewer, et al., Computational social science. Science **323**(5915), 721–723 (2009)

R. Kitchin, *The Data Revolution: Big Data, Open Data, Data Infrastructures and their Consequences* (Sage, 2014)

C.A. Lopes, S. Bailur, Gender Equality and Big Data. UN Women (2018). https://undg.org/wp-content/uploads/2018/02/Gender-equality-and-big-data-en.pdf. Accessed 25 Oct 2019

F. Mecatti, F. Crippa, P. Farina, A special gen (d) re of statistics: roots, development and methodological prospects of gender statistics. Int. Stat. Rev. **80**(3), 452–467 (2012)

I. Permanyer, On the measurement of gender equality and gender-related development levels. J. Hum. Dev. **9**(1), 87–108 (2008)

I. Permanyer, The measurement of multidimensional gender inequality: continuing the debate. Soc. Indic. Res. **95**(2), 181–198 (2010)

I. Permanyer, A critical assessment of the UNDP's gender inequality index. Fem. Econ. **19**(2), 1–32 (2013)

A.D. Sagar, A. Najam, The human development index: a critical review. Ecol. Econ. **25**(3), 249–264 (1998)

United Nations Member States, Millenium Development Goals and beyond 2015. https://www.un.org/millenniumgoals/. Accessed 25 Oct 2019
United Nations Member States, Sustainable Development Goals. https://sustainabledevelopment.un.org/?menu=1300. Accessed 25 Oct 2019
United Nations News, Despite progress, companies face gender equality backlash: UN business body (2019) https://news.un.org/en/story/2019/03/1034891. Accessed 25 Oct 2019
World Bank, Gender Data Portal (2019) https://datatopics.worldbank.org/gender/. Accessed 25 Oct 2019

Agent-Based Organizational Ecologies: A Generative Approach to Market Evolution

César García-Díaz

1 Introduction

Social researchers have long studied the evolution of social systems (Breslin 2011; Felin et al. 2014; Hannan and Freeman 1977; Hodgson and Knudsen 2010; Nelson and Winter 1982), conceived as either a population of organizational forms (Bogaert et al. 2016; Levinthal 1997), a group of firms (Chang 2015), or a set of individuals embedded in a communication structure (Jackson 2008; Young 2001).

Computational approaches have gradually been incorporated as methodological tools to study social and organizational structures in fields such as economics (e.g., Chang 2015), sociology (e.g., Harrison 2004), management (e.g., Levinthal 1997), and political science (e.g., Jung and Lake 2011). This chapter is devoted to the computational exploration of a specific social structure: markets.

An *organizational ecology* is a set of members (e.g., individuals, organizational forms, firms) in a population that results from the different interacting forces affecting them (Jung and Lake 2011). This chapter reviews the implications of using *agent-based computational modeling* (ABM) to study markets as emergent ecologies. In an ABM framework, markets emerge in a *generative process*; their structures result from the interplay between entry, exit, and adaptation of heterogeneous firms. I argue that agent-based computational methods have emerged naturally (i) to allow inclusion of realistic firm behavior, (ii) to properly capture dynamic market processes, (iii) to simultaneously inspect the effects of adaptive features of individual entities and selection forces (Levinthal 1997), and (iv) to explore (co)evolutionary processes between firm-level characteristics, consumer

C. García-Díaz (✉)
Department of Business Administration, School of Economics and Business, Pontificia Universidad Javeriana, Bogotá, Colombia
e-mail: ce.garciad@javeriana.edu.co

T. Rudas, G. Péli (eds.), *Pathways Between Social Science and Computational Social Science*, Computational Social Sciences,
https://doi.org/10.1007/978-3-030-54936-7_8

behavior, and market structure (Fatas-Villafranca et al. 2019). In particular, I highlight the potential of ABM in two aspects: one is investigating aggregation problems in market structure emergence by opening the door to inspect alternative *generative* paths to well-known statistical regularities (García-Díaz et al. 2015; Page 2012); the other is exploring how market competition interacts with different internal features of firms (e.g., learning and adaptation).

Using the potential of computational modeling comes at a price, however, posing new challenges on the analysis of market structure dynamics. There has been a strong tendency to confound statistically assessed macro-level outcomes of micro-level calibrated behavior with explanatory mechanisms. Thus, the mere task of computationally reproducing macro-level features from micro-level calibrated behavior does not guarantee an *understanding* of a generative process. ABM can well provide micro-level descriptions of agents and their interactions that produce macro-level outcomes (*"if you didn't grow it, you didn't explain its emergence,"* Epstein 1999, p. 43). But usually there are no systematic procedures to investigate the generative process, that is, the *operational description* of the chain of causal events (León-Medina 2017). Besides the *generative sufficiency* test, some researchers also call for a thorough understanding of the causal links that reproduce a given property (*"if you don't know how you grew it, you didn't explain it,"* Macy and Flache 2009, p. 263). Model traceability may be a potential way to overcome that.[1] I briefly review this particular issue later in this chapter.

The interplay between entry and exit of firms has been approached differently by economics, management, and sociology. Therefore, Sects. 2, 3, and 4 briefly review key contributions from economics' industrial organization, sociology's organizational ecology, and management's NK fitness landscape models, with special attention to their links to modern computational approaches.[2] In Sect. 5, I illustrate how the consideration of computational methods has emerged to complement and overcome some limitations in the empirical tradition of the abovementioned fields. I also argue that agent-based computational methods are best suited to integrate economic, managerial, and sociological perspectives through a generative approach. Section 6 reviews methodological issues regarding explanation in generative processes. Section 7 concludes the chapter.

2 Industrial Organization and Computation

The field of industrial organization (IO) has long focused on answering questions such as the following: "[W]hy are some industries dominated worldwide by a

[1]I thank one of the reviewers for calling this to my attention.

[2]The reader may find extensive accounts of the contributions of these fields, for example, at Baumann et al. (2019), Carroll and Khessina (2019), Caves (1998), Hannan (2005), Salimath and Jones III (2011), and Sutton (2007).

handful of firms? Why is the size distribution of firms within most industries highly skewed?" (Sutton 2007, p. 2304). The early contributions to empirical IO research had been based on the structure-conduct-performance paradigm (Bain 1951; Caves 2007; Sutton 2007), putting forward a causal relationship between market structure, firm behavior, and profits. Due to data limitations, these works used a cross-sectional approach (Chang 2011b) and were mainly focused on long-run (equilibrium) effects.

IO has produced an impressive amount of studies that have significantly advanced the understanding of empirical regularities (Caves 2007). Nevertheless, the particular focus on long-run equilibrium in empirical studies, poses methodological problems in regard of the proper understanding of dynamic market processes. As stated by Chang (2011a, p. 28–29), "[w]hile the theoretical perspective supporting these empirical studies typically defined long-run equilibrium either with or without entry barriers, the actual process of firm entry and exit through which an industry approaches the stable long-run equilibrium remained outside the purview of standard industrial organization literature." The focus on long-run equilibrium has also been present in mathematical models. Game theory has undoubtedly played a significant role in understanding industry dynamics (e.g., Sutton 2007), in spite of its limitation to incorporate sophisticated firm entry and exit patterns. These mathematical renderings have been always characterized by firms with perfect foresight (e.g., Jovanovic 1982) and with the ability to maximize profits as a result of their exhaustive calculations of the possible outcomes of strategic decisions.

The correlation between entry and exit rates (Chang 2011b; Dunne et al. 1988), the persistence of entry and exit in mature industries (Chang 2015, 2018), and the *shakeout* patterns in incipient markets are important regularities across industries (Chang 2009). Some markets with high entry rates may also have high exit rates, whereas some others with low entry rates may also have low exit rates (Chang 2015). Industries may experience market shakeouts: an initial rise in the number of firms is followed by a sharp decline. Models relying on the traditional assumptions of game-theoretical methods have been used to explain market shakeouts (e.g., Klepper and Graddy 1990).

But despite the significant contributions of equilibrium-based models (e.g., see Asplund and Nocke (2006) and their findings relating firm turnover with market size), a proper understanding of entry and exit dynamics should focus not only on long-run outcomes but also on the transient process of market histories. Market shakeouts are out-of-equilibrium processes (Chang 2011a) that reflect, for example, the impacts of overexpectations in business decisions. Such overexpectations lead firms first entering the market and then leaving. This limited managerial rationality is at odds with those game-theoretical models that assume firms having perfect knowledge about the consequences of their decisions (Chang 2009).

Computational models have come into play as a means of rendering dynamic processes in markets. Many computational studies simulated the stochastic processes of entry, growth, and exit without much emphasis on firm behavior (e.g., Dosi et al. 1995; McCloughan 1995). Nonetheless, the inclusion of firm-level representations of heterogeneity (e.g., in terms of expectation formation) has long been a concern (Witteloostuijn and Maks 1988). Some other works, like the Markov perfect equilibrium models, relied on game-theoretical assumptions like having full rationality and focused on equilibrium outcomes (e.g., Besanko et al. 2010; Doraszelski and Satterthwaite 2010; Ericson and Pakes 1995). The latter approach suffers from the "curse of dimensionality" (see Chang 2009, p. 77 and Chang 2011a, p. 30) because of the computational burden the model imposes by considering firms' optimizing behavior concerning all their potential future states.

Agent-based computational models take a totally different approach; they oftentimes depart from the game-theoretical perfect optimizing behavior. Markets are characterized by unanticipated disruptions like changes in technological regimes (Chang 2011b) or the evolution of demand (Valente 2012). Therefore, the assumption of perfect foresight, including the possibility of knowing the probability distribution of future states, is highly questionable.

In contrast, real firms face market complexities by thinking inductively. Inductive reasoning implies a backward-looking behavior that conceives belief formation (mental models) of the states of the world. Beliefs can change according to the experiences perceived from the environment. Positive experiences strengthen beliefs, whereas negative experiences weaken them. As Arthur (1994, p. 407) puts it, "[e]ach agent will normally keep track of the performance of a private collection of such belief-models. When it comes time to make choices, he acts upon his currently most credible (or possibly most profitable) one."

A clear advantage of agent-based models is that they require a lower computational capacity than that of Markov perfect models, still being also capable to reproduce well-known empirical regularities (Chang 2015). For instance, agent-based models have reproduced market shakeouts with inductively reasoning agents. (Chang 2009) considered entry and exit dynamics with Cournot competition and technological search through innovation and imitation (Fig. 1). García-Díaz and Witteloostuijn (2011) built an entry and exit model with adaptive firms; considering a diversity of production costs, they explored different consumer taste distributions, showing that low-cost diversity leads to a shakeout (Fig. 2). The abandoning of the perfect rationality assumption has opened the door to a new generation of computational models that take into account learning features (Dosi et al. 2016; Rouchier 2017), environmental turbulence (Chang 2011b), and coevolutionary settings (Fatas-Villafranca et al. 2019).

3 Organizational Ecology and Simulation Modeling

Sociology's organizational ecology (OE) has been concerned with the antecedents and consequences of organizational demographic diversity: "Why are there so many kinds of organizations?" (Hannan 2005, p. 51). OE has produced an impressive number of empirical studies (Carroll and Hannan 2000; Hannan 2005) that collect and analyze the entire life data of industry populations. OE studies have explored the determinants of firm entry and exit to markets in a variety of industries.

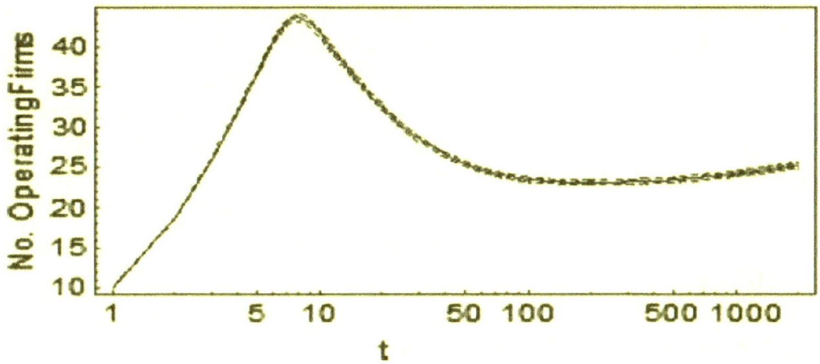

Fig. 1 The market shakeout phenomenon as reproduced by Chang (2009). The number of operating firms in the market first rises and then declines over time. (Reprinted by permission from Springer Nature Customer Service Centre GmbH: Chang 2009)

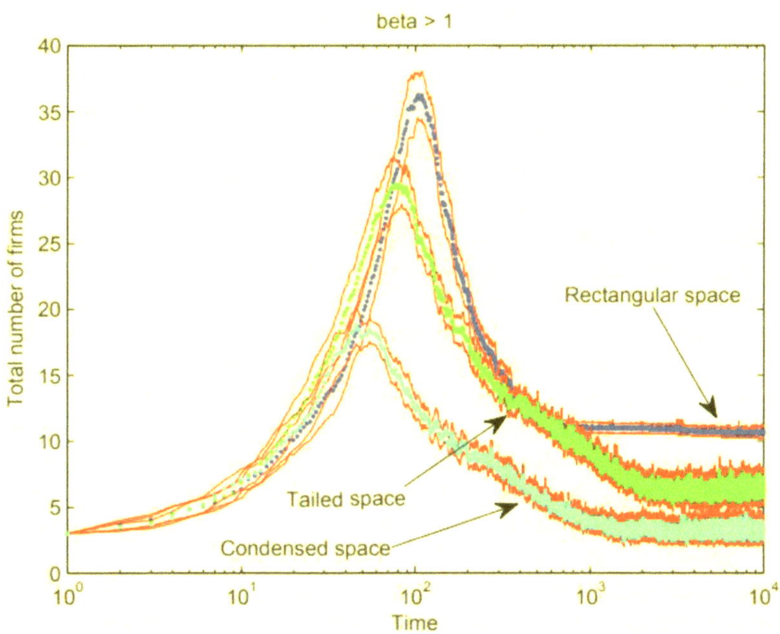

Fig. 2 The market shakeout phenomenon as reproduced by García-Díaz and van Witteloostuijn (2011). Firms with low-cost diversity – parameterized with a coefficient *beta* – populate markets with different consumer taste distributions: homogeneous (rectangular), tailed (unimodal), and condensed (with a few dominant consumer tastes) spaces. (Reprinted by permission from Springer Nature Customer Service Centre GmbH: García-Díaz and Witteloostuijn 2011)

A key assumption of OE is that adaptation takes place dominantly at the population level through the market inflow and outflow of firms. Selection forces are considered the main driver of industry evolution; firm-level adaptation plays a marginal role (Hannan and Freeman 1977). The view that population structures are mainly shaped by entry and exit was reinforced by arguments on the importance of organizational inertia: "Organizational ecology builds on a different assumption, that core structures of organizations are subject to strong inertial pressures and efforts at changing such structures substantially increase the chances of failure" (Hannan 2005, p. 59).

OE has studied ecological processes from a variety of angles. One stream explored the determinants of firm mortality with firms differing in scope (niche width theory; see Freeman and Hannan 1983). Another stream of OE research addressed the dynamics between founding rates, mortality rates, and firm counts mediated by the interacting forces of *legitimation* and *competition* (density dependence theory; Hannan and Carroll 1992). A third stream explained the effects of market concentration on the niche width-specific mortality of the so-called *generalist* and *specialist* organizations, with unimodal consumer distributions – resource spaces – in place (resource-partitioning theory, Carroll et al. 2002). Unlike economists, organizational ecologists have also studied nonproduct markets and have offered explanatory mechanisms incorporating concepts beyond production technologies, quantities, and prices (e.g., organizational identity; Hannan 2005).

Organizational ecologists quickly realized the methodological benefits of incorporating computer simulation tools to their research agendas. Their early computer simulation models addressed firm population growth trajectories based on the density dependence theoretical premises (Hannan and Carroll 1992). Barron (1999) investigated the joint effect of founding, mortality, and growth rates on the evolution of organizational populations, reproducing population trajectories that first rise and then decline after reaching a peak. Carroll and Harrison (1994) explored the reciprocal effects of two competing organizational populations. Harrison (2004) inspected four different modeling perspectives of organizational growth (i.e., Gribrat's law, Gibrat's law with effects of firm count and age, proportional growth, and Schumpeterian growth) and their effect on population dynamics. Harrison's work revealed the inherent complexity of market dynamics. He showed that each industry or organizational population had different distinctive features, so it is very difficult a single growth model would be able to capture all feasible population trajectories.

Despite the important contributions of simulation models in OE, these efforts have usually ignored the role of organization-level behavior in shaping markets. Structural changes have been rather characterized by stochastic renderings of entry and exit, reproducing population dynamics without the organization's ability to make strategic decisions. This orientation might be in part due to three issues: the early dominant Darwinian approach that downplayed the role of individual organizational adaptation; the lack of an empirical base to validate organizational-level behavior, especially when dealing with nonproduct markets; and the potential "risk" of falling in the realm of economics' strategic decision models when dealing with product markets.

Despite pleas to develop joint economic and sociological approaches to study industry dynamics (Boone and Witteloostuijn 1995; Witteloostuijn and Boone 2006), cross-fertilization could not really materialize. Rather, hybrid computational models were constructed, combining an ecological base approach with firm-level economic behavior. García-Díaz and Witteloostuijn (2006) inspected the preconditions of resource release. Concerning the resource release effect, resource-partitioning theory argues that competition at the center of the resource space caused by generalist firms leave peripheral spots unattended, which so become attractive ground to specialist firms. García-Díaz and Witteloostuijn (2006) concluded that resource release is not likely to occur without having some evolutionary changes in the shape of the resource space. García-Díaz et al. (2015) built an agent-based model that reproduced several statistical regularities observed in resource-partitioning studies; they also found that even non-scale-driven firms need a minimum scale of production to thrive in the market.

4 NK Models and Industry Dynamics

Management scholars have also studied industry dynamics using a model inspired in evolutionary biology (Baumann et al. 2019; Robertson 2019): the NK model. This model was first applied to management topics by Levinthal (1997). It has been used in a stream of papers in the computational literature in management, providing important contributions concerning organizational learning, the implications of problem decomposition, and the effects of different cognitive representations in organizational search problems, among others (see Baumann et al. 2019 for a detailed review of NK models in the management literature).

The NK rendering aims to model the easiness (or difficulty) of adaptation across several degrees of interdependence. Several variations of the NK model exist, the simplest assuming (N) different organizational sub-units, each facing a binary decision problem (Csaszar 2018). The performance (*fitness* value) of each of sub-unit might depend on the decisions of K other. Given some adaptation constraints (e.g., the organization walks along the fitness landscape by modifying only the choice of a single sub-unit at a time), the problem for the organization is to find an N-dimensional string $\mathbf{X}$ that locally maximizes its fitness. The performance contribution of each sub-unit x_i, $f_i(\cdot)$, depends on its own decisions and the choices of other K sub-units $x_{j_{i1}}, x_{j_{i3}}, \ldots, x_{j_{iK}}$, being j_{ik} the index of the kth sub-unit affecting x_i. Then, the performance contribution of x_i is $f_i(x_i; x_{j_{i1}}, x_{j_{i2}}, \ldots, x_{j_{iK}})$. The overall performance of the organization, $V(\mathbf{X})$, can be computed as:

$$V(\mathbf{X}) = \frac{1}{N} \sum_{i=1}^{N} f_i(x_i; x_{j_{i1}}, x_{j_{i2}}, \ldots, x_{j_{iK}}). \tag{1}$$

Let us assume that an organization is composed by $N = 3$ different sub-units, x_1, x_2, and x_3. Each sub-unit may opt for one out of two possible different design configurations, so the total number of organizational design possibilities is 2^3. Let us also assume that the organization wants to find which design brings about the highest performance of the organization, $V(\mathbf{X})$. That implies that the organization has to undertake a search of the highest peak on the fitness landscape.

As said, the fitness value of each sub-unit's decision depends on the choices of other K sub-units. The simple case is when fitness contributions are fully independent ($K = 0$). Table 1 illustrates an example of the fitness contribution of each ith sub-unit, $f_i(\cdot)$, when $K = 0$. The configuration [000] means that the organizational design choices are $x_1 = 0$, $x_2 = 0$, and $x_3 = 0$. Since the individual fitness contributions are $f_1(x_1 = 0) = 0.70$, $f_2(x_2 = 0) = 0.32$, and $f_3(x_3 = 0) = 0.23$, the overall fitness value for the organization is computed as $(0.70 + 0.32 + 0.23)/3 = 0.41$.

Notice that the highest fitness value is found at design configuration [011], with $V(\mathbf{X}) = 0.74$. This is equivalent to getting the highest possible fitness values of all sub-units simultaneously (0.70, 0.61, and 0.91). Thus, in absence of interdependence between sub-units, *maximizing the overall performance of the organization equates maximizing the performance of each sub-unit.*

In contrast, if the fitness contribution of each sub-unit depends on the remaining $K = N - 1$ sub-units (maximum interdependence), having each sub-unit maximizing its fitness value separately may bring about a suboptimal configuration for the organization. An example is shown in Table 2. The highest fitness value for the organization (0.84) is found at design configuration [100], which does not coincide with putting together the maximum fitness contributions of the three sub-units (0.98, 0.97, 0.82).

Adaptation takes place by assuming that an organization walks along the fitness landscape in search for the optimal design. Through successive gradual changes in the organizational design, the organizations test alternative configurations in the

Table 1 A numerical example of possible organizational design configurations with $N = 3$ organizational sub-units and an absence of performance interdependence between sub-units, $K = 0$. For instance, the performance of the first sub-unit x_1 when it takes the value of zero, $f_1(x_1 = 0)$, is always 0.70, independent of the arrangement of choices of the other two sub-units. The same logic applies to compute $f_1(x_1 = 1)$ and the fitness values of the remaining sub-units

Configuration	f_1	f_2	f_3	$V(\mathbf{X})$
[000]	0.70	0.32	0.23	0.41
[001]	0.70	0.32	0.91	0.64
[010]	0.70	0.61	0.23	0.51
[011]	0.70	0.61	0.91	0.74
[100]	0.27	0.32	0.23	0.27
[101]	0.27	0.32	0.91	0.50
[110]	0.27	0.61	0.23	0.37
[111]	0.27	0.61	0.91	0.60

Table 2 A numerical example of possible organizational design configurations with $N = 3$ organizational sub-units, where the performance of each sub-unit is dependent on the other two ($K = 2$). The performance $f_1(x_1 = 0)$ might be either 0.22, 0.97, 0.24, or 0.77, depending on the choices of the sub-units x_2 and x_3 ([00], [01], [10] and [11]). The same applies to calculate the performance of the sub-units x_2 and x_3

Configuration	f_1	f_2	f_3	$V(\mathbf{X})$
[000]	0.22	0.52	0.72	0.49
[001]	0.97	0.70	0.40	0.69
[010]	0.24	0.13	0.53	0.30
[011]	0.77	0.30	0.63	0.56
[100]	0.88	0.97	0.68	0.84
[101]	0.98	0.80	0.25	0.68
[110]	0.32	0.47	0.82	0.54
[111]	0.48	0.40	0.15	0.34

vicinity of the current configuration. Such a process is called *local search*. Local search is performed by changing one digit of the configuration string at a time.

A landscape with $K = 0$ is a *smooth* landscape with a clearly visible global optimum. An organization walking along a smooth landscape would eventually reach the global optimum. On the contrary, higher interdependence (e.g., $K = N - 1$) implies a more difficult adaptation process, since the chances that an agent (i.e., an organization) walking the landscape gets trapped in suboptimal values increase. Then, the visualization of the global optimum becomes difficult. In that case, the landscape is *rugged*. Examples of smooth and rugged spaces are seen in Figs. 3 and 4.

NK landscapes have also been used to model several aspects of industry dynamics. A key feature is the incorporation of internal aspects of firm functioning, in contrast with the economics and sociological traditions where the firm is the smallest unit of reference. NK renderings are purely theoretical. Unlike OE studies, they have no empirical links to any particular industry, and they are less focused on the emergence of specific market structures. The NK model literature is more concerned about the relationship between firm's adaptive capabilities, competition, and performance.

The seminal study of Levinthal (1997) explored the implications of having both adaptation and selection forces in a population of organizational forms, offering novel ways to the computational modeling of the effects of different degrees of organizational adaptation. Nevertheless, Levinthal (1997) included neither population dynamics nor competition features. Later, Lenox et al. (2006) proposed an agent-based model of firm competition to inspect the relationship between firm's adaptive capabilities (through the degree of interdependence of firm's production sub-units), population size, and industry-level profits. They applied two scenarios respectively based on undifferentiated (Cournot) competition and quality differentiation. Marginal costs (in the Cournot model) and product quality (in the quality differentiation model) were dependent on firm's internal organization of

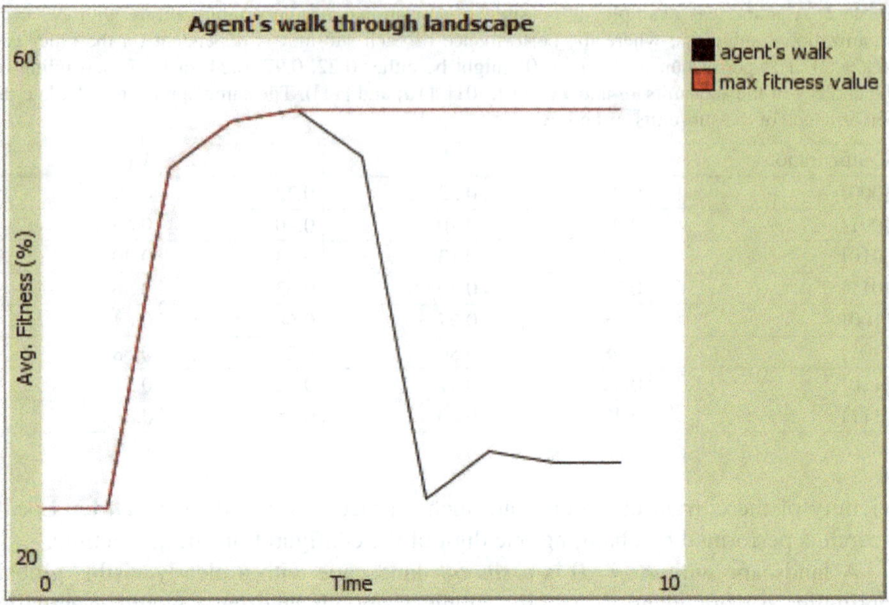

Fig. 3 An NK representation of a smooth landscape in NetLogo ($N = 3$, $K = 0$). The black line indicates an agent's walk through a fitness landscape, whereas the red line indicates the maximum fitness value found along the walk at time t. The agent moves to the unwalked spot with the highest fitness value in the neighborhood. In a smooth landscape, the agent is always able to locate the spot with maximum fitness value (global optimum) during the first hill climb

N activities. They found a non-monotonic relationship between average industry profits and the degree of interdependence K. When using the Cournot competition framework, Lenox et al. (2007) found that industries with a high K value had a high firm turnover and shakeout.

Following the works of Lenox et al. (2006, 2007), Knudsen et al. (2014) studied the effects of the expansion of firms' scale of operations; they combined the Cournot model with the NK fitness landscape framework to provide quantity and unit production cost possibilities for the firms. They reproduced observable market shakeout patterns, also finding that firm growth was a facilitating condition for the shakeout pattern to emerge.

5 Generative Processes of Markets

The previous sections illustrated how ABM has naturally emerged in different branches of the social sciences as a response to the need to incorporate bounded rationality into firm behavior, to investigate market dynamics through nonequilibrium market processes, and to understand the joint impact of firms' internal

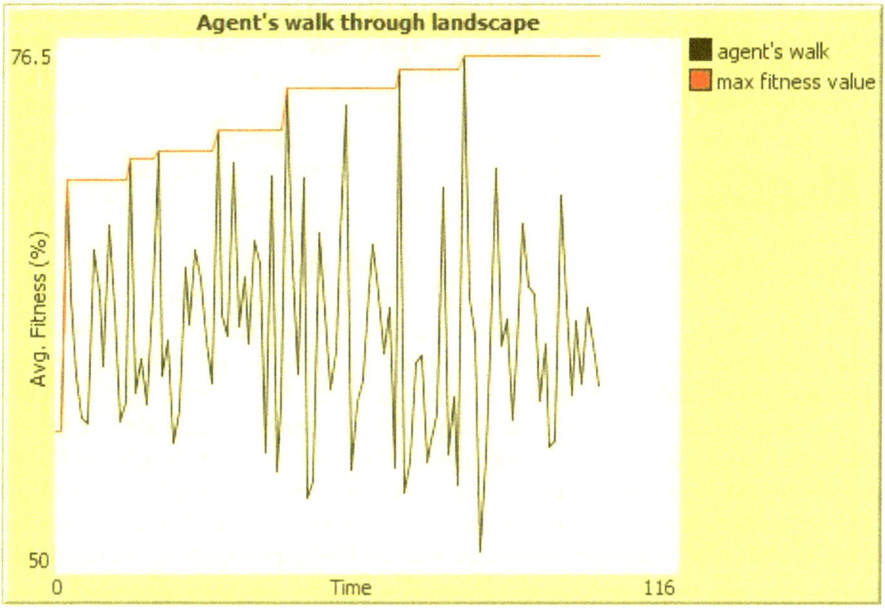

Fig. 4 An NK representation of a rugged landscape in NetLogo ($N = 10$, $K = 9$). The black line indicates an agent's walk, whereas the red line indicates the maximum fitness value found along the walk at time t. The agent moves to the unwalked spot with the highest fitness value in the neighborhood. In a rugged landscape, the agent always need to go downhill in order to find peaks with higher fitness values. If going downhill in the search process is not allowed, agents are eventually trapped in local optima

organization (i.e., adaptive capabilities) and competition on the emergence of market structures.

Considering economic behavior *as it is*, as opposed to *what it should be*, requires an alternative agenda for market studies (March 2014). This idea has long been around before the advent of ABM (Cyert and March 1963) and could finally be materialized due to the availability of modern computational methods. Abandoning the assumption of forward-looking agents endowed with perfect foresight has both technical and epistemic advantages. Assuming boundedly rational, backward-looking firms characterized by simple behavioral heuristics avoids unnecessary computational efforts, thus allowing to include features as cognitive biases, inductive reasoning, and learning. Different types of learning may trigger a variety of market structures (Vriend 2000). Bounded rationality also conforms with unexpected changes in technology regimes, with evolving consumer preferences, heterogeneous firm adaptive capabilities, and so forth.

The idea of reproducing market-level characteristics from firm-level behavior and interaction necessitates studying evolving industry dynamics and considering market change as a *generative process* (Chang 2018; Epstein 1999). This shift involves rethinking the ways we understood and modeled social systems so far, especially in economics (Arthur 2014). Approaches that explore statistical regularities help to choose between narratives for causal mechanisms; ABM can complement this by exploring temporal dynamics of these causal mechanisms. Understanding markets involves explaining multilevel phenomena, like the interaction between firms and market structures. ABM allows to study aggregation effects (Page 2012), which are not easy to perceive by mathematical or statistical methods. Doing so, ABM can provide alternative (i.e., generative) explanations to well-known statistical regularities.

ABM can handle different degrees of firm-level adaptation (e.g., local search or long jumps in the fitness landscape) along with market rivalry that forces the least profitable firms to leave. With these possibilities in place, ABM-based research can lessen the divide between the traditional equilibrium focus of IO, the selectionist approach of OE, and the adaptive perspective of the NK framework. It also offers new avenues to develop *agent-based ecological frameworks* that reflect both firm behavioral aspects and market competition dynamics. Robertson and Caldart (2009, p. 165) rightly argued that "[t]he conceptual divide between the dynamics of the interactions between firms and the dynamics within them is useful to emphasize the main features and challenges associated with each of these approaches. In the real world, however, firms face the challenge of managing their external and internal landscapes simultaneously."

Another important coevolutionary process in markets that ABM might deal with is the one between firm strategy and consumer preferences. A big deal of models either ignore the role of evolving demand in shaping market structures (Valente 2012) or consider it static. The power of ABM can be used to explore the interplay between the changing demand and the adaptive capabilities of the firm. Felin et al. (2014) note that bounded rationality is oftentimes seen as a search process in a predetermined fitness landscape configuration. Adopting a coevolutionary lens would allow studying agents' strategy changes as the landscape modifies, as a consequence of agents' actions.

6 Explanation and Generative Processes

Considering markets as generative processes by means of computational modeling also brings about interpretative and methodological issues. The generative perspective is different from the traditional deductive and inductive approaches (Epstein 1999). Finding a generative explanation through computational modeling basically follows an abductive process (Davis 2018; Miller 2015). In trying to explain a particular phenomenon, modelers aim to find a given specification through successive iterations with the model. That is, modelers learn from model outcomes, which in turn act as inputs to further refine the internal mechanisms of the model itself. This type of investigation process is in contrast with the standard natural science approach in economics (Davis 2018).

As a generative approach implies making the chain of events that relate micro-level behavior with macro-level outcomes explicit (León-Medina 2017), simply reproducing market-level properties out of firm-level interactions and relating the two in a statistical fashion is not sufficient for explanation (Davis 2018). Because of the pervasiveness of statistical methods in the social sciences, some researchers propose statistical hypotheses as the only way to think about expectations concerning computational model outcomes, as if models generated empirical data (Knudsen et al. 2019), also considering mere explored statistical regularities as generative explanations.

Agent-based modelers are sometimes criticized for building black box-like computational representations with inner workings hard to understand, so necessitating the use of narratives to explain the internal chain of events. Using narratives is not a problem. Empirical researchers use sophisticated statistical methods with results connected to verbally constructed theories. The point is that a genuine generative explanation has to provide a convincing micro-meso-macro-succession of events concerning the phenomenon in question. Narratives have to be based on *operational* descriptions of the linked events, instead of on accounts of *correlated* variables. Therefore, models would have to incorporate auxiliary meso-level constructs that relate the micro level with the macro level.

There is no single answer for the above concern, and probably a combination of different actions should be undertaken. A first possible way to develop a consensus around explanatory mechanisms is through model *replication* and *code sharing*. For those aims, agent-based modelers have proposed documentation standards that help to explicitly describe models' components (the ODD protocol; see Grimm et al. 2010). A complementary option is model *traceability*, as developed for policy simulations (Lotzmann and Wimmer 2012). Traceability schemes have been developed to help maintain communication between model developers and stakeholders. Model traceability concerns establishing documentation procedures and protocols to easily switch from the conceptual model to the simulation code and results. A third possibility – and a challenging one – is *docking*: the task of comparing two different modeling frameworks to inspect whether they provide similar behavior and results (Burton 2003).

A triangulation is necessary to investigate the generative processes of agent-based organizational ecologies: the combination of statistical, interpretative, and computational tools can help to identify the underlying causal mechanisms. The divide between economics, sociology, and management at studying industry evolution is artificial. The construction of diverse hybrid conceptual models that incorporate both internal and external firm aspects, adaptation and selection features, and economics-based behavioral drivers along with sociological and managerial aspects should be encouraged.

7 Conclusions

This chapter provided an overview of the emergent agent-based computational approach to industry evolution in the pertaining economics, sociology, and management literature. I argued that computational approaches, and particularly agent-based modeling, have endogenously emerged as a response to the need for representing firm-level behavior. Agent-based models represent a convenient way to include realistic, i.e., boundedly rational, firm behavior: to capture dynamic market processes without an extreme computational burden and to allow to simultaneously inspect joint effects of adaptation and selection forces in markets, as well as coevolutionary processes like those between firm strategy and consumer preferences.

Aiming at developing cross-disciplinary efforts, I envisage two fields where the computational perspective to market evolution might become highly relevant. The first is *entrepreneurship* (see Chang 2011a, for a similar argument). Entrepreneurship typically involves out-of-equilibrium processes that unfold over time (Lévesque and Stephan 2019; McMullen and Dimov 2013) where learning aspects (Cope 2005) and evolutionary perspectives (Breslin 2008) are crucial. Models can help to understand, for instance, the role of entrepreneurs in shaping demand through the creation of new product dimensions (Liu and Witteloostuijn 2018). These considerations make ABM suitable to build theoretical frameworks that help clarifying implications of different entrepreneurial choices (e.g., Yang and Chandra 2013). The literature on computational modeling in entrepreneurship is still scarce but growing (e.g., Caton 2017; Ganco and Agarwal 2009; Mauer et al. 2018; Welter and Kim 2018).

The second one concerns *competitive strategy* (Robertson 2019). ABM can explore the ecological dynamics in competitive landscapes (Lenox et al. 2006, 2007; Robertson and Caldart 2009), at the same time considering internal organizational processes (e.g., intraorganizational turnover; Haveman 1995). ABM models of strategy would allow to develop dynamic theories that represent the interplay between strategic decision-making and market developments.

Markets are complex adaptive systems (Arthur 2014) where the equilibrium focus has not sufficed to develop proper causal mechanisms: markets are *computational irreducible* in the sense that, in order to understand their dynamics, one has to explicitly model their out-of-equilibrium processes (Bookstaber 2017). ABM can effectively contribute to the cross-disciplinary efforts aiming at understanding markets' evolving character and the underlying integrative causal mechanisms.

Acknowledgments I have greatly benefited from the suggestions and comments of Gábor Péli, Tamás Rudas, Arjen van Witteloostuijn, and two anonymous reviewers. They have provided invaluable insights to this work. All errors remain my own.

References

W.B. Arthur, Inductive reasoning and bounded rationality. Am. Econ. Rev. **84**(2), 406–411 (1994)
W.B. Arthur, *Complexity and the Economy* (Oxford University Press, New York, 2014)
M. Asplund, V. Nocke, Firm turnover in imperfectly competitive markets. Rev. Econ. Stud. **73**(2), 295–327 (2006)
J.S. Bain, Relation of profit rate to industry concentration: American manufacturing, 1936–1940. Q. J. Econ. **65**(3), 293–324 (1951)
D.N. Barron, The structuring of organizational populations. Am. Sociol. Rev. **64**(3), 421–445 (1999)
O. Baumann, J. Schmidt, N. Stieglitz, Effective search in rugged performance landscapes: a review and outlook. J. Manag. **45**(1), 285–318 (2019)
D. Besanko, U. Doraszelski, Y. Kryukov, M. Satterthwaite, Learning-by-doing, organizational forgetting, and industry dynamics. Econometrica **78**(2), 453–508 (2010)
S. Bogaert, C. Boone, G. Negro, A. van Witteloostuijn, Organizational form emergence: a meta-analysis of the ecological theory of legitimation. J. Manag. **42**(5), 1344–1373 (2016)
R. Bookstaber, *The End of Theory: Financial Crises, the Failure of Economics, and the Sweep of Human Interaction* (Princeton University Press, Princeton, 2017)
C. Boone, A. van Witteloostuijn, Industrial organization and organizational ecology: the potentials for cross-fertilization. Organ. Stud. **16**(2), 265–298 (1995)
D. Breslin, A review of the evolutionary approach to the study of entrepreneurship. Int. J. Manag. Rev. **10**(4), 399–423 (2008)
D. Breslin, Reviewing a generalized darwinist approach to studying socio-economic change. Int. J. Manag. Rev. **13**(2), 218–235 (2011)
R.M. Burton, Computational laboratories for organization science: questions, validity and docking. Comput. Math. Organ. Theory **9**(2), 91–108 (2003)
G.R. Carroll, S.D. Dobrev, A. Swaminathan, Organizational processes of resource partitioning. Res. Organ. Behav. **24**, 1–40 (2002)
G.R. Carroll, M.T. Hannan, *The Demography of Corporations and Industries* (Princeton University Press, Princeton, 2000)
G.R. Carroll, J.R. Harrison, On the historical efficiency of competition between organizational populations. Am. J. Sociol. **100**(3), 720–749 (1994)
G.R. Carroll, O.M. Khessina, Organizational, product and corporate demography, in ed. by D.L. Jr Poston *Handbook of Population* (Springer, Cham, 2019), pp. 521–553
J. Caton, Entrepreneurship, search costs, and ecological rationality in an agent-based economy. Rev. Austrian Econ. **30**(1), 107–130 (2017)
R.E. Caves, Industrial organization and new findings on the turnover and mobility of firms. J. Econ. Lit. **36**(4), 1947–1982 (1998)
R.E. Caves, In praise of the old IO. Int. J. Ind. Organ. **25**(1), 1–12 (2007)
M.-H. Chang, Industry dynamics with knowledge-based competition: a computational study of entry and exit patterns. J. Econ. Interac. Coord. **4**(1), 73–114 (2009)
M.-H. Chang, Agent-based modeling and computational experiments in industrial organization: growing firms and industries in silico. East. Econ. J. **37**(1), 28–34 (2011a)
M.-H. Chang, Entry, exit, and the endogenous market structure in technologically turbulent industries. East. Econ. J. **37**(1), 51–84 (2011b)
M.-H. Chang, *A Computational Model of Industry Dynamics* (Routledge, London/New York, 2015)
M.-H. Chang, Computational industrial economics: a generative approach to dynamic analysis in industrial organization, in *The Oxford handbook of computational economics and finance*, ed. by S.-H. Chen, M. Kaboudan, Y.-R. Du (Oxford University Press, Oxford, 2018), pp. 594–634
J. Cope, Toward a dynamic learning perspective of entrepreneurship. Entrep. Theory Pract. **29**(4), 373–397 (2005)
F.A. Csaszar, A note on how NK landscapes work. J. Organ. Des. 7(1), 15 (2018)
R.M. Cyert, J.G. March, *A Behavioral Theory of the Firm* (Prentice Hall, Englewood Cliffs, 1963)

J.B. Davis, Agent-based modeling's open methodology approach: Simulation, reflexivity, and abduction. Œconomia. Hist. Methodol. Philos. **8**(4), 509–529 (2018)

U. Doraszelski, M. Satterthwaite, Computable markov-perfect industry dynamics. RAND J. Econ. **41**(2), 215–243 (2010)

G. Dosi, O. Marsili, L. Orsenigo, R. Salvatore, Learning, market selection and the evolution of industrial structures. Small Bus. Econ. **7**(6), 411–436 (1995)

G. Dosi, M.C. Pereira, M.E. Virgillito, The footprint of evolutionary processes of learning and selection upon the statistical properties of industrial dynamics. Ind. Corp. Chang. **26**(2), 187–210 (2016)

T. Dunne, M.J. Roberts, L. Samuelson, Patterns of firm entry and exit in us manufacturing industries. RAND J. Econ. **19**, 495–515 (1988)

J.M. Epstein, Agent-based computational models and generative social science. Complexity **4**(5), 41–60 (1999)

R. Ericson, A. Pakes, Markov-perfect industry dynamics: a framework for empirical work. Rev. Econ. Stud. **62**(1), 53–82 (1995)

F. Fatas-Villafranca, C.M. Fernández-Márquez, F.J. Vázquez, Consumer social learning and industrial dynamics. Econ. Innov. New Technol. **28**(2), 119–141 (2019)

T. Felin, S. Kauffman, R. Koppl, G. Longo, Economic opportunity and evolution: beyond landscapes and bounded rationality. Strateg. Entrep. J. **8**(4), 269–282 (2014)

J. Freeman, M.T. Hannan, Niche width and the dynamics of organizational populations. Am. J. Sociol. **88**(6), 1116–1145 (1983)

M. Ganco, R. Agarwal, Performance differentials between diversifying entrants and entrepreneurial start-ups: a complexity approach. Acad. Manag. Rev. **34**(2), 228–252 (2009)

C. García-Díaz, A. van Witteloostuijn, Co-evolutionary market dynamics in a peaked resource space, in *Advances in Artificial Economics*, ed. by C. Bruun (Springer, Berlin, 2006), pp. 93–104

C. García-Díaz, A. van Witteloostuijn, Firm entry diversity, resource space heterogeneity and market structure, in *Emergent Results of Artificial Economics*, ed. by S. Osinga, G.J. Hofstede, T. Verwaart (Springer, Berlin, 2011), pp. 153–164

C. García-Díaz, A. van Witteloostuijn, G. Péli, Micro-level adaptation, macro-level selection, and the dynamics of market partitioning. PLoS ONE **10**(12), e0144574 (2015)

V. Grimm, U. Berger, D.L. DeAngelis, J.G. Polhill, J. Giske, S.F. Railsback, The ODD protocol: a review and first update. Ecol. Model. **221**(23), 2760–2768 (2010)

M.T. Hannan, Ecologies of organizations: diversity and identity. J. Econ. Perspect. **19**(1), 51–70 (2005)

M.T. Hannan, G.R. Carroll, *Dynamics of Organizational Populations: Density, Legitimation, and Competition* (Oxford University Press, Oxford, 1992)

M.T. Hannan, J. Freeman, The population ecology of organizations. Am. J. Sociol. **82**(5), 929–964 (1977)

J.R. Harrison, Models of growth in organizational ecology: a simulation assessment. Ind. Corp. Chang. **13**(1), 243–261 (2004)

H.A. Haveman, The demographic metabolism of organizations: industry dynamics, turnover, and tenure distributions. Adm. Sci. Q. **40**, 586–618 (1995)

G.M. Hodgson, T. Knudsen, *Darwin's Conjecture: The Search for General Principles of Social and Economic Evolution* (University of Chicago Press, Chicago, 2010)

M.O. Jackson, *Social and Economic Networks* (Princeton University Press, Princeton, 2008)

B. Jovanovic, Selection and the evolution of industry. Econometrica J Econ. Soc. **50**, 649–670 (1982)

D.F. Jung, D.A. Lake, Markets, hierarchies, and networks: an agent-based organizational ecology. Am. J. Polit. Sci. **55**(4), 972–990 (2011)

S. Klepper, E. Graddy, The evolution of new industries and the determinants of market structure. RAND J. Econ. **21**, 27–44 (1990)

T. Knudsen, D.A. Levinthal, P. Puranam, A model is a model. Strategy Sci. **4**(1), 1–3 (2019)

T. Knudsen, D.A. Levinthal, S.G. Winter, Hidden but in plain sight: the role of scale adjustment in industry dynamics. Strateg. Manag. J. **35**(11), 1569–1584 (2014)

M.J. Lenox, S.F. Rockart, A.Y. Lewin, Interdependency, competition, and the distribution of firm and industry profits. Manag. Sci. **52**(5), 757–772 (2006)

M.J. Lenox, S.F. Rockart, A.Y. Lewin, Interdependency, competition, and industry dynamics. Manag. Sci. **53**(4), 599–615 (2007)

F.J. León-Medina, Analytical sociology and agent-based modeling: is generative sufficiency sufficient? Sociol. Theory **35**(3), 157–178 (2017)

M. Lévesque, U. Stephan, It's time we talk about time in entrepreneurship. *Entrepreneurship Theory and Practice, Advanced online publication* (2019). https://doi.org/10.1177/1042258719839711

D.A. Levinthal, Adaptation on rugged landscapes. Manag. Sci. **43**(7), 934–950 (1997)

M. Liu, A. van Witteloostuijn, Emergence of entrepreneurial populations: a feature dimensionality approach. Small Bus. Econ. **54**, 1–19 (2018)

U. Lotzmann, M.A. Wimmer, Provenance and traceability in agent-based policy simulation, in *ESM'2012–The European Simulation and Modelling Conference: Modelling and Simulation*, 2012, pp. 203–207

M.W. Macy, A. Flache, Social dynamics from the bottom up: agent-based models of social interaction, in *The Oxford Handbook of Analytical Sociology*, ed. by P. Hedström, P. Bearman (Oxford University Press, Oxford, 2009), pp. 245–268

J.G. March, The two projects of microeconomics. Ind. Corp. Chang. **23**(2), 609–612 (2014)

R. Mauer, R. Wuebker, J. Schlüter, M. Brettel, Prediction and control: an agent-based simulation of search processes in the entrepreneurial problem space. Strateg. Entrep. J. **12**(2), 237–260 (2018)

P. McCloughan, Simulation of concentration development from modified gibrat growth-entry-exit processes. J. Ind. Econ. **43**(4), 405–433 (1995)

J.S. McMullen, D. Dimov, Time and the entrepreneurial journey: the problems and promise of studying entrepreneurship as a process. J. Manag. Stud. **50**(8), 1481–1512 (2013)

K.D. Miller, Agent-based modeling and organization studies: a critical realist perspective. Organ. Stud. **36**(2), 175–196 (2015)

R.R. Nelson, S.G. Winter, *An Evolutionary Theory of Economic Change* (Harvard University Press, Cambridge, 1982)

S.E. Page, Aggregation in agent-based models of economies. Knowl. Eng. Rev. **27**(2), 151–162 (2012)

D.A. Robertson, *Agent-Based Strategizing* (Cambridge University Press, Cambridge, 2019)

D.A. Robertson, A.A. Caldart, *The Dynamics of Strategy: Mastering Strategic Landscapes of the Firm* (Oxford University Press, Oxford, 2009)

J. Rouchier, Agent-based simulation as a useful tool for the study of markets, in *Simulating Social Complexity (Understanding Complex Systems)*, ed. by B. Edmonds, R. Meyer (Springer, Cham, 2017), pp. 671–704

M.S. Salimath, R. Jones III, Population ecology theory: implications for sustainability. Manag. Decis. **49**(6), 874–910 (2011)

J. Sutton, Market structure: theory and evidence, in *Handbook of Industrial Organization*, ed. by M. Armstrong, R. Porter, vol. 3 (North-Holland, Amsterdam, 2007), pp. 2301–2368

M. Valente, Evolutionary demand: a model for boundedly rational consumers. J. Evol. Econ. **22**(5), 1029–1080 (2012)

N.J. Vriend, (2000). An illustration of the essential difference between individual and social learning, and its consequences for computational analyses. J. Econ. Dyn. Control. **24**(1), 1–19

C. Welter, S. Kim, Effectuation under risk and uncertainty: a simulation model. J. Bus. Ventur. **33**(1), 100–116 (2018)

A. van Witteloostuijn, C. Boone, A resource-based theory of market structure and organizational form. Acad. Manag. Rev. **31**(2), 409–426 (2006)

A. van Witteloostuijn, J. Maks, Workable competition and the barrier market. Eur. J. Polit. Econ. **4**(1), 117–135 (1988)

S.-J.S. Yang, Y. Chandra, Growing artificial entrepreneurs: advancing entrepreneurship research using agent-based simulation approach. Int. J. Entrep. Behav. Res. **19**(2), 210–237 (2013)

H.P. Young, *Individual Strategy and Social Structure: An Evolutionary Theory of Institutions* (Princeton University Press, Princeton, 2001)

Networks of the Political Elite and Political Agenda Topics: Creation and Analysis of Historical Corpora Using NLP and SNA Methods

Attila Gulyás, Martina Katalin Szabó, Orsolya Ring, László Kiss, and István Boros

1 Introduction

In this chapter we present the methodologies and results of the recent interdisciplinary research project focusing on elite networks in the early communist era in Hungary. We combine methods and approaches of social network analysis, computational linguistics, elite, and historical research. Such interdisciplinary approach introduces quantitative methods to fields where currently qualitative analysis is mainly used.

Our study is centered around a specific era in Hungarian history (between 1948 and 1956), during which Mátyás Rákosi was the first communist leader of the Stalinist dictatorship in Hungary. This era – usually referred to as *Rákosi era* – was characterized by centralization and strong party leadership coupled with an almost constant internal struggle for power. These years laid down the foundations for the subsequent 34 years of the socialism in Hungary.

Another characteristic feature of this and the following era was the strong influence of personal (social) networks at every level of political decision-making and related processes (Sík 2001). Up until now no research has been conducted to uncover the networks of the Hungarian political elite. Here, we uncover the latent network of the political elite in this era. While doing so, we also take a novel approach of elite research of this era and on this group.

Here, we analyze two written sources. Text network analysis is applied to the committee meeting reports of key political bodies, while a classification-based

---

A. Gulyás (✉) · M. K. Szabó · O. Ring · L. Kiss · I. Boros
RECENS group, Centre for Social Sciences, Hungarian Academy of Sciences Centre of Excellence, Budapest, Hungary
e-mail: gulyas.attila@tk.mta.hu; szabo.martina@tk.mta.hu; ring.orsolya@tk.mta.hu; kiss.laszlo@tk.mta.hu; boros.istvan@tk.mta.hu

T. Rudas, G. Péli (eds.), *Pathways Between Social Science and Computational Social Science*, Computational Social Sciences,
https://doi.org/10.1007/978-3-030-54936-7_9

approach is applied to the collection of agenda items and speakers of committee meetings of the same political bodies. Based on the insights obtained, we propose a novel approach to the elite definitions used for this era and stress the added value of CSS to history.

2 Data and Analysis

The Stalinist dictatorship of the period under investigation in Hungary ruthlessly punished resistance at any level starting from everyday people to the highest levels in the political hierarchy (using show trials in numerous cases).

Here, political elite means the "positional" elite in our perspective: "the communist power elite contains those who nominated the nominators of the nominees at lower levels" – essentially those present at the top of the nomenclature (Gyarmati 2009).

The positional elite consists of members of two specific party committees of the Hungarian Workers' Party (HWP): the Political Committee and the Secretary of the party. These were the major decision-making bodies of the governing party, and their members were viewed as the most powerful political actors. Also, as Table 1 shows, even a high-level executive position was not enough for one to become a member of these committees and join the circle of the positional elite.

Table 1 shows a few high-ranking party members with diverse key positions, also stating if they were considered as being in the positional elite. We assume that the formation of political power goes beyond the simple concept of positionality.

Table 1 High-ranking party members and their positions

Name	Hierarchical position	Positional elite
Mátyás Rákosi	General secretary of the Party, prime minister	Yes
Lajos Ács	Secretary of the CC[a], leader of the Department of Party and Mass Organizations	Yes
István Kovács	Secretary of the CC, leader of the Cadre Department	Yes
Ernő Gerő	Secretary of the CC, deputy chairman of the Council of Ministers	Yes
Márton Horváth	Director of the Department of Agitation and Propaganda	Yes
Zoltán Vas	Director of the National Planning Office	Yes
István Friss	Director of the Plan, Financial and Commercial Department	No
József Köböl	General secretary of the Union of Construction, Wood, and Building Material Workers	No
Mihály Farkas	Secretary of the CC, Minister of Defense	Yes
Ferenc Erdei	Minister of Agriculture, Minister of Justice	No
Zoltán Komócsin	Director of the Department of Agitation and Propaganda	No

[a]The Central Committee (referred to as CC in the table) was a key administrative body of the HWP. Being a member of this body was considered the highest party recognition

In the following we attempt to describe the political elite with a network defined by joint political action, modeled with the co-occurrence of the names of party members in written historical sources. We will assume that joint political action or cooperation in political matters can be interpreted as certain relationship between party members.

2.1 Textual Sources and Analysis Approaches

Text corpora play a key role in many fields of scientific research, such as in sociology, history, and linguistics. A digital corpus contains a huge amount of language data, and the digital form allows one to use algorithms to quickly and cost-effectively process and query the dataset, from quantitative and qualitative points of view, respectively (McEnery 2005:449).

The sources needed to map networks of the elite in the era investigated are quite rare, since most of their interactions are not documented. The main source used for this study is a set of official notes of the committee meetings of the Hungarian Workers' Party (HWP) produced between 1949 and 1956. These reports contain enough information to identify activities of various party members in relevant political topics, thus allowing the construction of a multimodal social network. Also, these reports are among the few sources preserved in a usable condition from that era.

Besides trying to find answers to historical and sociological research questions using this source, we seek to create text bases that can also be analyzed using semi-automated methods for further studies.[1]

The two main textual sources used in this research are compared in Table 2: the so-called decree reports of HWP committee meetings and the collection of agenda topics of the committee meetings.

The decree reports of HWP committee meetings are a set of reports describing meetings of various bodies. The HWP controlled the state between 1948 and 1956 as a party state, thus its leadership meetings are among the most important historical sources of that era. The edited or stenographic reports are not preserved in the HWP archives; only the decree reports are available. These are not edited reports; they rather resemble memos, not having a rigid structure save for some formal headers. These headers contain the date of the meeting; a list of attendees (from the party and

[1]In recent decades the volume of digitized text (originally paper-based sources) that permitted corpora-building in Hungarian has grown dramatically (Simon and Sass 2012:1–2). However, to the best of our knowledge, there is only one large and freely available Hungarian historical corpus, the so-called Magyar Történeti Szövegtár (MTSZ, Hungarian Historical Text Collection), containing also texts produced during the era in question. Though this corpus has a number of advantages (for instance, it consists of a vast volume of documents, representing hundreds of years of language usage), it also has some notable disadvantages concerning the query opportunities of the corpus. In addition, querying of the semantic features of the database is not possible.

Table 2 Basic features of the textual sources used

HWP decree reports	HWP meeting agenda
Available as scanned images	Available as digitized text
Not well structured	Well structured
Considerable preprocessing and corrections required	Some preprocessing and corrections required
Detailed descriptions of items and participating actors	Very short item descriptions with only the names of topic presenters
Covers meetings between 1949 and 1956	Covers meetings between 1949 and 1956

invited speakers); agenda points with their speakers; and finally the resolution or the decision of the body. Usually these reports span 3–4 pages per meeting. Overall, 4409 pages were added to this text base, composed of Political Committee and Secretary Committee reports. These reports are available in a digital text format after having executed optical character recognition (OCR). The digital text was in general of bad quality due to the age and condition of the scanned documents. Hence before use, they had to be corrected in a semi-automated way, focusing on having the correct names of the political actors properly identified in the text. For example, some frequent Hungarian family names referred to different people in the text (e.g., István Kovács, Sándor Kovács); here, individual tags are assigned to these individuals based on their given names. Although the main goal is to correct the names, it also allows us to limit the set of names used in the analysis. In the full text over 6000 names are mentioned, but in our analysis (with any political relevance) we will consider only 200 of them.

Our second source, just containing the agenda items of these meetings was constructed from the agenda collection published by the National Archives of Hungary (Varga 2007), containing the agenda and the speakers of the meetings of leading bodies of the HWP in chronological order. This source is noticeably smaller than the previous source, but its quality was superior. For this source no error correction was needed at all. Furthermore, this is an edited source; hence, there are no literal inaccuracies, and the agenda descriptions are much clearer and to the point. The preprocessing steps carried out on our second corpus were far less time-consuming and much more cost-effective than that for the first corpus since the documents were available in a PDF format. Consequently, the format processing from PDF to plain text did not cause frequent character problems in the texts. Also, in these sources common names do not appear frequently (e.g., István Kovács was also unique, although it refers to multiple members in the full meeting minutes).

The two methods we apply here create social networks from the information in the corpora in characteristically different ways. Words constitute the nodes in the text network analysis method. The edges between members of the elite are given by the co-occurrences of their names in the text. The classification methods compress the textual information into categories. There, the network between political actors is produced using a multimodal approach; the network connections between actors indicate occurrences in similar categories of the different text sources we analyze.

2.2 Text Network Analysis (TNA)

This is the method we use on the first corpus, the decree reports of HWP committee meetings. The general purpose of TNA is to describe the structure of the text, to identify the central concepts of it, and to visualize their relationships. The text network is created based on simple co-occurrences. Probability and proportion – based on the likelihood of the occurrence of a given word – are affected by cultural and linguistic factors, such as the form of the textual data and the vocabulary of the authors (Paranyushkin 2011). Words of the text are mapped into nodes, and they get connected when two words co-occur within the defined range of a text. This range or unit of text might either be a document, a paragraph, a sentence, or a given range of words as well. When the co-occurrence is defined as the distance between words, it is a common practice to use stop lists that contain words that should not be considered when calculating the distance. Usually these stop lists include articles (definite/indefinite), prefixes and suffixes, or other words specific to a corpus. Their main purpose is to prevent distortions from words that do not fundamentally change the information content of a sentence.

The names of the political actors are defined as nodes of the network – each name receiving a unique tag as described above to avoid confusing people with identical names. We assume that when the names of multiple political actors occur in the same sentence, they are involved in actions that are related to each other. Co-occurrence is defined as having a maximal distance of three words to each other (not counting words on the stop list as intervening words) and appearing in such a way at least three times in the source. In the network we connect nodes with co-occurrence conditions like these.

Figure 1 shows the network constructed from the committee meeting minutes between 1949 and 1956. The size of an individual node is proportional to its frequency of co-occurrence with others. The weight of the nodes is defined as the number of connections it has – hence people with larger corresponding nodes are well embedded in the network. Clusters are represented in different colors. The basic network parameters are given in the following table.

The network parameter values in Table 3 tell us that the network is sparse (low density) but well clustered (positive medium modularity). We expect to obtain clusters that correspond to groups of party members responsible for individual governmental areas: agriculture, industry, foreign policy, etc. Our expectation is observing intense internal communications between people working in the same area. However, the clusters obtained do not seem to correspond to such areas.

Table 3 Network parameter values

Parameter	Value
Avg. degree	6.081
Density (network)	0.012
Modularity (network)	0.486

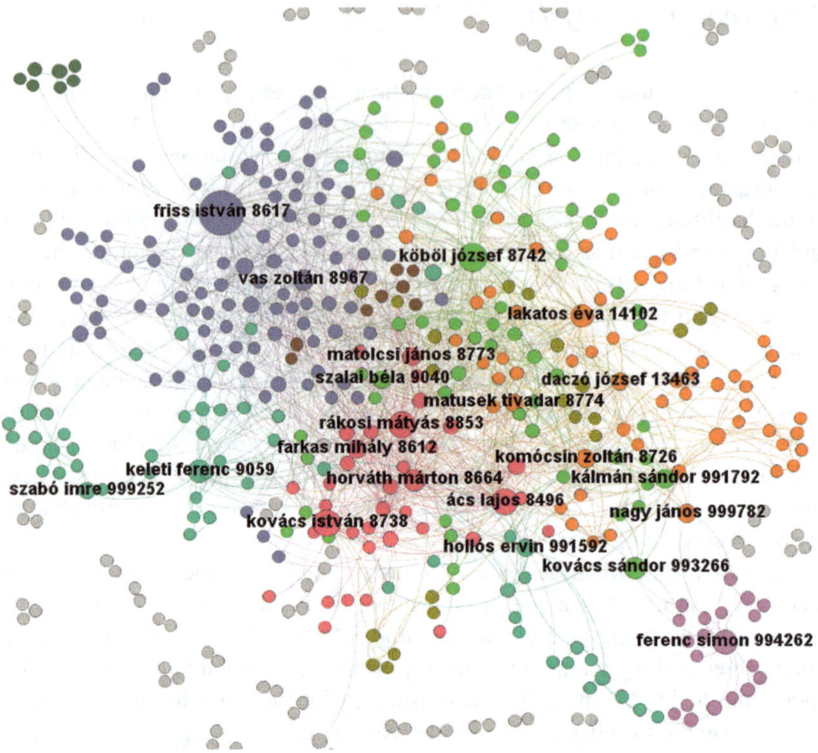

Fig. 1 The text-based network of name co-occurrences in the HWP committee meeting reports (Party Committee and Secretary 1949–1956)

Figure 1 shows the unique tags of the most specific actors (numeric postfixes are used to separate actors with similar names).[2] Most of the party members in Fig. 1 are also present in Table 1. Interestingly, in several cases members of the positional elite do not have considerable weights in the network. For instance, one of the major figures of the positional elite was Ernő Gerő, holding various ministry and executive positions (industry and economy); but he does not have a strong weight in this network, whereas we would expect him to be a central figure. In contrast, István Friss[3] is perhaps the most prominent person in the network, yet he was not a member of the positional elite, and his actual power certainly did not match that of Gerő. But Friss basically acted as the right-hand man of Gerő; hence his strong position in the network of Fig. 1 indicates his actual importance in the general political discourse. Another actor was István Kovács, the secretary of the party with the responsibility of

[2]The numerical suffix of these actors corresponds to their identifier in the personnel database used for their identification.

[3]István Friss (1903–1978) was the executive of the State Economics Department then later the Planning, Finance, and Trade Department.

organizing committee meetings and taking notes. His membership of the positional elite is confirmed in Fig. 1. But in contrast to István Friss, his practical role was rather administrative, explaining his frequent appearance in the communication.

As we said earlier, we expected that the TNA clustering would organize party members into groups according to governmental areas. However, the fact that the clusters of Fig. 1 do not reflect such structures suggests that the edges (co-occurrences) do not necessarily connect party members discussing topics from the same governmental area. A random verification of the sources also shows that our actors are often mentioned in contexts irrelevant to their actual political roles, like the enumeration of party members receiving awards or traveling abroad.

2.3 Classification Methods

Given that co-occurrences alone may not necessarily indicate joint political action, we decided to broaden our approach. Instead of simply observing textual co-occurrences, we also investigate co-occurrences in activities related to similar topics. We have to determine the topics in which our actors were actively involved. We construct a multimodal network connecting people to agenda items, and also agenda items to governmental categories (Fig. 2).

As our first textual source proved to be unsuitable for this approach, we created a second corpus of agenda items, containing the main speakers and participants of the HWP committee meetings. In this corpus there are 2241 agenda items, each item having a party member as the assigned presenter of the item during committee meetings.

We opt for a coding developed in the Comparative Agendas Project (CAP, Baumgartner et al. 2019). The codebook is adapted to the topics of Hungarian public policy (Boda and Sebők 2018). The CAP codes are assigned via double-blind coding; after determining coding reliability (using intercoder reliability measures), the doubtful cases are revisited and received consensus codes.

In the CAP coding system, every agenda item is categorized as a main topic and has one subtopic. We only use the main topic categories, since our goal is to connect political actors communicating in the same broad areas. Below, we present the two

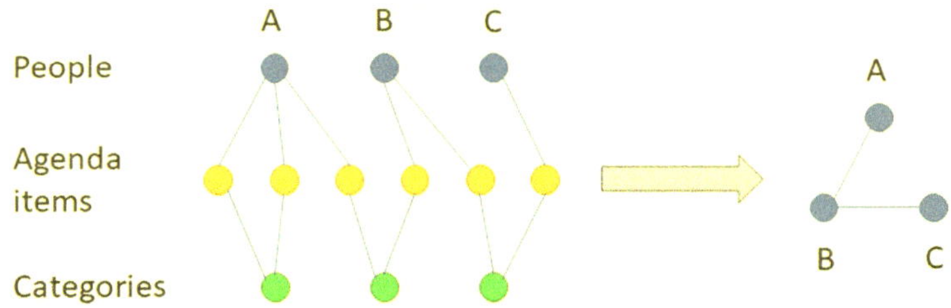

Fig. 2 Transforming the multimodal network

approaches we apply for assigning CAP codes to the agenda items: a dictionary-based approach using a specific ontology created for this corpus and a generic AI approach using a linear classifier.

2.4 Ontology-Based Classification

The first approach classifying agenda topics by CAP codes is based on dictionaries. We build an ontology, a dictionary that contains certain language elements (keywords) grouped into various classes, topics, and subtopics. By also incorporating the semantic relations among these elements, ontologies store entities in a hierarchical way (Noy and McGuinness 2001; Sure et al. 2002).[4] The agenda items receive CAP codes by searching for the keywords in them (Fig. 3). Here, we assign a single CAP code to each individual agenda item.

The categories of our ontology are the main topics in the CAP coding design. All the concepts identified in the agenda items are assigned to CAP categories. A word in the corpus is treated as a keyword, and it is included in the ontology if its presence associates an agenda item with a single topic. Words corresponding to two or more topics are excluded.

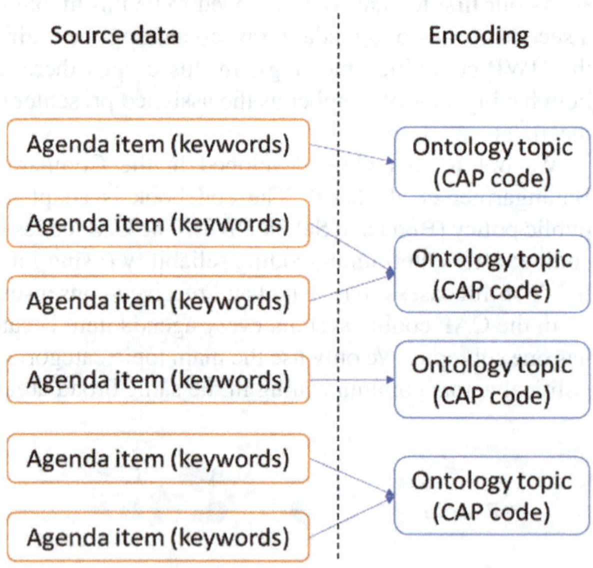

Fig. 3 Assigning CAP codes to the agenda items

[4]For this study the Protege tool is used. It is a free, open-source ontology editor that serves as a framework for building intelligent systems. This program was chosen after considering easy accessibility and operability and because it fully supports the latest c(OWL 2).

Table 4 The main statistics of the ontology

CAP-code	Main topic	N (keywords)	CAP-code	Main topic	N (keywords)
1	Macroeconomic policy	28	13	Social policy	14
2	Human rights	7	14	Area policy, housing	9
3	Health policy	10	15	Financial regulation	11
4	Agricultural policy	42	16	National defense	13
5	Employment policy	22	17	Science policy	9
6	Education policy	17	18	Foreign trade policy	9
7	Environmental policy	0	19	Foreign policy	14
8	Energy policy	10	20	Government operation	37
9	Migration policy	6	21	Land policy	7
10	Transport policy	9	22	Cultural policy	30
12	Justice	24	SUM		328

To avoid overfitting – i.e., determining too many or possibly non-relevant keywords for a single CAP code – we use stop lists containing the most frequent non-content words (articles, affixes, etc.). The first version of our ontology is created by the manual double encoding of the 20% of the corpus. As only a fraction of items can be categorized with this method, we resort to learning ontology creation[5] in order to enrich the codes assigned to the agenda items. Many more agenda items could be categorized this way than by manual encoding; but the efficacy of the method is still quite low. There are several too generic keywords that could be reasonably associated with multiple topics. Therefore, extended stop lists and a posteriori manual filtering have to be employed to produce reliable results.

The structure of the ontology obtained is relatively simple: it consisted of 21 topics (each matching a given CAP code) and 328 keywords semantically connecting to these topics. Table 4 shows the CAP codes and the number of keywords connected to them in the ontology.

Although Table 4 only lists the number of unique keywords associated to the topics, Table 4 also suggests the importance of topics indirectly. For instance, governmental operations, agriculture, and macroeconomic policies are the most important topics, whereas environmental policies are not present in the agenda items at all.

[5]During the encoding process, the code table was enriched by the words of a given agenda item that is assigned a given topic in the ontology. Of course as more items are included, the association list (word codes) increases significantly and results in overfitting.

Table 5 Manual classification results of 2241 items

CAP code	Freq	CAP code	Freq
1	196	14	30
2	16	15	58
3	20	16	94
4	240	17	49
5	210	18	37
6	135	19	119
8	2	20	606
9	11	21	25
10	21	22	145
12	57	99	115

There are 2241 agenda items we need to classify, each of them being assigned a specific CAP code, as described in Table 5. The 12 uncategorizable and 115 off-topic items are coded as 99.

The results of this classification are suitable for creating the multimodal network outlined above. However, in order to provide a method that could handle a much higher volume of data (and to verify the manual encoding), we also implement an AI-driven classifier.

2.5 *Classification Using an AI Classifier*

A major drawback of the ontology-based classification is its dependence on how topics change in the agenda items. There may be cases where the keywords are very specific or become obsolete. The latter means that as more and more items get processed, some keywords simply do not occur anymore due to the change of language or the change of topics. While automatic encoding techniques exclude such words, they still may remain classified as keywords with manual encoding. In practice this means that manual classifications may become unstable: different encoders might classify an item with different CAP codes or would misclassify some. We assume that the topics in the agenda items remain sufficiently homogenous over the given time period hence the keywords remain relevant.

In order to provide stability for processing further data in the continuation of this study, robust, automated processing has to be used. We apply an AI-based classification to the currently available data as a preparation for the future analysis of higher data volumes. Due to the low number of cases, we only present the methodology and some preliminary results. We use the fastText linear classifier implementation to produce a classifier for the agenda items. This is easy to use, and because its main unit for training is a set of character n-grams, not words, it creates a good vector representation of complete words (Joulin et al. 2016).

Classifiers in general require some preprocessing work on the corpus. For the fastText classifier, the complete corpus is transformed into a tidy text format – all

Table 6 Performance (classification efficiency) of binary classifiers for the codes 4, 5, and 20

Accuracy	Code 4	Code 5	Code 20
Iteration 1	0.9063	0.8928	0.8348
Iteration 2	0.9018	0.9152	0.8214
Iteration 3	0.8929	0.9375	0.8080
Iteration 4	0.9018	0.8839	0.8571
Iteration 5	0.8929	0.9018	0.8259

words are turned into lower case; glyphs and special characters are removed from the agenda items; and accented characters are exchanged for regular characters in the English alphabet. We use a very limited stop list to filter out only the most common words (mainly articles and conjunctions). We deliberately avoid using lemmatization as it may change the semantic relationship of a word with other words in a sentence (e.g., a simple accusative case suffix can significantly affect its meaning).

Initially we train the fastText classifier with 2000 samples from the dataset and used the remaining samples (~200–10%) to test the classifier. The target categories of the classifier are the complete set of CAP codes. Due to the small number of samples, the classification accuracy of this model is low (<10%); hence we take a different approach, using individual binary classifiers for each CAP code. Table 5 indicates that there are several codes with only a few occurrences; hence no classifier was trained for these cases. The three most relevant codes are 4,5 and 20; in the following we will present only their results.

The classifiers are trained in 25 epochs, with words occurring at least twice included in the training set. To obtain more meaningful results, five iterations are ran for each classifier.

The accuracy scores in Table 6 for three CAP codes indicate that the binary classifiers provide better results, but looking at the confusion matrices of each classifier tells us that the false-positive rate of these classifiers is rather high (sometimes exceeding 50%). Simply put, the high accuracy comes from accurately predicting when a text does not belong to a given category, but in most cases a text corresponding to a given category is misclassified as a negative.

We conclude that the currently available sample size of about 2000 is clearly insufficient for preparing a robust, automated classification. Therefore, we decide to rely on manual classification.

2.6 *The Resulting Relational Network*

As shown in Fig. 2, the CAP encoding is used to create a multimodal network connecting people to different CAP codes through agenda items. This network is then transformed into a relational network of party members through their connections with different CAP codes. Thus, when two people are connected by

Table 7 Network parameter values

Parameter	Value
Avg. degree	8.105
Density (network)	0.022
Modularity (network)	0.303

at least three CAP codes, they are connected in the network. The strength of their ties is determined by the number of CAP codes they are connected through.[6]

Now we are able to model the informal structure using this method. However, this network does not necessarily contain actual existing ties; rather, it indicates potential ones. Hence our model is not an explicit description of the real network of the elite but rather an approximation of it. An edge between two people is created only if they are connected through at least three topics to filter out noise. Applying a higher number than that results in losing ties known to exist based on historical knowledge, while a lower number results in including insignificant ties.

Also, it should be mentioned that a formal relationship (an interaction concerning the discussion of a proposal or the submission of a joint proposal on a meeting) does not necessarily mean that there is an existing personal relationship among the interacting people. The clusters formulated from this network nevertheless show the extent of the authorities of various high party executive positions. As opposed to the case of the network presented in Fig. 1, here the clusters do indeed map to certain policy areas. Through this network we can analyze the manifestation of the "leading role" of the party, the operation of various controlling bodies and their authority, and the connections among the party and various sectors of politics (public policy) as well.

The network parameters presented in Table 7 have similar characteristics to those shown in Fig. 1. This second network is also sparse, but it has a higher density; the modularity indicates that the network is also clustered (the clustering coefficient of 0.864 further emphasizes this). The size of an individual node is proportional to its betweenness centrality. This measure shows how many nodes are connected to each other via this node in their shortest connecting paths. We interpret this measure as the importance or embeddedness of a person. The clustering in this second network (Fig. 4) clearly reflects political domains: red, macroeconomic policy; purple, agricultural policy; blue, government operation; green, cultural policy; and yellow, national defense.

Some actors who have a strong presence in Fig. 1 are also significant in Fig. 4; see, for example, István Friss, Mátyás Rákosi, and Lajos Ács. Interestingly, all the important party members listed in Table 1 are present in Fig. 4. Besides the central actors, others act as bridges between clusters. The bridging role has multiple meanings. If party members are active in various topics, then we assumed that their

[6]A similar method was applied by Leo S. Gregory during the network analysis of the Political Committee of the Chinese Communist Party. There political topics were defined not through CAP codes but the public political statements (Gregory 2013).

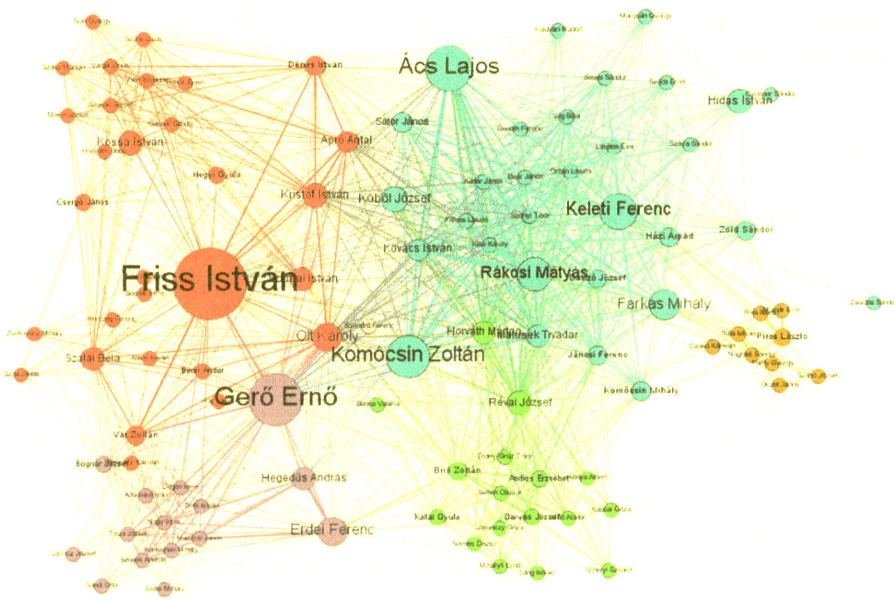

Fig. 4 The relational network based on joint political participation in similar topics

political power might be greater than their actual formal political positions would suggest. Therefore, bridges are not only relevant by connecting clusters; they also indicate personal significance. For example, the network position of Márton Horváth (executive of the Department of Agitation and Propaganda and then later chief editor of Free People[7]) demonstrates a key characteristic of the era: the extensive influence of agitation and propaganda issues on cultural, scientific, educational, agricultural, foreign, and economic policies. We find István Friss in a similar position being connected to multiple topics as in Fig. 1; but now his importance is even more emphasized. The same holds for Ernő Gerő: he has much more connections in Fig. 4 than in Fig. 1.

Bridges also provide information about the clusters themselves. The breadth and increasing influence of economic policy management is also highlighted by the position of Károly Olt (Minister of Finance between 1950 and 1956) and Zoltán Vas (responsible for economic affairs in the party leadership and head of the National Planning Institute), both acting as bridges among numerous clusters.

Figure 4 also shows that the members of the "Troika[8]" dominate distinct areas. Mátyás Rákosi is central to a broad range of governmental and party

---

[7]Free People ("Szabad Nép") was the daily newspaper of the communist parties (first the MKP then the HWP) between 1942 and 1956.

[8]Troika corresponds to the three key politicians of this era: Mátyás Rákosi, Ernő Gerő, and Mihály Farkas. "Trojka" is best translated as "trio."

operations, Mihály Farkas dominates state defense, and Ernő Gerő is a key figure in agriculture (through weak ties) and – through the strong connection to István Friss – macroeconomics and employment policy. Unexpectedly, connections among these members are relatively weak. Both Farkas and Gerő have only weaker ties to Rákosi, and there is no connection between Farkas and Gerő at all. This does not mean that there was no cooperation between them; it only shows that they did not act jointly in the open political scene.

The connection between Gerő and Friss is interesting from another aspect as well. Gerő is connected to macroeconomics and employment policy through Friss, and he is also the one with the most connections to agriculture. The cluster of agriculture highlights the strong connection between agriculture and macroeconomics, which is further stressed by the connections of Gerő to the highest executives of agricultural policy: András Hegedűs, János Matolcsi, and Ferenc Erdei.

The politicians representing the trade unions have a strong presence in the cluster of macroeconomics and employment policy, as the unions themselves are related to employment policy. The two most important politicians, Antal Apró and István Kristóf, are surrounded by others active in employment policy who are also tied to other members in the cluster through István Friss, alongside Apró and Kristóf.

The agriculture, macroeconomic, and governmental function clusters are dense clusters with many bridges among them. Cultural policy and state defense however are much less dense and also largely separated from other clusters. Their members are mostly active in professional topics, being connected to other clusters through József Révai, the most important member of the cultural politics cluster.

The clusters cover the key topics mentioned in the committee meetings and for each person there is a dominant topic. However, the bridges between clusters also demonstrate that the politicians are involved in many other topics at the same time. Besides the strong central position of key party members (Rákosi and Gerő), this network provides some interesting information about the individual roles of other players. A good example is, again, István Friss, who was clearly the most active member in this segment of the political discourse. He had an efficient bridging role, having connections to most clusters. Another person worth mentioning is István Kovács, the secretary of the party who was responsible for the preparation (and record keeping) of the committee meeting reports. Though he has many connections in this network, it is a historical fact that he did not have any actual political power.

All these findings indicate that besides discussing the elite purely from a political position perspective, there is another possible way to conceptualize the political elite. Since both corpora consist of meeting notes that essentially record communications of party members and discourses covering a wide range of topics, we decided to investigate the elite from the standpoint of the political discourse and arrived at a notion we denoted as *discursive elite*.

2.7 A New Approach to the Political Elite of This Era

Discursive political theory describes politics as discursive reality. The actors entering a discourse become the part of the reality being constructed. The purpose of discursive action – besides utilizing political power – is to define the political community outward and to articulate it inward as well (Szűcs 2008:11).

Here we assume that the extent of political power (the socioeconomic domains politics aims to influence and control) can be analyzed to a great extent by the topics occurring in the discourse of leading party bodies. Furthermore, we assume that the inner political structure could be mapped by using the expressions and actions of political bodies, organizations, and individuals. The choice of topics of a given person in a given body – what social system it addresses – indicates what the specific body or person seeks to control.

If we apply the discursive political approach to the higher party management of the Rákosi era, then basically we should analyze the discursive space of the leading bodies of the party. The exchanges during the committee meetings of the leading bodies of the leading communist party (the proposals and comments found in the reports of these meetings) might be interpreted as the discursive space. In the earlier stage of the reported research, we collected all the persons that participated in these meetings and also those who were mentioned in individual agenda items. We found that some active participants did not have formal positions in these bodies, while some members of these leading bodies did not actively participate in the meetings. Also, those who submitted proposals included people with a wide range of positions – they might be officials of the Central Committee, directors in state organizations, ministry officials, or even leaders of mass organizations. Lastly, some politicians never actually participated at any meeting, but they were still mentioned during the meetings.

Analyzing the networks in Figs. 1 and 4 we can see that using pure positionality to categorize somebody as a member of the political elite might be misleading. Accordingly, we propose a new concept called the *discursive elite*. The members of the discursive elite are the political actors who take part in the political discourse of the leading party and governmental bodies. These people are the key influencers of politics interpreted as discursive reality. Clearly, holding a formal political position associated with the positional elite – in this era this means either having Political Committee or Secretariat membership – was not a prerequisite for belonging to the discursive elite. Thus, the discursive elite is a larger and much more heterogenous group than the positional elite.

There are only a few party members of the positional elite who cannot be considered to be in the discursive elite. The most important among them is János Tausz, the minister of internal commerce. From our historical knowledge, he was not among the most powerful members of the party despite his high position. This is underlined by the fact that he did not actively participate in the committee meetings that discussed the most relevant political and state-governance topics.

While there are people normally not considered to be the members of the political elite in general, our results indicate that their political participation was considerable and extensive. Good examples are – as Table 1 also mentions – István Friss, József Köböl, or Zoltán Komócsin, who held executive positions either in state control or in the unions. These members of the discursive elite held professional state-government positions and were very active participants of the political discourse. In Fig. 4, we see members of the positional elite in key positions, but the political emergence of other party members tells us that participating in the political discourse – and thereby strongly influencing politics – was not the privilege of the positional elite. The concept of discursive elite suggests the dynamics of human interactions influencing actual political decisions as opposed to restricting elite membership to static political positions.

3 Conclusions

In this study the main goal was to identify the power networks of a characteristic era in Hungarian history, namely, the Rákosi era, which began after the stabilization following World War 2 in 1948 and finally led to the Hungarian revolution in 1956. We sought to uncover how the political (power) elite had been formed through informal networks manifesting in joint political actions and political participation.

We conducted our research work using computational social science methodologies mostly centered around social network analysis (SNA) and natural language processing (NLP) methods. First, we constructed the informal network of the actors discussing various topics based on their co-mentions in committee meeting reports using text network analysis (TNA). This required significant preprocessing realized in a semi-automated way. As a second methodology, we created a multimodal network using classification methods (both manual and AI-driven methods). We were able to construct a discursive network of party members by transforming this multimodal network into a relational network. In that network we formulated connections among the actors based on common topics they discussed. Here, a restricted source was used, not requiring any substantial amount of data preprocessing, namely, the agenda items of the committee meetings. We used a semi-automated coding algorithm to classify the agenda items, and then we drew up a multimodal network of party members, agenda items, and topics. This was transformed into a relational network of people jointly participating in the topics discussed on the meetings. This methodology could be further improved by using more advanced techniques based on artificial intelligence (for coding), but for our current purposes, the semi-automated coding was sufficient. In other experiments we wanted to use an AI classification method for this task using a linear classifier, but our sample size did not allow us to rely on this classifier in the end.

Overall, then, we found that joint political action confirms our intuitions about the interconnected nature of party politics in that era. At the same time we obtained valuable information on well-known individual politicians. Key party members like

Mátyás Rákosi and Ernő Gerő were much less exposed at the party meetings than expected, while others, like István Friss, assumed surprisingly important network positions. Most party members participating in meetings were viewed as elite by the positional approach to elite. Still, we found party members who could not be classified as positional elite, but their presence in these networks was considerable – like Zoltán Komócsin, Ferenc Erdei, and István Friss.

These findings are in broad agreement with our current historical knowledge of that era, yet it also requires to change our view on the elite. We introduced the term discursive elite, to extend the current elite definition and to highlight features of this group that had not been clear hitherto. With the methodologies used in our study, we extended the toolkit of historians to be able to reveal latent power networks. While these results only cover this specific and short political era, the experience gained can be applied to investigate more complex political systems on a wider timeframe.

The efficiency of the methods presented strongly depends on the quality of the sources. This has led us to use two sources in this study. Yet from other historical sources, it is clear that our analysis could not adequately map discursive elite. For example, the existence of the "Troika" – an inner circle of the three most powerful party members – shows the role of other informal relationships in shaping political discourse and political outcomes. Therefore, it would be useful to also investigate the idiosyncratic historical background with the traditional tools of history and political science in every case.

Alongside the corpora we produced, we were able to create dictionaries specific to this era that could be used in automated processing of other documents from the examined era or decades of the communist regime that followed. The coding process could be greatly improved by the use of proven AI-driven classification methods that would allow us to apply our methodologies to big data as well.

References

F.R. Baumgartner, C. Breunig, E. Grossman, *Comparative Policy Agendas: Theory, Tools, Data* (Oxford University Press, Oxford, 2019)

Z. Boda, M. Sebők (eds.), *A magyar közpolitikai napirend: Elméleti alapok, empirikus eredmények* (MTA TK Politikatudományi Intézet, Budapest, 2018)

L.S. Gregory, A Social Network Analysis of the Chinese Communist Party's Politburo. Thesis and Dissertation Collection, Naval Postgraduate School, Montrerey, 2013

G. Gyarmati, Hatalmi Elit Magyarországon a 20. Század Második Felében. Korunk **20**(3) (2009), 75–88

A. Joulin, E. Grave, P. Bojanowski, T. Mikolov, Bag of Tricks for Efficient Text Classification (2016). ArXiv:1607.01759 [Cs]

T. McEnery, Corpus linguistics, in *The Oxford Handbook of Computational Linguistics* (Oxford University Press, Oxford, 2005), 448–463

N.F. Noy, D.L. McGuinness, A guide to creating your first ontology, in *Ontology Development*, vol 101 (Stanford University, Stanford, 2001)

D. Paranyushkin, Identifying the Pathways for Meaning Circulation Using Text Network Analysis (2011). Retrieved 23 Nov 2017. http://noduslabs.com/research/pathways-meaning-circulation-text-network-analysis/

E. Sík, Aczélhálóban ("In the Net of Aczél. Contribution to understanding of the operation of social capital"). Szociológiai Szemle **3**, 64–77 (2001)

E. Simon, B. Sass, Nyelvtechnológia És Kulturális Örökség, Avagy Korpusépítés Ómagyar Kódexekből. Általános Nyelvészeti Tanulmányok **XXIV**, 243–264 (2012)

Y. Sure, J. Angele, S. Staab, R. Struder, D. Wenke, OntoEdit: collaborative ontology development for the semantic web, in *Proceedings of the 1st International Semantic Web Conference (ISWC2002)* (Springer, 2002)

Z.G. Szűcs, *A Diszkurzív Politikatudomány Mint Politikai Diszkurzustörténet. Tudományelméleti És Módszertani Esszé* (Miskolci Egyetem Politikatudományi Intézet, Miskolc, 2008)

G.T. Varga, *Az MDP Központi Vezetősége, Politikai Bizottsága És Titkársága Üléseinek Napirendi Jegyzékei II. Kötet 1953–1956 (A Magyar Országos Levéltár Segédletei, 18/2. Budapest, 2007) | Könyvtár | Hungaricana*, vol 2 (Magyar Országos Levéltár, Budapest, 2007)

Participatory Budgeting: Models and Approaches

Haris Aziz and Nisarg Shah

1 Introduction

Participatory budgeting (PB) is a paradigm which empowers residents to directly decide how a portion of the public budget is spent. Specifically, residents deliberate over spending priorities and vote over how the budget should be allocated to different public projects. Projects which receive broad support from the community are then funded through the process.

This process was initiated as a radical democratic project in the city of Porto Alegre, Brazil, led by the leftist Workers' Party (Cabannes 2004). Over the next few decades, it quickly spread across the world; it has been implemented by over 1,500 municipalities (Röcke 2014) and in locations as diverse as Guatemala, Peru, Romania, and South Africa (Shah 2007). The nonprofit organization *Participatory Budgeting Project*[1] alone has helped allocate more than US$300 million of public budget in 29 cities across the USA and Canada. In Europe, the effort is led by Paris and Madrid, each spending at least $100 million a year on public projects through participatory budgeting (Gutiérrez 2017; Legendre et al. 2017). In particular, Madrid developed an online open-source platform, *Decide Madrid*, which has been used by more than 30 local governments (Samir 2018). PB is still spreading in more regions

[1]http://www.participatorybudgeting.org

H. Aziz (✉)
UNSW Sydney and Data61 CSIRO, Sydney, NSW, Australia
e-mail: haris.aziz@unsw.edu.au

N. Shah (✉)
University of Toronto, Toronto, ON, Canada
e-mail: nisarg@cs.toronto.edu

T. Rudas, G. Péli (eds.), *Pathways Between Social Science and Computational Social Science*, Computational Social Sciences,
https://doi.org/10.1007/978-3-030-54936-7_10

across the world. For example, Toronto recently completed its 3-year pilot of PB (Murray 2019), and the state of New South Wales in Australia recently started PB under the name *My Community Project*.[2]

Participatory budgeting is typically a long process; in many municipalities, one PB cycle takes 1 full year. While the exact implementation details differ from one PB instance to another, at a high level, the process is composed of the following stages, which allow residents to come up with effective project proposals and provide their preferences over budget allocation (Shah 2007).

- First, the geographical region may be divided into several subregions (e.g., districts), and one PB process may be conducted in each district separately. The goal of this stage is to allow residents to focus on the projects in their own neighborhood and community. The total available budget for each district is also typically decided at this stage.
- Next, residents share and discuss ideas through neighborhood meetings and online tools. This allows them to come up with preliminary project proposals.
- These initial proposals are then developed into feasible proposals by focus groups and vetted by experts. Often, this is the stage where a proposal may be broken into distinct stages of implementation, and a cost estimate may be derived for each implementation stage. Projects may also be classified into categories such as recreation, infrastructure, health, education, transportation, etc.
- This may be followed by several rounds of deliberation through which residents finalize a small number of proposals to be included in the final vote.
- The final stage of the PB process is the voting stage. In this stage, eligible residents vote over how the public budget should be spent across the finalized proposals, and these votes are aggregated into a final budget allocation.

Additional effort is often required to make PB a success. For example, advertisement and promotion through various channels, including social media, can be the key to increasing civic engagement. This can be crucial in encouraging various minorities to participate and ensuring that their preferences are incorporated into the decision-making. See the edited volume by Shah (2007) for a detailed survey of the entire PB process.

In this chapter, we limit our attention to the final *voting* stage of the process. That is, we assume that the project proposals have been detailed and finalized and projects to be included on the final ballot have been filtered. We also assume that project costs have been estimated and the total available budget is known. Research on the voting stage of the PB process focuses on four important considerations:

(i) Decision space: Is the space of possible outcomes discrete (e.g., because a project can only be "funded" or "not funded")? Or is it continuous (e.g., because a project can be allocated different amounts of funds to implement it to different degrees of effectiveness)?

[2]https://www.nsw.gov.au/improving-nsw/projects-and-initiatives/my-community-project/

(ii) Preference modeling: How will residents' preferences over the projects be represented for the purpose of mathematical analysis?
(iii) Ballot design (aka preference elicitation): It is often desirable for the modeling to allow complex preferences. However, it may be infeasible to ask residents to report such complex preferences. What should the ballot ask from residents which will serve as a proxy for their preferences and allow them to effectively convey their preferences?
(iv) Vote aggregation: How will the votes cast by residents be aggregated into a final allocation of the available public budget? How should this process fairly incorporate aggregation of the preferences of different communities of residents and efficiently allocate the public budget?

The need for systematic modeling and study of these considerations has recently inspired a body of research in the computational social choice literature (Brandt et al. 2016; Endriss 2017). This body of research sits at the intersection of computer science, social science, and economics and aims to use mathematical modeling and algorithmic techniques to design different approaches to the voting stage of the PB process.[3] The goal of this chapter is to present a comprehensive overview of the research on PB in computational social choice. Some of the original contributions of this chapter include providing a unifying theoretical framework which allows viewing the different research works through a single lens and providing a taxonomy of the different PB models which highlights their unique modeling choices.

1.1 Outline

The organization of this chapter is as follows. In Sect. 2, we present a general mathematical formulation of PB and list several prominent features that distinguish different implementations of PB. Next, we focus on popular PB models that make specific design choices in terms of these features and present a taxonomy of these models. We pay special attention to the representation and elicitation of preferences and popular desiderata underlying the vote aggregation methods.

In Sect. 3, we survey the research on the "integral model" of PB, in which each project can only be fully funded or not funded. In Sect. 4, we survey the research on the "continuous model" of PB, in which the funding level of each project can lie on a continuum. We highlight the differences between motivations and results under both discrete and continuous models. Finally, in Sect. 5, we discuss possible extensions and directions for future research.

[3] In several PB programs, the final decision of budgets is entirely done by deliberation by a small subset of residents. Deliberation has its advantages and disadvantages. While it allows an in-depth discussion of possible outcomes from different perspectives, it also poses the danger of the process being dominated by an unrepresentative subset of the residents, thus thwarting the democratic objective.

2 Mathematical Formulation

We begin by reviewing various parameters which play a key role in formulating the participatory budgeting problem. Later, we consider specific choices of these parameters, which give rise to popular PB models.

- *Residents:* There is a set of residents (aka agents or voters) $N = \{1, \ldots, n\}$. In some applications of PB, residents are divided across different geographical regions (e.g., districts or wards), and each region conducts a separate PB election, in which residents of that region vote.
- *Resources and budgets:* There is a set of d resources, denoted R. The available budget is $\mathbf{B} = (B_1, \ldots, B_d)$, where B_r denotes the amount of resource r available. In most applications of PB, money is the only limiting resource. Thus, much of the work on PB in computational social choice has focused on the case of a single resource, in which case the available budget is denoted by the scalar B.
- *Projects:* There is a set of projects $P = \{p_1, \ldots, p_m\}$. In some applications, projects are categorized into various domains (e.g., infrastructure, security, education, health and wellness, etc.).
- *Degree of completion:* Some projects can be completed to different degrees. For example, completing 65% of a project proposing neighborhood clean-up may mean a part of the neighborhood being cleaned up. Some projects may have discrete milestones, and only the first few milestones may be achieved; e.g., a project proposing the creation of a public park may include construction of the park as the first milestone and construction of a children's playground within the park as the second milestone. For other projects, the degree of completion may be binary; they must be either fully implemented or not implemented at all (e.g., installing a fountain in a park). For project p, let $\mathcal{X}_p$ denote the set of its possible degrees of completion and x_p denote its actual degree of completion in an outcome. In some models of PB, which we refer to as *bounded models*, there is an upper bound (aka *cap*) on the degree of completion of each project p, denoted by q_p. In *unbounded models*, there are no such caps.[4] We discuss popular choices of $\mathcal{X}_p$ in Sect. 2.1. We assume $0 \in \mathcal{X}_p$ for all projects p. Let $\mathcal{X} = \mathcal{X}_{p_1} \times \ldots \mathcal{X}_{p_m}$. Various models of PB studied in the literature crucially differ in this key choice; we elaborate on this in Sect. 2.1.
- *Costs:* Each project $p \in P$ has an associated cost function $c_p : \mathcal{X}_p \to \mathbb{R}^d$, where the vector $c_p(x_p)$ gives the amount of each resource required to complete project p to degree x_p. We assume that $c_p(0) = \mathbf{0}$ and c_p is monotonically non-decreasing.
- *Budget Allocations:* An outcome $\mathbf{x} = (x_p)_{p \in P} \in \mathcal{X}$ is characterized by the degree of completion of each project p. Note that this also specifies the amounts of different resources that will be devoted to each project. The outcome is feasible

[4] Note that the cost function of a project and the total available budget may still induce an effective upper bound on its degree of completion.

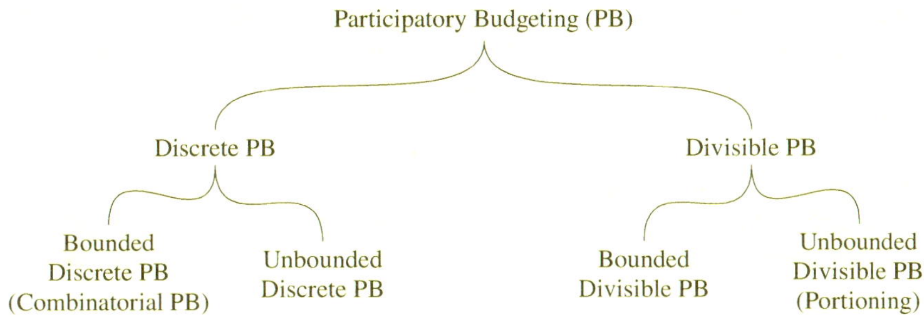

Fig. 1 A taxonomy of participatory budgeting models

if $\sum_{p \in P} c_p(x_p) \leq \mathbf{B}$, where addition is component-wise. We refer to feasible outcomes as *(budget) allocations*. Let $\mathcal{A}$ denote the space of allocations.
- *Resident Preferences:* Each resident $i \in N$ has preferences over which projects should be implemented and to what degree. These can be represented through an ordinal preference relation $\succsim_i$ or a cardinal utility function u_i over the space of allocations $\mathcal{A}$. We elaborate on this later in Sect. 2.2.1.

2.1 Decision Space and Popular PB Models

Figure 1 presents a taxonomy of the popular PB models in the literature, which crucially differ in their modeling of the possible degrees of completion of projects. We review each model in further detail.

2.1.1 Bounded Discrete PB (Combinatorial PB)

This is perhaps the most widely studied and applied model of PB. In this model, projects must be either fully implemented or not implemented at all. Hence, the set of possible degrees of completion is $\mathcal{X}_p = \{0, 1\}$ for each project $p \in P$. Note that this model has unit caps ($q_p = 1$). Consequently, the cost function c_p of project p is effectively a vector, which indicates the amounts of different resources needed to fully implement project p. Feasible allocations in $\mathcal{A}$ are subsets of projects which can be implemented simultaneously subject to budget constraints. In essence, this is a multi-agent variant of a multi-dimensional knapsack problem.

2.1.2 Discrete PB

In this model, each project p has discrete possible degrees of completion. However, unlike combinatorial PB, there may be more than two possible degrees of completion. These degrees of completion can also mathematically capture funding of multiple of units of the same project. For example, installing ten units of public toilets can be viewed as having a single project with 10 different degrees of completion. Typically, we assume that $\mathcal{X}_p = \mathbb{N} \cup \{0\}$ for each project p. Note that this model has no cap on the degree of completion, but limited availability of resources may still place a natural upper bound on the degree of completion of each project in any feasible allocation. This model is applicable when each project is ambitious, and its full implementation can potentially use up all the available resources.

2.1.3 Divisible PB

In this model, it is assumed that projects can be implemented to any fractional degree. In the version with caps, we can assume without loss of generality that the cap is $q_p = 1$, i.e., $\mathcal{X}_p = [0, 1]$ for each project $p \in P$; in this case, x_p denotes the fraction of project p that is completed. In practice, projects often cannot be executed to arbitrary fractional degrees. However, in cases of sufficient granularity, assuming divisible PB can help for computational reasons; more details are provided in Sect. 4.

2.1.4 Unbounded Divisible PB (Portioning)

In this model of divisible PB, there are no caps on the degrees of completion. Thus, $\mathcal{X}_p = \mathbb{R}_+$ for each project p. When there is a single resource type (e.g., money), under mild assumptions,[5] deciding the degrees of completion of different projects is equivalent to deciding how the available budget will be divided among the projects. This setting is known as *portioning* in the literature (Airiau et al. 2019).

Example Suppose there is a set of 3000 residents $N = \{1, \ldots, 3000\}$, a set of four projects $P = \{A, B, C, D\}$, and a single resource (money) with a total budget of \$7 million. Suppose the cost functions of the projects are as follows:

- $c_A(x_A) = \$3 \text{ million} \times x_A$
- $c_B(x_B) = \$3 \text{ million} \times x_B$
- $c_C(x_C) = \$2 \text{ million} \times x_C$
- $c_D(x_D) = \$2 \text{ million} \times x_D$

Suppose 2000 residents like projects A and B equally but derive no value from C or D. 500 residents derive value only from C and the remaining 500 only from D.

In the divisible PB model with unit caps (i.e., $\mathcal{X}_p = [0, 1]$ for each project p), we have numerous choices. For instance, we could implement 7/10 fraction of each

[5] E.g., if the cost function of each project is strictly increasing, then there is a one-to-one correspondence between the amount of money allocated to the project and its degree of completion.

project. Or we could implement 1/2 fraction of A and B each and fully implement C and D.

In the combinatorial PB model (i.e., $\mathcal{X}_p = \{0, 1\}$ for each project p), given the budget of $7 million, we can implement A and B, which would make 2000 residents very happy but 1000 residents very unhappy, or we can implement one of A and B together with both C and D, which would make 2000 residents partially happy and 1000 residents very happy.

How do we quantify how happy the residents are? How do we make trade-offs between such decisions? To understand this better, we need to understand modeling of resident preferences and goals of the PB process, which we review below.

2.2 *Preference Modeling and Ballot Design*

Two important decisions in the PB process are the modeling of residents' preferences (which helps in the mathematical analysis of the effectiveness of the process) and the format in which the residents cast their votes.

2.2.1 Preference Modeling

Recall that $\mathcal{A}$ is the set of feasible allocations. In an expressive model, each resident i has a cardinal utility function $u_i : \mathcal{A} \to \mathbb{R}$.[6] If only comparisons among allocations are required, one may work instead with an ordinal preference relation $\succsim_i$ over $\mathcal{A}$.[7]

While such an expressive modeling allows capturing all possible preferences, it does not utilize the structure that is often present in the preferences. Below, we discuss three common approaches to modeling such structure.

The first approach is to directly impose a structural assumption on the utility function or the preference relation. For example, in combinatorial PB, where allocations are effectively feasible subsets of projects and a utility function can be represented as $u_i : 2^P \to \mathbb{R}$, one may impose that u_i satisfies subadditivity or submodularity (when projects are substitutes of each other) or superadditivity or supermodularity (when projects are complements of each other).

The second approach is to use spatial models, where allocations are embedded in a metric space, each resident has a preferred allocation, and her utility for another

[6]We note that in this case, much of the existing research on PB in computational social choice, inspired from classical research on voting, implicitly assumes that $u_i(\mathbf{0}) = 0$ and u_i satisfies monotonicity (i.e., $u_i(\mathbf{x}) \geq u_i(\mathbf{x}')$ for $\mathbf{x} \geq \mathbf{x}'$). In words, implementing more projects or implementing projects to a greater degree cannot bring less happiness. In the PB context, implementing projects requires spending costly resources, which makes this assumption questionable. See Sect. 5 for further discussion.

[7]The ordinal preference relation is typically assumed to be transitive and complete.

allocation is a (typically non-increasing) function of its distance to her preferred allocation. For example, Garg et al. (2019) study ℓ_p normed utilities, under which the utility of resident i (with preferred allocation $\mathbf{x}_i^*$) for allocation $\mathbf{x}$ is $u_i(\mathbf{x}) = -\|\mathbf{x} - \mathbf{x}_i^*\|_p$, where $\|\cdot\|_p$ is the ℓ_p norm.

The third—and perhaps the most popular—approach is to model residents' preferences over individual projects and use a natural rule to extend them to preferences over allocations. Below, we review common ways to achieve this for both cardinal utility functions and ordinal preference relations.

Cardinal Extensions

Fain et al. (2016a) study the class of *scalar separable* utility functions, whereby resident i derives utility $u_{i,p} \cdot f_p(x_p)$ from each project p,[8] and her utility for an allocation $\mathbf{x}$ is simply additive across projects, i.e., $u_i(\mathbf{x}) = \sum_{p \in P} u_{i,p} \cdot f_p(x_p)$. This is a reasonable assumption when the different project proposals are completely independent of each other. In case of combinatorial PB, where $x_p \in \{0, 1\}$ for each project p, this effectively reduces to $u_i(S) = \sum_{p \in S} u_{i,p}$ for each $S \subseteq P$; this is commonly known as an *additive* utility function (Benade et al. 2017; Bhaskar et al. 2018; Fain et al. 2018). A further restriction of $u_{i,p} \in \{0, 1\}$ for each resident i and project p gives rise to *dichotomous* preferences (Aziz et al. 2018; Faliszewski and Talmon 2019), under which each resident approves or disapproves each project and only cares about the number of her approved projects that are implemented. In combinatorial PB, another common extension is the *max set extension* (Caragiannis et al. 2017), whereby the utility of a resident for an allocation is her utility for her most favorite project that is funded: $u_i(S) = \max_{p \in S} u_{i,p}$ for each $S \subseteq P$. This is applicable when the projects are substitutes of each other, and the resident will derive value from only one of the implemented projects.

Ordinal Extensions

When residents are assumed to have ordinal preferences over projects, we use the following notation. Each resident i has a weak order preference relation $\succsim_i$ over P. We denote by $\succ_i$ the strict part and by $\sim_i$ the indifference part of the relation $\succsim_i$. We denote by $E_i^1, \ldots, E_i^{k_i}$ the equivalence classes of the relation $\succsim_i$ in decreasing order of preferences; thus, each set E_i^j is a maximal equivalence class of objects among which agent i is indifferent, and k_i is the number of equivalence classes. Given an ordinal preference relation $\succsim_i$, one can induce resident i's preference relation over the set of allocations $\mathcal{A}$ in several natural ways.

[8] Note that $f_p : \mathcal{X}_p \rightarrow \mathbb{R}$ is a resident-independent function. Typically, it is assumed to be non-decreasing and, when $\mathcal{X}_p$ is continuous, smooth and concave as well.

The (first-order) stochastic dominance (SD) extension is defined as follows (see, e.g., Brandl et al. 2016). For allocations $\mathbf{x}, \mathbf{y} \in \mathcal{A}$, we say that $\mathbf{x} \succsim_i^{SD} \mathbf{y}$ if and only if $\sum_{j=1}^{l} \sum_{p \in E_i^j} x_p \geq \sum_{j=1}^{l} \sum_{p \in E_i^j} y_p$ for all $l \in \{1, \ldots, k_i\}$. Since this extension requires adding the degrees of completion of different projects, it makes more sense in models where the degrees of completion of different projects are comparable (e.g., in combinatorial PB or divisible PB with unit caps). The desirable aspect of the SD extension is that $\mathbf{x} \succsim_i^{SD} \mathbf{y}$ is equivalent to saying that $\mathbf{x}$ yields at least as much utility to resident i as $\mathbf{y}$ does, for all additive utilities over projects that are consistent with preference relation $\succsim_i$. However, the SD relation is not complete (i.e., it does not compare every pair of allocations).

One popular complete extension is the lexicographic extension $\succsim_i^{lex}$, under which resident i is assumed to care significantly more about project p than about project p' whenever $p \succ_i p'$. Formally, $\mathbf{x} \succ_i^{lex} \mathbf{y}$ if and only if for the smallest (if any) l with $\sum_{p \in E_i^l} x_p \neq \sum_{p \in E_i^l} y_p$, we have $\sum_{p \in E_i^l} x_p > \sum_{p \in E_i^l} y_p$. Once again, since this requires adding the degrees of completion of different projects, it makes more sense when the degrees of completion of different projects are comparable. However, if each resident submits a strict ordering over the projects, the equivalence classes become singletons, and the definition makes sense for other models of PB too. For combinatorial PB, under the lexicographic extension, resident i compares two allocations by comparing her most favorite project that implemented in each allocation, breaking ties by her next most favorite project, and so on. This is similar to the max set extension mentioned above for cardinal utility functions and is a reasonable assumption when projects are substitutes of each other.[9] Klamler et al. (2012) study a broader class of preference extensions that includes the lexicographic extension.

Another complete extension is derived by converting ordinal preferences to cardinal preferences using *scoring rules*. A scoring vector is denoted by $\mathbf{s} = (s_1, \ldots, s_m)$, where $s_1 \geq \ldots \geq s_m \geq 0$. Given a ranking $\succ_i$ over projects, it is assumed that resident i has utility $u_{i,p} = s_k$ for project p, where k is the rank of project p under $\succ_i$. Then, any of the cardinal extensions mentioned above can be used to induce the resident's preferences over allocations.

2.2.2 Ballot Design

Even in the simplest case of combinatorial PB, there can be exponentially many allocations. This makes it infeasible to ask residents to communicate their full preferences, thus motivating the use of more restrictive preference elicitation techniques.

9For instance, the projects may propose to build public parks in different locations, but a resident may only be interested in using a single park that is built closest to her home.

For example, even when residents' preferences over allocations are induced from their ranked preferences over individual projects, asking residents to rank as many as 20 projects[10] can be tiresome, and the cognitive burden can lead to fewer residents voting or residents making poor choices (Iyengar and Lepper 2000). Hence, most real-world applications of PB use easier ballot formats such as k-approval voting (where residents approve the k projects they like the most), approval voting (where residents approve all projects that they like), range voting (where residents rate projects), or knapsack voting (where residents provide the ideal allocation according to their preferences) (Goel et al. 2019).

2.3 *Vote Aggregation*

Once the representation of residents' preferences and the format in which they cast their votes are decided, the next challenge is to decide how to aggregate their votes into a collective outcome, which can be a single allocation or a distribution over allocations (if randomization is permitted).

This is perhaps the most well-studied aspect of the entire PB process. We review several prominent approaches to vote aggregation in computational social choice.

2.3.1 Welfare Maximization

This approach is applicable when individual preferences are represented as cardinal utility functions. It uses a *social welfare function*, which combines individual utility functions of residents into a societal utility function and finds an allocation maximizing this societal utility function. There are three popular social welfare functions:

- The *utilitarian welfare* of an allocation is the sum of utilities it gives to residents: $UW(\mathbf{x}) = \sum_{i \in N} u_i(\mathbf{x})$ for $\mathbf{x} \in \mathcal{A}$.
- The *egalitarian welfare* of an allocation is the minimum utility it gives to any resident: $EW(\mathbf{x}) = \min_{i \in N} u_i(\mathbf{x})$ for $\mathbf{x} \in \mathcal{A}$. Maximizing this welfare function is seen as an extreme form of one interpretation of fairness.
- The *Nash welfare* of an allocation is the product of utilities it gives to residents: $NW(\mathbf{x}) = \prod_{i \in N} u_i(\mathbf{x})$ for $\mathbf{x} \in \mathcal{A}$. Maximizing this welfare function is seen as a compromise between utilitarianism and egalitarianism.

[10]The 2018 PB in Cambridge, USA involved 20 projects: https://pb.cambridgema.gov/pbcycle5

2.3.2 The Axiomatic Approach

This approach entails specifying intuitive normative axioms that the vote aggregation rule must satisfy and searching for rules that satisfy as many of the axioms as possible. Many compelling axioms have been proposed for PB (see, e.g., Aziz et al. 2018; Faliszewski and Talmon 2019); the two examples given below are selected because they apply to all the models of PB we described above.

Definition 1 (Exhaustiveness (Aziz et al. 2018)) A feasible outcome (i.e., allocation) $\mathbf{x}$ is called exhaustive if an outcome $\mathbf{y}$ is not feasible whenever $y_p \geq x_p$ for all projects p and a strict inequality holds for at least one project. In words, it should not be possible to implement any more projects or any projects to a greater degree with the leftover budget. A vote aggregation rule is exhaustive if it always outputs an exhaustive allocation. A similar axiom is termed budget monotonicity by Faliszewski and Talmon (2019).

Definition 2 (Discount Monotonicity (Faliszewski and Talmon 2019)) Suppose a vote aggregation rule outputs allocation $\mathbf{x}$. Suppose project p receives a revised cost function c'_p such that $c'_p(x_p) \leq c_p(x_p)$ for all $x_p \in \mathcal{X}_p$, and, all else being equal, the vote aggregation rule now outputs allocation $\mathbf{y}$. Then, $y_p \geq x_p$. In words, if a project becomes more affordable, it should not be implemented to a lesser degree.

Faliszewski and Talmon (2019) study a number of additional axioms that are specifically applicable to settings where residents have dichotomous preferences.

2.3.3 Fairness

Finally, an important consideration in democratic decision-making is to fairly take into account the preferences of all residents when making the collective decision. We review one fairness axiom that is applicable in our general PB framework. For other axioms that are applicable in more specific environments (e.g., with dichotomous preferences), we refer interested readers to the work of Aziz et al. (2018).

Definition 3 (Core (Fain et al. 2016a, 2018)) An allocation $\mathbf{x}$ is said to be in the core if no subset of residents $S \subseteq N$ can find an outcome $\mathbf{x}'$ that is feasible given a budget of $|S|/n \cdot \mathbf{B}$ such that $u_i(\mathbf{x}') \geq u_i(\mathbf{x})$ for every resident $i \in S$ and a strict inequality holds for at least one $i \in S$.

The notion of the core is based on the idea that a group of residents S collectively deserve at least $|S|/n$ fraction of the budget spent on their needs. This is formalized by requiring that they not be able to allocate this fraction of the budget in a way that they prefer more.

3 Discrete Participatory Budgeting

We are now ready to review how these approaches have been applied to different models of PB. Recall that in the discrete model, the decision space is discrete. This model is applicable when projects can only be executed up to discrete levels. For example, a project p that proposes to build public toilets in the community can have degree of completion $x_p \in \mathbb{N} \cup \{0\}$, where x_p may denote the number of public toilets that are actually built; the assumption here is that each individual toilet is either built fully or not built at all. This discrete aspect affects the modeling of residents' preferences as well as the design of ballots and vote aggregation procedures.

This model is a natural generalization of *multi-winner voting* (alternatively known as *committee selection*), which has been widely studied in social choice theory (Aziz et al. 2017; Faliszewski et al. 2017). In multi-winner voting, there are m candidates, and k of them are to be selected based on voters' preferences. This is a special case of discrete PB (more specifically, of combinatorial PB) where each candidate p is a project with unit cap ($q_p = 1$) and unit cost ($c_p = 1$), and the budget limit is k. We first review prior literature on multi-winner voting and other similar models of decision-making and then provide an overview of various approaches to discrete PB.

3.1 *Review of the Literature on Settings Related to Discrete PB*

The simplest special case of combinatorial PB is multi-winner voting. As explained above, this is where each project has unit cost. A natural goal in this setting is to maximize the utilitarian welfare. In this case, each voter approves a subset of candidates, and the goal is to select k candidates with the highest number of total approvals. This can be accomplished efficiently by a greedy algorithm which selects candidates one by one in the decreasing order of the number of approvals they receive. However, when we replace the selection of k candidates by a knapsack constraint to extend this to combinatorial PB with dichotomous preferences, the problem becomes NP-hard even for a single agent (Karp 1972). The greedy algorithm is still well-defined in this setting; it selects projects one by one in the decreasing order of the number of approvals they receive, skipping any project if its inclusion exceeds the budget limit. This aggregation rule is often used in practice.[11] The literature also focuses on several other objectives in multi-winner voting; an excellent overview is given by Faliszewski et al. (2017).

Klamler et al. (2012) also focus on committee selection under knapsack constraints. However, in their model, there is a single agent with a ranking over individual candidates (projects). They explore different ways to extend this to a

[11] See, for example, the PB process in Toronto (Murray 2019) and in Cambridge (Mundt 2017).

ranking over sets of candidates and then select a committee of k candidates subject to a knapsack constraint. A similar model has been considered by Delort et al. (2011).

Another model related to PB is *combinatorial public projects* (CPP) (Papadimitriou et al. 2008). In this model, a set of public projects are to be selected subject to a resource constraint, similarly to PB. However, research on CPP focuses on designing truthful mechanisms by charging payments to agents (Dughmi et al. 2016), which is unrealistic in the PB setting.

Conitzer et al. (2017) propose the *public decision-making* model, in which there are a finite number of *issues*, and for each issue, there is a set of alternatives which can be implemented. A feasible outcome consists of choosing a single alternative corresponding to each issue. One can view this as a special case of discrete PB, where each issue corresponds to a unique type of resource of which one unit is available and implementing any alternative of an issue requires the full one unit of the corresponding resource. One of the fairness definitions they propose is *proportionality up to one issue*, which is a relaxation of the core defined in Sect. 2.3.3.

Lu and Boutilier (2011) introduce *budgeted social choice*. In this framework, each alternative has a cost, and the goal is to select a set of alternatives subject to budget constraint. This framework generalizes the PB model we discuss as it allows the cost of an alternative to be dependent on the number of voters who derive utility from that alternative. However, they work with a limited modeling of preferences wherein there is a common *positional scoring rule* which maps an alternative's rank in a voter's preference ranking to the voter's utility for the alternative. This is also a common approach in the resource allocation literature (Bouveret and Lang 2011).

3.2 *Approaches to Discrete PB*

We now provide an overview of various approaches to PB in the discrete model.

3.2.1 Welfare Maximization

This is a natural approach when residents have cardinal utilities. Maximizing the utilitarian welfare subject to the budget constraint is the classic knapsack problem. While the problem is NP-hard (Karp 1972), there exists a pseudo-polynomial time dynamic programming algorithm and a fully polynomial time approximation scheme (FPTAS) (Vazirani 2013).

Fluschnik et al. (2017) consider the combinatorial PB model and study the computational complexity of maximizing (1) the utilitarian social welfare with respect to additive utilities, (2) the utilitarian social welfare with respect to the max extension, and (3) the Nash social welfare with respect to additive utilities. All of these problems are NP-hard except under severe restrictions on residents' utility functions.

3.2.2 Elicitation

Eliciting cardinal utilities in the PB context can be challenging and may impose high cognitive burden on the voters, potentially even deterring them from voting in the first place (Garg et al. 2019). This has inspired a line of research that aims to propose simpler ballot formats. Goel et al. (2019) introduce *knapsack voting*, under which residents simply report their favorite allocation. They suggest aggregating knapsack votes using a greedy approach. In combinatorial PB, the number of "approvals" that a project receives is the number of residents who include the project in their favorite allocation, and projects are selected in the decreasing order of their number of approvals subject to the budget constraint. In discrete PB, the algorithm starts by setting the degree of completion of each project to zero. Then, at each step, it considers increasing the degree of completion of each project to its next possible value and, among the feasible improvements, selects the one which is part of the favorite allocation of most voters. They show that this approach has compelling welfare and incentive guarantees under restrictive models of resident preferences.

Benade et al. (2017) compare knapsack voting to three other ballot formats: ranking projects by value, ranking projects by value-for-money, and a new format that they call *threshold approval voting*. They use the *implicit utilitarian voting* framework (Boutilier et al. 2015; Procaccia and Rosenschein 2006), in which residents are assumed to have underlying cardinal utilities and the information they provide on the ballot is treated as a proxy for their underlying utilities. They show that in the worst case, knapsack voting leads to an exponentially bad approximation of utilitarian welfare, ranking projects by value or value-for-money leads to a polynomially bad approximation, and threshold approval voting leads to a logarithmic approximation.

3.2.3 Incentives

Another line of research studies residents' incentive to misreport their preferences in order to induce an outcome that is better than what would be implemented had they reported their honest preferences. While all reasonable deterministic aggregation rules are susceptible to such manipulations due to the classic impossibility result by Gibbard (1973) and Satterthwaite (1975), Bhaskar et al. (2018) show that when residents rank projects by value, there exist randomized aggregation rules that are *strategyproof* (i.e., provide no incentive to residents to misreport) and achieve nearly optimal approximation of utilitarian welfare among all (even non-strategyproof) aggregation rules, implying that, at least when randomization is allowed, strategyproofness does not impose a significant burden in this setting.

3.2.4 Axiomatic Desiderata

Aziz et al. (2018) begin the formal study of defining axioms for PB that encode principles of proportional representation. They design algorithms for combinatorial PB with dichotomous preferences which satisfy such axioms. Their central result is an algorithm called the *generalized Phragmen's sequential rule*, which computes an allocation satisfying a reasonable proportional representation property. An alternative simple method, which is motivated by proportional representation concerns, is discussed by Aziz (2019). Faliszewski and Talmon (2018), in the same model, focus on monotonicity-style axioms of a range of aggregation rules. The common approach in this line of work is to focus on axioms and algorithms for multi-winner voting (Aziz et al. 2017; Faliszewski et al. 2017) and extend them to combinatorial PB.

Since combinatorial PB is more general than multi-winner voting, the negative existential and computational results from multi-winner voting carry over directly. For example, strategyproofness and any weak notion of proportional representation are inherently incompatible (Peters 2018), and finding Pareto optimal allocations is typically NP-hard (Aziz et al. 2016).

Shapiro and Talmon (2017) generalize Condorcet's principle (also referred to as popularity in allocation settings) to PB and devise an algorithm to compute allocations satisfying this principle. Since Condorcet's principle is based on majoritarian comparisons, the approach is not well-suited for proportional representation of minorities, which is a prime concern in real-world applications of PB (Ganuza and Francés 2012; World Bank 2008).

3.2.5 Fairness

Fain et al. (2018) investigate fairness in a setting that generalizes discrete PB. In particular, they show that allocations in the core may not exist in discrete PB with additive utilities, but allocations that provide a logarithmic approximation of the core are guaranteed to exist and can be computed efficiently. For the special case of discrete PB with binary additive utilities (where each resident has utility 1 or 0 for each project and utilities are additive across projects), it is an open question whether allocations in the core always exist. Other work on proportional representation for the combinatorial PB model (see, e.g., Aziz et al. 2018) can also be viewed as focusing on fairness.

3.2.6 Other Approaches

Rios and Insua (2008) study the combinatorial PB model with additive utilities. They propose to use a generalization of the Kalai-Smorodinsky Kalai and Smorodinsky (1975) cooperative game solution to find a desirable allocation. A similar approach is taken by Rios et al. (2005).

4 Divisible Participatory Budgeting

The divisible model of PB applies to scenarios in which funding for a project can lie on a continuum. For instance, recall the example from Sect. 3, which involved a project proposing to build five public toilets in the neighborhood. We argued that this falls under the discrete model of PB, as one can only provide funding to build an integral number of toilets. In contrast, a project on maintaining the cleanliness of the toilets may be funded partially without any integral restrictions and thus may fall under the divisible model of PB.

While a divisible model can often be approximated by a discrete model, the divisible model is interesting in its own right as it often leads to better existential and computational results (see, e.g., the work of Fain et al. 2018).

4.1 Review of the Literature on Settings Related to Divisible PB

A strand of the literature focuses on a multi-dimensional continuous space, where the dimensions correspond to categories of projects such as defense, health, and education (Freeman et al. 2019; Garg et al. 2019). The model is especially relevant when deciding the composition of funding across different categories is more important than deciding particular projects within a category. This literature typically works with the spatial model of resident preferences, in which each resident has an ideal point in the space. The model is closely related to spatial voting models studied in political science (Arrow 1990; Enelow and Hinich 1984).

Recall that in bounded divisible PB, each project has a cap of $q_p = 1$ (without loss of generality). One might consider a special case of this setting, where the cost function of each project p is given by $c_p(x_p) = x_p$ and the total budget is $B = 1$. Thus, the set of feasible allocations is given by $\{\mathbf{x} : \sum_{p \in P} x_p \leq 1\}$. This can be thought of as portioning (also called fair mixing) (Aziz et al. 2019), where x_p can be thought of as the fraction of budget devoted to project p (after appropriate renormalization of cost functions). As noted in Sect. 2.1, this is equivalent to the unbounded divisible PB model. Alternatively, one can also think of this model as representing probabilistic voting (Gibbard 1977), where each project p is a candidate and x_p denotes the probability of choosing candidate p as the winner of the election.

4.2 Approaches to Divisible PB

4.2.1 Welfare Maximization

Goel et al. (2019) study the divisible model of PB with unit caps. They consider cardinal additive utilities as well as a spatial model of utilities. They point out that a simple greedy algorithm finds an allocation maximizing the utilitarian welfare.

This algorithm sorts the projects in decreasing order of their value-for-money ($\sum_{i \in N} u_i(p)/c_p$) and fully funds projects in this order until no more projects can be fully funded. The remaining budget is allocated to partially fund the next project in the list.

Garg et al. (2019) consider a model of decision-making in continuous spaces that is more general than PB. In their model, each resident i has a favorite point $\mathbf{x_i}$ in the space, and her utility for an allocation $\mathbf{x}$ is based on the ℓ_p norm, specifically, $u_i(\mathbf{x}) = -\|\mathbf{x} - \mathbf{x_i}\|_p$. They study a class of algorithms called iterative local voting (ILV), in which residents are iteratively asked to modify the current allocation by moving it toward their favorite allocation within a local neighborhood until convergence. This approach adopts the classic stochastic gradient descent (SGD) method from optimization to multi-agent decision-making.

4.2.2 Fairness

Fain et al. (2016b) study proportional representation in the divisible model of PB with scalar separable utility functions. They argue that a variant of the classic game-theoretic notion of core captures fairness in the setting. Recall that an allocation $\mathbf{x}$ is in the core if *no* subset of voters S can use $|S|/n$ of the budget to find an allocation which is no less appealing than $\mathbf{x}$ to any member of S and strictly more appealing than $\mathbf{x}$ to some member of S. They show that in the divisible model, there always exists an allocation in the core and present a polynomial-time algorithm for computing one.

Recall that in portioning (i.e., unbounded divisible PB), there are no caps on the degrees of completion. Hence, as discussed in Sect. 2.1, one can normalize the model such that feasible allocations satisfy $\sum_{p \in P} x_p \leq 1$, and x_p can be thought of as the fraction of budget devoted to project p (Airiau et al. 2019). Aziz et al. (2019) consider portioning—they refer to it as fair mixing—with dichotomous preferences, which was originally studied by Bogomolnaia et al. (2005). They consider the relative merits of several rules. They show that maximizing Nash welfare satisfies several strong fairness properties including the fact that it finds an allocation in the core. It also satisfies the *average fair share* (AFS) guarantee, which requires that for any set of residents S, the average utility of residents in S is at least $|S|/n$. Another rule with several desirable properties that emerges from their study is the *conditional utilitarian rule* (CUT). In CUT, each resident is assumed to have a "personal budget" that is $1/n$ fraction of the total budget.[12] Each resident spreads her personal budget among the projects that she likes that are liked by the most number of residents. Unlike the maximum Nash welfare solution, CUT does not satisfy Pareto optimality, AFS, or core fairness.

Airiau et al. (2019) consider the portioning problem where voters express preferences over individual projects and use the stochastic dominance (SD) relation

[12]Note that CUT easily generalizes to the case where agents have unequal personal budgets.

to reason about their preferences over probability distributions over projects. They find that using a scoring rule as a proxy for utilities consistent with the ordinal preferences and then maximizing the Nash social welfare with respect to these proxy utilities achieves desirable notions of fairness. Another notable example of algorithms for portioning is the egalitarian simultaneous reservation algorithm (Aziz and Stursberg 2014) for agents with weak and transitive ordinal preferences over projects.

4.2.3 Incentives

Aziz et al. (2019) prove that for the portioning problem with dichotomous preferences, both CUT and the Nash welfare maximizing rule provide strict incentives for residents to participate in the voting process. Additionally, CUT is strategyproof (i.e., provides voters no incentive to misreport), while maximizing Nash welfare is not.

Freeman et al. (2019) consider the spatial voting model and focus on the setting in which a resident's disutility for an allocation is its ℓ_1 distance from the resident's ideal allocation (aka bliss point). They present the independent market mechanism, which is both strategyproof and satisfies a basic notion of proportional representation.

5 Extensions and Future Directions

We presented a survey of research on participatory budgeting. Specifically, we presented a mathematical model which classifies existing research across different axes and surveyed computational and axiomatic approaches proposed in the literature.

Research on participatory budgeting within computational social choice is still in its infancy. As such, only limited models of resident preferences and decision-making processes have been explored. Further research is required to design better theoretical models and approaches for participatory budgeting and bridge the gap between theory and practice. To that end, it is crucial to leverage insights from various disciplines such as computer science, social science, microeconomics, and public policy. In this section, we review several directions in which these models can (and should) be extended to bring them closer to reality and for them to inform real-world implementations of PB.

- *Multi-dimensional constraints:* Much of the literature focuses on a single, knapsack-style constraint which stems from a limit on the available money. While current real-world implementations of PB also only deal with money, this is one dimension where further research on practical approaches to conducting PB with multi-dimensional constraints that capture other types of costs (e.g., costs to the environment) can lead the frontier of novel PB implementations.

- *Voter or voter group entitlements:* In the classic PB setting, giving equal consideration (or weight) to all voters is a normative desideratum. However, one can imagine a more general setting in which different voters or groups of voters bring different resources to the system and their entitlements need to reflect their contributions. In other words, different voters or groups of voters may have a claim over different parts of the universal budget.
- *Distributional constraints:* Some applications of PB impose lower and upper bounds on the amount of funding which can be allocated to each project or each category of projects (e.g., education or healthcare). Studying the effects of such constraints on the welfare, fairness, and incentives is an interesting direction.
- *Hybrid models:* While our taxonomy (and the literature) partitions PB models into discrete and divisible, there can be hybrid models allowing some projects to be funded only at discrete levels while others on a continuous scale.
- *Complex resident preferences:* Most of the positive axiomatic and algorithmic results in the literature rely on stylized modeling of resident preferences. In practice, residents have complex preferences which stem from intricate synergies between different projects. An important research direction is to extend the existing results to more general classes of utility functions. For example, most utility functions considered in the literature treat projects as independent or substitutes; tackling complementarity in projects remains an interesting challenge.
- *Initial endowments:* In reality, some projects may already have some funds allocated to them (e.g., through previous iterations of PB). The choice between extending funding to existing projects versus funding new projects can give rise to novel challenges.
- *Pledges:* In some cases, private organizations may be willing to pledge financial support to a project conditional on certain projects receiving at least a minimum level of funding. This can affect the choice of vote aggregation methods.
- *Market-based approaches:* In recent years, market-based approaches have been investigated for public decision-making (Garg et al. 2018). Extending this line of approach to PB can lead to intuitive mechanisms for PB.
- *Strategic agent models:* The work on incentives in PB focuses on strategyproofness, which aims to prevent manipulations by rational agents. However, people often do not act like rational agents. An interesting direction is to use insights from behavioral game theory, develop models of realistic manipulations that residents may use in PB, and design algorithms to prevent such manipulations.
- *The role of information:* It requires significant effort to inform residents of the costs, benefits, and complexities of different projects. The manner in which this information is communicated can have significant effect on the preferences of the residents; this is a complex issue which requires a detailed study.
- *An end-to-end model:* Finally, as we mentioned at the beginning of this chapter, we only focus on the final stage of voting within the entire PB pipeline. However, the initial stages have a direct impact on the final outcome. For example, the agenda-setting phase where projects are proposed by the residents themselves crucially affects the latter stages. Formally modeling the entire PB process and designing end-to-end solutions is a complex challenge of paramount importance.

To conclude, participatory budgeting is an important grassroots approach to democracy. Research on models, methods, and axioms for PB will provide insights that will be valuable to both the theory and practice of PB.

Acknowledgements Haris Aziz is supported by a UNSW Scientia Fellowship. Nisarg Shah is supported by an NSERC Discovery grant. The authors thank Aditya Ganguly, Barton Lee, and Dominik Peters for very helpful feedback.

References

S. Airiau, H. Aziz, I. Caragiannis, J. Kruger, J. Lang, D. Peters, Portioning using ordinal preferences: fairness and efficiency, in *Proceedings of the 28h International Joint Conference on Artificial Intelligence (IJCAI)*, 2019

K. Arrow, *Advances in the Spatial Theory of Voting* (Cambridge University Press, 1990)

H. Aziz, Participatory budgeting: are we really giving a voice to everyone? 2019. https://medium.com

H. Aziz, P. Stursberg, A generalization of probabilistic serial to randomized social choice, in *Proceedings of the 28th AAAI Conference on Artificial Intelligence (AAAI)* (AAAI Press, 2014), pp. 559–565

H. Aziz, J. Lang, J. Monnot, Computing Pareto optimal committees, in *Twenty-Fifth International Joint Conference on Artificial Intelligence, IJCAI 2016*, pp. 60–66, 2016

H. Aziz, M. Brill, V. Conitzer, E. Elkind, R. Freeman, T. Walsh, Justified representation in approval-based committee voting. Soc. Choice Welf. **48**(2), 461–485 (2017)

H. Aziz, B.E. Lee, N. Talmon, Proportionally representative participatory budgeting: axioms and algorithms, in *Proceedings of the 17th International Conference on Autonomous Agents and Multi-Agent Systems (AAMAS)* (IFAAMAS, 2018), pp. 23–31

H. Aziz, A. Bogomolnaia, H. Moulin, Fair mixing: the case of dichotomous preferences, in *Proceedings of the 20th*, pp. 753–781, 2019

G. Benade, S. Nath, A.D. Procaccia, N. Shah, Preference elicitation for participatory budgeting, in *Proceedings of the 31st AAAI Conference on Artificial Intelligence (AAAI)* (AAAI Press, 2017), pp. 376–382

U. Bhaskar, V. Dani, A. Ghosh, Truthful and near-optimal mechanisms for welfare maximization in multi-winner elections, in *Proceedings of the 32nd AAAI Conference on Artificial Intelligence (AAAI)*, pp. 925–932, 2018

A. Bogomolnaia, H. Moulin, R. Stong, Collective choice under dichotomous preferences. J. Econ. Theory **122**(2), 165–184 (2005)

C. Boutilier, I. Caragiannis, S. Haber, T. Lu, A.D. Procaccia, O. Sheffet, Optimal social choice functions: a utilitarian view. Artif. Intell. **227**, 190–213 (2015)

S. Bouveret, J. Lang, A general elicitation-free protocol for allocating indivisible goods, in *Proceedings of the 22nd International Joint Conference on Artificial Intelligence (IJCAI)* (AAAI Press, 2011), pp. 73–78

F. Brandl, F. Brandt, W. Suksompong, The impossibility of extending random dictatorship to weak preferences. Econ. Lett. **141**, 44–47 (2016)

F. Brandt, V. Conitzer, U. Endriss, J. Lang, A. Procaccia (eds.), *Handbook of Computational Social Choice* (Cambridge University Press, 2016)

Y. Cabannes, Participatory budgeting: a significant contribution to participatory democracy. Environ. Urban. **16**(1), 27–46 (2004)

I. Caragiannis, S. Nath, A.D. Procaccia, N. Shah, Subset selection via implicit utilitarian voting. J. Artif. Intell. Res. **58**, 123–152 (2017)

V. Conitzer, R. Freeman, N. Shah, Fair public decision making, in *Proceedings of the 18th ACM Conference on Electronic Commerce (EC '17)*, 2017

C. Delort, O. Spanjaard, P. Weng, Committee selection with a weight constraint based on a pairwise dominance relation, in *Proceedings of the Second International Conference on Algorithmic Decision Theory (ADT)*, pp. 28–41, 2011

S. Dughmi, T. Roughgarden, Q. Yan, Optimal mechanisms for combinatorial auctions and combinatorial public projects via convex rounding. J. ACM **63**(4), 30 (2016)

U. Endriss (ed.), *Trends in Computational Social Choice* (AI Access, 2017)

J.M. Enelow, M.J. Hinich, *The Spatial Theory of Voting: An Introduction* (CUP Archive, 1984)

B. Fain, A. Goel, K. Munagala, The core of the participatory budgeting problem, in *Proceedings of the 12th International Workshop on Internet and Network Economics (WINE)*, Lecture Notes in Computer Science (LNCS), pp. 384–399, 2016a

B. Fain, A. Goel, K. Munagala, The core of the participatory budgeting problem, in *Proceedings of the 12th International Conference on Web and Internet Economics (WINE '16)*, pp. 384–399, 2016b

B. Fain, K. Munagala, N. Shah, Fair allocation of indivisible public goods, in *Proceedings of the 19th ACM Conference on Economics and Computation (ACM-EC)*, pp. 575–592, 2018

P. Faliszewski, N. Talmon, A framework for approval-based budgeting methods. CoRR, abs/1809.04382, 2018

P. Faliszewski, N. Talmon, A framework for approval-based budgeting methods, in *Proceedings of the 33rd AAAI Conference on Artificial Intelligence (AAAI)*, 2019

P. Faliszewski, P. Skowron, A. Slinko, N. Talmon, Multiwinner voting: a new challenge for social choice theory, chapter 2, in *Trends in Computational Social Choice*, ed. by U. Endriss (AI Access, 2017)

T. Fluschnik, P. Skowron, M. Triphaus, K. Wilker, Fair knapsack. CoRR, abs/1711.04520, 2017

R. Freeman, D.M. Pennock, D. Peters, J.W. Vaughan, Truthful aggregation of budget proposals, in *Proceedings of the 20th ACM Conference on Electronic Commerce (EC '19)*, 2019

E. Ganuza, F. Francés, The deliberative turn in participation: the problem of inclusion and deliberative opportunities in participatory budgeting. Eur. Polit. Sci. Rev. **4**(2), 283–302 (2012)

N. Garg, A. Goel, B. Plaut, Markets for public decision-making. arXiv:1807.10836, 2018

N. Garg, V. Kamble, A. Goel, D. Marn, K. Munagala, Iterative local voting for collective decision-making in continuous spaces. J. Artif. Intell. Res. **64**, 315–355 (2019)

A. Gibbard, Manipulation of voting schemes: a general result. Econometrica **41**(4), 587–601 (1973)

A. Gibbard, Manipulation of schemes that mix voting with chance. Econometrica **45**(3), 665–681 (1977)

A. Goel, K.K. Anilesh, S. Sakshuwong, T. Aitamurto, Knapsack voting for participatory budgeting. ACM Trans. Econ. Comput. **7**(2), 8:1–8:27 (2019). ISSN 2167-8375

B. Gutiérrez, Madrid as a democracy lab, 2017. https://www.opendemocracy.net/democraciaabierta/bernardo-guti-rrez/madrid-as-democracy-lab. Accessed 02 May 2018

S.S. Iyengar, M.R. Lepper, When choice is demotivating: can one desire too much of a good thing? J. Pers. Soc. Psychol. **79**(6), 995 (2000)

E. Kalai, M. Smorodinsky, Other solutions to nash's bargaining problem. Econometrica **43**(3), 513–518 (1975)

R.M. Karp, Reducibility among combinatorial problems, in *Complexity of Computer Computations* (Springer, 1972), pp. 85–103

C. Klamler, U. Pferschy, S. Ruzika, Committee selection under weight constraints. Math. Soc. Sci. **64**(1), 48–56 (2012)

J. Legendre, H. Madénian, P.L. Scully, Participatory budgeting in Paris, France, 2017. http://participedia.net/en/cases/participatory-budgeting-paris-france. Accessed 02 May 2018

T. Lu, C. Boutilier, Budgeted social choice: from consensus to personalized decision making, in *Proceedings of the 22nd Joint Conference on Artifical Intelligence (IJCAI '11)*, pp. 280–286, 2011

M. Mundt, Participatory budgeting evaluation report. https://pb.cambridgema.gov/read_the_pb3_evaluation, 2017. Last Accessed 5 Aug 2019

C. Murray, Toronto's participatory budgeting pilot evaluation. https://www.toronto.ca/legdocs/mmis/2019/bu/bgrd/backgroundfile-124370.pdf, 2019. Last Accessed 5 Aug 2019

C. Papadimitriou, M. Schapira, Y. Singer, On the hardness of being truthful, in *Proceedings of the 49th Symposium on Foundations of Computer Science (FOCS)* (IEEE, 2008), pp. 250–259

D. Peters, Proportionality and strategyproofness in multiwinner elections, in *Proceedings of the 17th International Conference on Autonomous Agents and Multi-Agent Systems (AAMAS)*, vol. 1549–1557, 2018

A.D. Procaccia, J.S. Rosenschein, The distortion of cardinal preferences in voting, in *Proceedings of the 10th*, pp. 317–331, 2006

J. Rios, D. Rios Insua, A framework for participatory budget elaboration support. J. Oper. Res. Soc. **59**(2), 203–212 (2008)

J. Rios, D.R. Insua, E. Fernandez, J.A. Rivero, Participatory budget formation through the web, in *E-Government: Towards Electronic Democracy, International Conference, TCGOV 2005, Bolzano, 2–4 Mar 2005, Proceedings*, pp 268–276, 2005

A. Röcke, *Framing Citizen Participation: Participatory Budgeting in France, Germany and the United Kingdom* (Springer, 2014)

R. Samir, Madrid, part 1: e-democracy hub of Europe, 2018. http://netdem.nl/en/articles/madrid-e-democracy-hub-europe/. Accessed 28 Apr 2019

M.A. Satterthwaite, Strategy-proofness and Arrow's conditions: existence and correspondence theorems for voting procedures and social welfare functions. J. Econ. Theory **10**(2), 187–217 (1975)

A. Shah, *Participatory Budgeting*. Public sector governance and accountability series. The World Bank, 2007

E. Shapiro, N. Talmon, A condorcet-optimal participatory budgeting algorithm. arXiv preprint arXiv:1709.05839, 2017

V.V. Vazirani, *Approximation Algorithms* (Springer Science & Business Media, 2013)

World Bank, *Brazil: Toward a More Inclusive and Effective Participatory Budget in Porto Alegre* (World Bank, 2008)

From Durkheim to Machine Learning: Finding the Relevant Sociological Content in Depression and Suicide-Related Social Media Discourses

Júlia Koltai, Zoltán Kmetty, and Károly Bozsonyi

1 Introduction

The phenomenon of suicide is in the focus of social scientists since Durkheim (1897, 2002). In his classic work "Suicide" – he analyzed the spatial and temporal variation of suicide. He argued that in most cases, social causes could be found behind suicide variation. The topic was always in the center of scientific discourse, but after his work, even more attention was generated.

The work of Zoltán Kmetty was funded by the Premium Postdoctoral Grant of the Hungarian Academy of Sciences.

The work of Julia Koltai was funded by the Premium Postdoctoral Grant of the Hungarian Academy of Sciences.

J. Koltai (✉)
Centre for Social Sciences, Hungarian Academy of Sciences Centre of Excellence, Budapest, Hungary

Faculty of Social Sciences, Eötvös Loránd University, Budapest, Hungary
e-mail: koltai.julia@tk.mta.hu

Z. Kmetty
Centre for Social Sciences, Hungarian Academy of Sciences Centre of Excellence, Budapest, Hungary

Faculty of Social Sciences, Eötvös Loránd University, Budapest, Hungary
e-mail: kmetty.zoltan@tk.mta.hu

K. Bozsonyi
Faculty of Humanities and Social Sciences, Károli Gáspár University of the Reformed Church, Budapest, Hungary
e-mail: bozsonyi.karoly@kre.hu

T. Rudas, G. Péli (eds.), *Pathways Between Social Science and Computational Social Science*, Computational Social Sciences,
https://doi.org/10.1007/978-3-030-54936-7_11

In recent years, an increasing tendency of suicide has been observed in some western countries. In 2015, the female suicide rate in the UK reached its highest point of a decade.[1] Suicide rates also increased in the USA, especially in the young age cohorts. These tendencies drew the attention of social and medical scientists, and several studies were published, which attempted to provide an explanation for this phenomenon. One of the research teams who picked up on the increasing trends analyzed the online footprints of suicide and depression. Internet and social media sites provide new ways for people to express their positive feelings, but they are also platforms to express suicide ideation or depressed thoughts. Most of these contents are not notes about real suicides, but some of them are a cry for help. Suicide- and depression-related content varies among platforms, and it is not evident how a researcher can find these contents in mass data of social media. Accordingly, most of the studies, which focused on suicide, and used digital data for the research, used small sample sizes – big data approach is still uncommon in this field.

Digital data recording, social media, IoT, and digitalization, in general, produce such amount of data that was unimaginable in the past. Traditional surveys produced nicely structured and research question-related data, which was easy to analyze. These datasets are in sharp contrast with digital data, which are usually stored in highly unstructured formats and contain a lot of text data.

The noisiness, unstructuredness, and qualitative nature of this kind of data bring up new challenges for social researchers. It is far from evident how we can retrieve the relevant information from this vast amount of data. In our paper, we present possible solutions for this problem, applying a complex methodology of natural language processing, neural networks, and machine-learning algorithms.

We used the corpus of more than four million Instagram posts, related to mental health problems. However, this data is very diverse in content. It includes posts beyond those relevant from our theoretical point of view – trends of depression- and suicide-related content. Thus, after defining the initial corpus, we present two different strategies to find the relevant sociological content in the noisy environment of social media. The first approach starts with a topic modeling (Latent Dirichlet Allocation), resulting in an output that serves as the basis of a supervised classification method based on advanced machine-learning techniques. The other strategy is built on an artificial neural network-based word embedding language model. We primarily decided to use a word embedding model to validate the results of the topic model. Moreover, a word embedding approach can also help to discover the deeper understanding of main semantic patterns behind the posts we analyzed. The combination of the two methods provides a powerful framework for analyzing and interpreting the digital footprints of depression and suicide.

In the next part of the paper, we will present the most relevant findings of the studies, which deal with the digital representation of depression and self-harm. In the data and methods section, we will give an overview of the methods we used and

[1]Suicides in the UK: 2015 registrations https://www.ons.gov.uk/peoplepopulationandcommunity/birthsdeathsandmarriages/deaths/bulletins/suicidesintheunitedkingdom/2015registrations

the data we collected for this study. In the empirical part, a very detailed explanation of the analysis process will be presented alongside the analysis of the results. In the final section, in addition to our conclusions, we will discuss the limitations of the paper and the possible application of the results in social media moderation practice.

2 Recent Studies of the Field

Recently several studies have been published investigating the possible relationships between the spread of suicides/mental problems in society and Internet use. Here we focus on those, which used social media data. In recent years, Twitter emerged as a valuable platform for examining the possible interrelationships of social media and suicide. A study conducted by Sueki (2015) examined the association between suicide-related tweets and the suicidal behavior of youngsters in Japan. The statistical results confirmed that tweeting "want to die" and "want to commit suicide" was significantly related to suicidal ideation and self-harming behavior. Although the findings are significant and relevant, the generalizability of the results is questionable, as the sample used in the study was not representative.

Braithwaite et al. (2016) created significant classifications for identifying participants at risk of suicide using a machine-learning method. However, their study can be criticized as it was based on a very low sample size. They used the data of only 17 at-risk participants for fitting their suicide-risk classification model. Kmetty et al. (2017) studied the temporal relationships between the moods of tweets and the number of committed suicides. They filtered more than 626 million tweets between February 1, 2012, and August 31, 2016, geolocated for the USA to identify those containing one of the following words, "depression," "depressed," "suicide," "Prozac," or "Zoloft." The "bad mood index" created on the basis of the relative frequency of these words was found to be related to the distribution of actual suicides across the different days of the week. However, the monthly distribution of the index was unrelated to the seasonal pattern of suicides across the different months.

Besides Twitter, Instagram is also a very widely used social media platform. Brown et al. (2018) conducted their study on Instagram data. They analyzed German language Instagram posts starting from April 2016. After filtering out the hashtags, relevant for the purposes of the study, they selected 2826 pictures and related messages from 1154 users in total, portraying non-suicidal self-injuries (NSSI). They found that the number of such posts increased significantly in the evening. No differences were found between different days of the week, although on Sundays, there was a noticeable, but non-significant increase. They established that the number of messages had a low but significant positive correlation with the apparent severity of the injury pictured.

In June 2014, Moreno et al. (2016) collected 225 Instagram posts containing the "self-injury" hashtag. They found that because of the peculiar language habits of the users, the system used by Instagram for monitoring dangerous or harmful content

("Content Advisory warning") could not be trusted.[2] Reece and Danforth (2017) recruited 166 volunteers from Amazon's Mechanical Turk crowd-working system, 71 of whom were diagnosed with depression. The researchers analyzed 13,184 Instagram photographs using both automatic image processing algorithms and human coders. In addition to the variables created on the basis of the characteristics of the pictures, the participants' activities on Instagram, their popularity, and their social embeddedness were also taken into consideration. A statistical model was set up for classifying healthy/depressed and healthy/non-diagnosed depressed groups, which proved to be successful as they were able to classify users to the abovementioned groups based on their Instagram pictures and other related characteristics.

Andalibi et al. (2017) analyzed 788 Instagram posts and pictures, which included the hashtag #depression. They also examined the 1741 comments connected to these posts. The authors established a complex coding system to handle the emotional content of the posts, and they found that while posts express negative feelings and experiences, commenters often respond to those with positive feedback and support. Comments included more positive emotions than negative ones, whereas occurrences of negative emotion outnumbered positive ones in the posts.

Cheng et al. (2017) conducted an online survey among Weibo users ($n = 974$), where beyond a classic questionnaire participants' Weibo posts published in the public domain were also downloaded. The associations between word counts of posts and suicide risk factors were examined and found that those with high suicide probability were marked by a greater total word count and higher usage of pronouns, prepend words, multifunction words, and lower usage of verbs. This latter finding is conflicting with the one of De Choudhury et al. (2016) as they found a contradictory relationship between the usage of verbs and suicidal ideation.

Although these studies used social media data to understand suicide, most of them were built on a small sample size. Thus, they could not provide an overview of how suicide and depression are represented overall in the online space. Our study takes one step forward and shows the main patterns of suicide- and depression-related contents of Instagram using a large amount of textual data.

3 Data and Methods

We used Instagram data collected by SentiOne, a web-based text analytics platform.[3] The first post of the dataset was created on January 16, 2016, and the last one

[2]However, it is important to mention, that Instagram developed its system since 2014. See: https://instagram-press.com/blog/2019/10/27/taking-more-steps-to-keep-the-people-who-use-instagram-safe/ (downloaded: November 8, 2019).

[3]SentiOne (www.sentione.hu) is an international social listening software, a content-based web analytics platform that covers and recognizes 30 languages all across Europe. It gathers, indexes, and analyses public online contents published anywhere on the web, starting with social media channels through blogs, forums, and websites. The interactive platform is built upon user-provided

on January 31, 2019. During this three-year period, Instagram data was collected every day with the same algorithm, through the application's API. The dataset contains all English language posts of this data collection.

From this database, we only used those posts, which include at least one hashtag and removed all other content, like simple texts and emojis from the posts. The approach is reasonable as post using hashtags become automatically searchable and can be accessed through the Instagram API. Posts without hashtags will only be searchable if the user set it public. It means that most of the Instagram posts, which are in the data warehouse of Sentione, are posts with hashtags, and there are only a few of that does not contain any. Hashtags are widely used on Instagram. People use hashtags to connect their pictures with different ideas or concepts; they are the simplified versions of associations and thoughts. One post can contain several hashtags and thus can provide a good overview of what words are linked in people's mind. Focusing only on hashtags simplifies the research as there is no need for text pre-processing, like stemming or lemmatization, or for deletion of articles, punctuation marks, and conjunctions as they do not appear in hashtags. Overall, using only hashtags does not only simplify the content but also makes the data more structured, and the linkages of words are still researchable.

In the first step, we selected those Instagram posts where specific hashtags appeared. We started with three keywords #depression, #depressed, and #suicide. Then, based on frequencies of the different words, which appeared in the selected posts, we extended the list of keywords with the following hashtags: #sucidal, #anxiety, #mentalhealth, #mentalillness, #selfharm, #sadness, #lonely, #broken, #cutting, #ptsd, #bipolar.

At this point, we did not use any stopwords to filter out the non-relevant content. Although stopwords could help us to clear out some of the non-relevant posts from the database, in this phase, it would have only served as a partial solution, because we could not define all the requisite stopwords in advance. That is why we decided to use a topic model approach at first to detect the relevant depression- and suicide-related contents.

The dataset, which was selected by the abovementioned hashtags, contained more than eight million Instagram posts. Some of these posts were duplicated in the dataset, as they contained more than one selected hashtag; thus these duplications had to be removed. We also deleted those users' posts, who had more than 100 posts in the corpus as we wanted to concentrate on the private users and filter out the institutional ones, who regularly posts about the same content. After the deletion of the duplications and institutional users, 4,190,031 posts remained in the initial corpus.

---

keywords and key phrases to look for the specific mentions that either in themselves or within their context contain those pre-given phrases that interest the user. The system gathers data almost real-time yet has a historical database that can go back up to 3,5 years. For quantitative research, data is structured by different focus points and research parameters and is visualized on an interactive dashboard. This technology also supports deep qualitative analysis as it enables thorough analysis and categorization of all the indexed web contents such as articles, posts, comments, etc., one by one.

We used two main methods: Latent Dirichlet Allocation and a neural network-based word embedding model to find the theory-relevant posts from the different contents, which remained in our corpora after the data cleaning process mentioned earlier. These methods were not initially created for the detection of content-specification, but our usage shows a possible application for this problem.

Probability-based topic models help us understand big text corpora (Liu 2015), as they are capable of exploring the hidden structure of the corpus. The Latent Dirichlet Allocation (LDA – see (Blei et al. 2003) is the most frequently used technique of topic modeling. LDA hypothesizes that words have strong semantic information, and similar documents (here, posts) use similar sets of words. Thus, LDA is a bag-of-words method (Mohr and Bogdanov 2013), where the order of the words does not matter, only their lack or existence in a given document (post) – which concept fits the use of hashtags as corpora. The concept of LDA is close to factor analysis, as its main objective is to find the latent structure that lies behind the data; however, LDA deals with texts and not with numbers. As in the case of factor analysis, each variable is assigned to each factor with some factor weight. In LDA each word is assigned to each topic with different weights. However, the algorithms behind LDA are based on a hierarchical Bayesian probabilistic model, where the topics and the distribution of topics among documents are hidden random variables, and the estimation of their conditional distribution is based on the observed collection of documents (DiMaggio et al. 2013). At the end of the process, each document (post) can be assigned to several topics with different probabilities. As LDA belongs to the unsupervised type of topic models, only the number of topics has to be given in advance, the interpretation of the final topics has to be done at the end of the process. We run the analysis in R using the *topicmodels* (Grün and Hornik 2011), *tm* (Feinerer 2018), *tidytext* (Silge and Robinson 2016), *dyplr* (Wickham et al. 2019), and *fastTextR* (Schwendinger 2017) packages.

Word embedding models can also help researchers in the understanding of large corpora. These models are called distributional semantic models as they can show the semantic representation of words in given corpora. Semantic representation, namely, the meaning of a word, is described by other words, which are the closest to a given word in a multi-dimensional (generally, a couple of hundred) vector space. Distance is originally based on the co-presence of words in a given frame, e.g., in a document or a given n-gram. We can define the width of this frame, e.g., the number of words and so decide, if we focus on the one-, two-, three-, or more words environment of a document or even on the whole document. Each type of word embedding method is based on the co-presence of words within the defined frame, and each of them results a multi-dimensional vector space. The goal is to reduce the number of dimensions (which is originally the number of words in the corpus) but preserve the distances between words. In the reduced dimensional vector space, two words will be close to each other if they share a similar environment in the corpus. Word embedding is a general term for these methods. There are different algorithms for word embedding with different advantages. One of the earliest word embedding models is the Latent Semantic Analysis (LSA), which initially took a document-word matrix created from the corpora and applied singular-value decomposition as a data reduction technique for the creation of a lower dimensional vector space (Evans and Aceves 2016). Later models (such as word2vec, GloVe and Fasttext) apply artificial neural networks to create the vector space. They apply a minimum one hidden layer, where reduced dimensions of the vector space are the dependent variables. With these algorithms, it is possible to train bigger corpora and create higher-dimensional vector spaces (Mikolov et al. 2013). Another advantage of these newer methods is that the distribution of explained variance between the reduced dimensions is more balanced. The analysis of the corpora is realized in this vector space. Each word is represented by a vector in the vector space and distances of words are defined as cosine of the angle between vectors. For the analysis, normalization of the word vectors is important, as it means that all the words will lie on the surface of a hypersphere (same dimensions as the vector space), and the angles will be proportionate to the distances on the surface (Kozlowski et al. 2018). As word embedding methods are able to provide information about the association of words, they are useful for analyzing several problems and research questions. Some of the application of word embedding methods are about the longitudinal examination of meaning of words (Kulkarni et al. 2015) and the cultural changes (Kozlowski et al. 2018) and stereotypes (Garg et al. 2018) behind them; but these models can also identify hard-to-detect styles like sarcasm (Joshi et al. 2016). Nevertheless, word embedding methods can be useful also for the detection of word clusters, which can help to discover the latent meanings and topics of a corpus. We trained the word embedding model in R using the text2vec (Selivanov and Wang 2018) package.

As our goal is not only to show how depression- and suicide-related sociological trends appear in the digital space of Instagram but also to present, how to retrieve the theory relevant content from the wide range of suicide-related posts, some of the researcher-related decisions of the abovementioned methods will be detailed in the next chapter.

4 Ways to Find and Analyze the Relevant Content

4.1 The Application of Topic Model

As described in Data and Methods, altogether, 14 hashtags were used to collect the posts for our corpus, which contained more than four million posts. We applied the unsupervised topic model Latent Dirichlet Allocation, to identify the explicitly suicide-related posts, and also the main discourses of this topic. Using LDA presents two main challenges, which have to be addressed in order to fit the model.

One of the challenges of LDA is the *selection of the number of topics*. The goal is to find the right number of topics and to make sure that the chosen solution is stable. There are several ways to specify the number of topics. We chose the ldatuning package of R, which calculates different metrics to estimate the most preferable number of topics for an LDA model. We selected 5 percent random samples (200,000 posts per sample) from the original corpus and repeated this

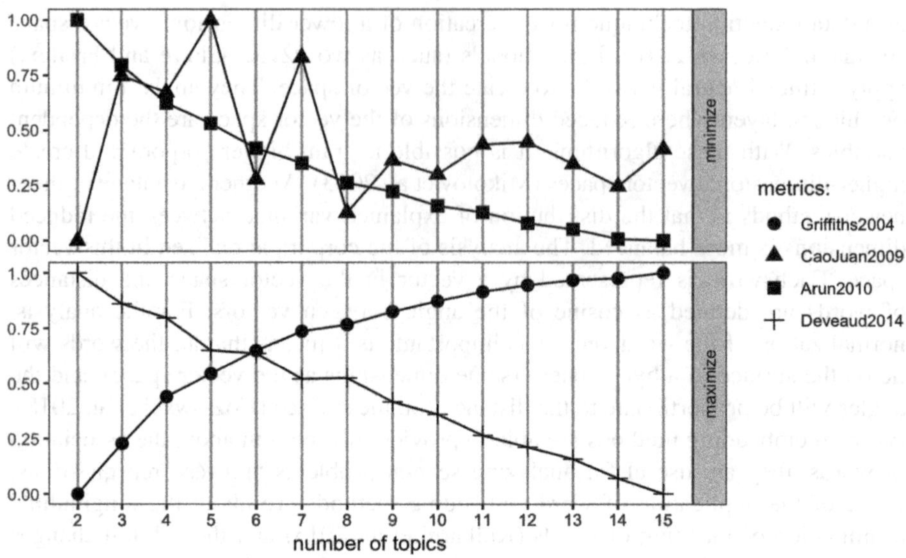

Fig. 1 Estimating the ideal topic number through different metrics

sampling process 5 times to check the stability of the structure. The 5 percent sample approach was chosen to minimalize the run time of the models. On each sample, we run the ldatuning package with parameters from 2 to 15 topics. We did not want to go above 15 topics, because we thought we would lose the interpretability of results. Based on the results of the different models, we decided to use the Gibbs sampling method for the estimation of LDA, as the alternative VEM algorithm provided fast, but more unstable solutions. Figure 1 presents one of the results of the five samples. The results of the five samples did not show high deviations from each other.

We can interpret the different metrics plotted in Fig. 1 as the following: in the case of the metrics of Cao-Juan2009 (Cao et al. 2009) and Arun2010 (Arun et al. 2010), low values indicate good solution; in the case of Griffiths2004 (Griffiths and Steyvers 2004) and Deveaud2014 (Deveaud et al. 2014), high value indicates the good choice. Based on this figure, there is no ultimate solution, but extracting at least six topics is a reasonable choice.

In order to decide on the final number of topics, we ran several LDA models on the 5 percent samples, with 6, 8, and 10 topics. We wanted to test the stability of the results with different topic numbers. Just as in the traditional method of K-means clustering, LDA only finds local optimums; thus different runs give different results. Accordingly, to *find the most stable solution for the number of topics*, we ran two models with the same number of topics at one time and repeated this process several times. Then, we calculated the ratio of similarly classified topics for each pair of runs. The result of this calculation was around the average of 65 percent for the 6 topics approach, 60 percent for the 8 topics approach, and the ratio was much lower for higher topic numbers. Thus, regarding stability, lower topic numbers provided a better fit.

However, overall stability was far from perfect (65 percent), even in the case of the six topic approach. Therefore, we ran the LDA 5 times, but now on the full corpus, with the six topics approach. We found that less than 50 percent of the posts were assigned to the same six topics in all of the five runs. Beyond the six base topics, there were some quite big topics, which also seemed coherent and interpretable. In the end, we decided to keep all the extracted topics, which contained at least two percent of the posts. Below 2 percent, only really small topics remained. To choose the approach of 13 topics instead of the originally selected 6 topics may seem odd; nevertheless, it was the direct result of the LDA algorithm. Like the K-means cluster, LDA also tries to avoid finding small topics. For example, if we change the number of topics from 6 to 10, LDA will not split the original six topics to smaller ones but re-run the whole analysis and tries to find 10 topics with approximately similar sizes. This is why higher topic numbers are more unstable.

To evaluate our approach, we calculated a *topic coherence* statistic, the UMass coherence measure (Mimno et al. 2011). The values of UMass are always negative, and the closer the value is to zero, the more coherent the topic is. The average topic coherence was −1.69, but there was a high variance between the topics: from −0.81 to −3.61. The average topic coherence improved compared to the original 6 topics models (where the value of UMass was around −2) and was similar to the simple 13 topics solution (where it was −1.7). Therefore, the 13 topics solution appeared to be an acceptable choice.

Though the number of topics was chosen, around 30 percent of the *posts were not stably classified* at this stage. In order to classify these texts, we applied the word embedding-based supervised classifier of fastText (Joulin et al. 2016).[4] We used all stably classified posts, as a training dataset and predicted the topics of the not stably classified posts. The average value of topic coherence decreased to −1.84, mostly because of one quite incoherent topic with a value of −3.45.

Taking a closer look at this topic, it did not contain any relevant depression- and suicide-related content, only posts about kitchen and grooming. Although #cutting is widely used in depression- and suicide-related posts, it is used in other domains as well, for example when people post about cutting their hair. As this topic contained no relevant information from the perspective of our research question, we decided to omit it from further analysis.[5] After the omission of this topic, the coherence value increased to −1.7.

The preliminary analysis also showed that two topics had very similar content about mental illnesses. The list of the most important keywords is also overlapping

[4]Just for testing the accuracy of the fastText classifier, first we selected 400,000 random posts from the ones, which were stably classified by the LDA model as a training dataset. Then, we selected another 100,000 random posts from the stably classified posts as a test dataset and predicted the topic classification. The accuracy was 95 percent.

[5]The topic contained 3.5 percent of the posts of our whole sample, thus it was a relatively small one.

Table 1 Most frequent 15 words of topics of the LDA analysis

Topic name	Words
1. Self-harm (15,8%)	Depression, sad, depressed, suicide, anxiety, suicidal, alone, broken, sadness, selfharm, cutting, lonely, love, worthless, anorexia
2. Memes (13%)	Memes, depression, meme, dankmemes, dank, funny, anime, edgy, lol, depressed, suicide, cringe, lmao, autism, filthyfrank
3. Art, photography (14,1%)	Art, lonely, broken, photography, depression, mentalhealth, love, anxiety, nature, artist, drawing, instagood, photooftheday, sad, travel
4. Fitness, diet (9,8%)	Mentalhealth, fitness, cutting, gym, health, motivation, bodybuilding, fitfam, anxiety, workout, fit, healthy, depression, weightloss, nutrition
5. Mental health awareness (8,4%)	Mentalhealth, love, anxiety, selflove, depression, motivation, selfcare, inspiration, mindfulness, meditation, health, life, happiness, healing, positivevibes
6. Illness and mental consequences (7,4%)	Anxiety, depression, ptsd, mentalhealth, stress, chronicpain, chronicillness, health, cbd, spoonie, slime, asmr, fibromyalgia, cannabis, satisfying
7. Mental illness, recovery (10,1%)	Mentalhealth, anxiety, depression, mentalhealthawareness, mentalillness, ptsd, recovery, bipolar, love, therapy, mentalhealthmatters, suicide, bpd, support, stress
8. Grunge, emo (6,6%)	Depression, grunge, depressed, aesthetic, sad, emo, tumblr, love, alternative, anxiety, dark, goth, music, suicide, scene
9. LikeforLike (7,9%)	Love, broken, sad, lonely, quotes, depression, follow, life, like4like, depressed, heart, like, happy, likeforlike, followforfollow
10. Poetry (3,3%)	Love, poetry, broken, quotes, writersofinstagram, depression, poem, writer, writing, poetrycommunity, poetsofinstagram, life, poet, words, sad
11. Religion (3,6%)	Love, depression, life, mentalhealth, anxiety, broken, quotes, hope, faith, god, peace, truth, motivation, jesus, inspiration

(like #mentalhealthmatters, #recovery, #mentalhealthawareness). Based on the values of topic coherence, it seemed reasonable to merge these topics: coherence values of the original topics were -1.74 and -1.45 and -1.67 for the merged one. Based on the coherence statistics and the similar keywords, we decided to merge the two topics.

The appearance of fairly similar topics also highlighted that it would not be worth creating more topics. Ultimately, we chose an 11 topics solution. Table 1 presents the most frequent 15 words of each topic in descending order.

Topic 1 is the darkest one. It shows high probability for all the hashtags we used for the selection of the posts and contains additional disturbing words like helpme and selfhate. This is the largest topic, and it contains around 16 percent of the posts.

Topic 2 is also large, but the posts, which are assigned to it, have completely different content. This topic covers the posts about memes and meme sites like filthyfrank. It also contains those posts, which use a wide range of hashtags to reach a broader audience. Topic 9 is similar to this one as it also contains "clickbait" posts, with hashtags, such as likeforlike and followforfollow. Nevertheless, topic 9 is more about heartbreak and loneliness.

Table 2 The ratio of positive and negative words in the different topics

	Positive	Negative
Self-harm	3,5%	30,1%
Memes	4,3%	10,2%
Art, photography	8,5%	8,0%
Fitness, diet	7,7%	4,3%
Mental health awareness	16,7%	5,0%
Illness and mental consequences	5,0%	8,0%
Mental illness, recovery	7,9%	9,8%
Grunge, Emo	8,1%	12,7%
LikeforLike	11,1%	10,5%
Poetry	10,0%	12,9%
Religion	18,0%	10,1%

Topic 3 is about photos and depression related art. This is the second largest topic, with more than 14 percent of the posts. Topic 10 (which is much smaller, only 3 percent of the posts) is similar to topic 3, as it is also connected to art, but while topic 3 is more about the visual representation, topic 10 is more about literary content. We can also include topic 8 in the art-related topics, as its posts are about specific dark music, such as grunge or goth, but hashtags like #emo or #alternative also appear here.

Topic 4 contains posts about fitness, gym, and diet associating mental health with exercising and doing sports, and their positive effect on mental health and depression. Topic 5 is also about positive messages for dealing with mental health problems, but with a focus on spirituality and positive emotions. Topic 11 covers similar perspectives, but from the point of religiosity and faith, emphasizing God and Jesus.

Topic 7 covers posts with more medical expression about mental illness than other topics. However, this topic is not exclusively negative, as it contains expressions like recovery or support as well. Medical expressions also appear in topic 6; however, it focuses on mental health issues as the consequence of other types of diseases.

To confirm our statement, that topic 1 is the darkest of all; we calculated the *sentiment value of each topic (see Table* 2 for the results). For this analysis, we used a sentiment dictionary called NRC Word Emotion Association Lexicon (Mohammad and Turney 2013). As we expected, the self-harm topic contained the most negative words: 30 percent of the words assigned to this topic were negative, and only 3.5 percent were positive. Also, this topic contained the most words of any kind of emotion (either positive or negative). Some topics, such as fitness, mental health awareness, or religion, contained more positive words than negative ones, which is also a clear sign that significantly different drivers can be identified behind the posts related to mental health or depression.

We also examined the *average number of likes and comments* of each topic (see appendix Table 3). It is not surprising that topic 9 – likeforlike – attracts the most likes and comments. However, the median value of likes is only 7, which is much lower compared to topic 4, where the median value is 12. Thus, we can conclude that a small number of posts with a large number of likes and comments in topic 9 cause this difference. Overall, topic 1 (self-harm) has the smallest number of likes and comments (and it is true even if we use the median instead of the average). It may indicate that people respond less actively to these types of contents. However, we cannot investigate this presumption, as we do not have information about the number of followers of users who posted about raw selfharm and selfhate.

The *temporal dynamics of posts* also presents interesting results. As we do not know the number of all Instagram posts created in the different periods of the data collection, we could only calculate the relative frequency of the different topics compared to all collected posts. Thus, the number of posts in a topic was compared to all downloaded mental health-related posts, which definitely makes our analysis limited. In this section, we primarily focus on topic 1, which covers self-harm posts because we have long-time temporal trends about the number of committed suicides, which seems independent from geographical factors. Thus, if we select the topic, which is the closest to the sociological and theoretical definition of our object of interest, we can compare these trends with the ones we found on our new type of data. Figure 2 presents the yearly, monthly, and days of week trends of suicide- and depression-related posts.

A general increase can be observed in the ratio of this topic during the examined period. In 2016, 12 percent of all the collected posts belonged to this topic, which increased to 15 percent in 2017 and above 17 percent in 2018. Thus, within this wide, mental health-related space, posts about suicide, and self-harm ideation increased significantly in the examined 3 years.

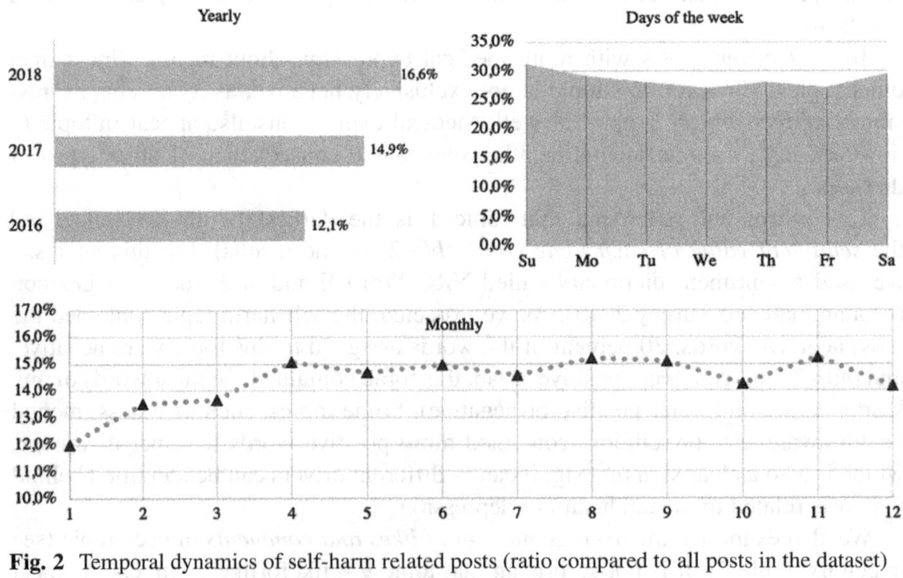

Fig. 2 Temporal dynamics of self-harm related posts (ratio compared to all posts in the dataset)

As a monthly trend, we observed that in the first 3 months of the year, the proportion of topic 1 is around 12–13 percent, then it increases to 15 percent and remains at that level.

There is another long-standing and spatially independent trend about the days of the week; thus, it is worth examining the weekly dynamic of the ratio of topic 1 among all the collected posts. The proportion of topic 1 is the highest on the weekends, and the lowest in the middle of the week (Wednesday).

Finally, we examined the ratio of topic 1 on some special days, like Christmas, New Year's Eve, and Valentine's day, just as (Beauchamp et al. 2014; Jessen et al. 1999; Zonda et al. 2009) examined these special days. However, we did not find any remarkable pattern in our data.

4.2 The Application of Word-Embedding

Our second approach for discovering the different discourses behind mental health-related Instagram posts is based on an artificial neural network word embedding model. We used the Glove algorithm (Pennington et al. 2014), with a 30-length window size, which means that the environment of 30 number of words was taken into account when pro-occurrences of words were calculated. This window size is larger than usual, but we wanted to make sure that all the hashtags that appeared in a post are assigned to the same window. The reason for this is that we hypothesized that the distance of two hashtags within the same post does not matter, only that they appear together in the same post. With this size of a window, each post served as one window. For the training of the vector space, we only used those words, which occurred at least 30 times in the corpus. Words with low occurrence could increase the instability of the word embedding models. Overall, more than 150,000 unique words were used in the training of the model. The trained vector space had 300 dimensions.

After we trained the model, we selected the 14 hashtags we applied in the data collection stage of the project and detected the closest words in the vector space for each of them. We used the threshold of 0.3 cosine similarity in the selection of the closest words; thus there were some initial hashtags, where less than 50 words were detected. The 0.3 value was based on our previous experiences on these word embedding models. Finally, we had 275 words in the close environment of the initial hashtags, from which we created a similarity matrix based on cosine similarities. To identify latent topics behind the corpus, we built a hierarchal cluster based on this similarity matrix.[6] As we wanted to group the words, and we could calculate the distance between the words, clustering seemed to be a reasonable solution for this task. Based on the interpretation of the dendrogram of the hierarchical cluster

[6]As the method and logic of clustering are based on distances of the elements, we used the complement of the similarity matrix, namely dissimilarity matrix.

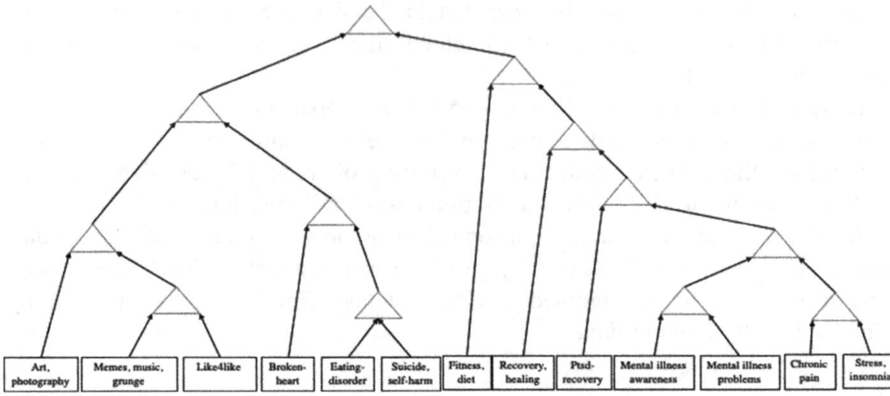

Fig. 3 Dendrogram of the hierarchical cluster of words in the vector space of the word embed-ding model

analysis, we decided to keep the solution of 13 word clusters. Here, the selection of the number of clusters was also not straightforward, but we could rely on the figure of the dendrogram. Figure 3 presents the simplified version of this dendrogram and thus show the distances of clusters with some interpretative labels of those. The list of all words assigned to each cluster is available in Table 4 in the appendix.

As expected, the word cluster of *chronic pain* and *stress/insomnia* are close to each other, as is *mental illness* and *mental illness awareness*. These four clusters merge before *ptsd recovery* joins them (right side of the figure). Words associated with *recovery and healing* create their own cluster, but on a higher level it merges to the first five. In the other part of the dendrogram, *suicide and self-harm*-related hashtags are close to *eating disorder*, and these two clusters are close to the cluster labeled as *broken heart*. Additionally, the cluster of *memes and music (grunge)* is close to the *likeforlike* and *followforfollow* clusters, which are also not that far from the *art and photography*-related cluster (left side of the figure). Last, but not least, we can find a distinct cluster about *fitness and diet*. This cluster is close to the recovery and mental illness-related clusters.

The main question is how similar are the results of word embedding clustering to the topics we got from the topic modeling approach. Most of the topics of the topic model can be identified in the results of the word embedding clustering, but there are some slight differences. The two main differences are the following. In the topic model, religion-related words shape a distinct topic; however, in the word embedding cluster, this topic does not appear separately. At the same time, eating disorder forms a separate cluster in the word embedding analysis, which does not appear in the topic model. Nevertheless, these are rather small differences; on the whole, the results of the two methodological approaches are exceedingly similar.

5 Discussion

People post/share/create billions of contents in social media every day, providing valuable data for researchers. However, social media data is always noisy, and it is hard to identify content germane to a particular research project. This paper had two main goals, a methodological and an analytical one.

Our methodological objective was to present a novel way to find the research related content in a large digital corpus, like the one we used from Instagram data. The combination of topic modeling and neural network word embedding methods seems to be a promising way to realize this goal. LDA topic modeling has been widely used in text mining research for a long time. However, the instability of the model and the problems regarding the selection of the correct topic number always create some uncertainty about the generalizability of the results. To solve this problem, we used a multi-step robust method to find the correct number of topics and validated the results by analysis of a word embedding model. In our approach, this neural network-based word embedding model played a limited but important role. First, it helped us to validate the results of the topic model. Second, the word embedding approach also discovered a deeper understanding of the main semantic patterns behind the posts we analyzed. The order of the application of topic modeling and word embedding is not straightforward. Reversely to our solution, there are examples in earlier researches, where word embedding results used as an input for topic modeling (see, e.g., Moody 2016). Utilized in either order, the combination of the two model families could give us more robust, more interpretable, and more valid results, which are all important for social science research.

Our analytical aim was the analysis of digital footprints of depression and suicide, on a platform, which is widely used by youngsters – Instagram. The results of the topic model showed that only 16 percent of the examined Instagram posts are connected to real suicidal ideation and/or depression. (For similar results on Twitter data, see O'Dea et al. 2015). People use depression and mental health-related hashtags in very different ways. Some people post about their (mental) illness, but it is more common to post about health (see also Muralidhara and Paul 2018), recovery, and mental health awareness (including diet and fitness). Many posts can be linked to art and poetry but also to music and music styles, such as grunge or black rock. In addition, plenty of clickbait posts and posts about memes are also present in the digital space of depression and suicide.

After identifying posts most connected to our research interest of suicide and self-harm, we analyzed the temporal dynamics of these contents. We found that it a striking result that the ratio of the self-harm-related posts increased heavily in the examined period. We do not have information about the socio-demographic composition of the users; thus, it is not possible to separate the effects, when the same user group starts to share more negative content; and when new users, who represent a different population with more negative feelings to share appear in the digital space. As the ratio of committed suicide is higher among older people, it is also possible that what we observe here is a process, where the overall age composition of the Instagram users becomes older in the analyzed time period, resulting in the changing content of the posts. Nevertheless, as the algorithm, based on which the API of Instagram provides data, is not transparent, change in this algorithm can also be a cause of these trends.

As we do not have information about the geolocation of the data, we do not have exact population numbers about the committed suicides in this time period, to which we could compare the temporal distribution of the posts. However, in the last few years, an increasing tendency was observed in the number of committed suicides in English-speaking countries, such as the UK and the USA. In the UK, female suicide rates in 2015 reached their highest point of the decade.[7] Overall, the suicide rates also increased in the USA, especially in the young age cohorts. Suicide rate was 10 per 100,000 inhabitants within the 15–24 age group in 2008, and 13.15 per 100,000 inhabitants within the same age group in 2016, which presents a 30 percent increase in this age cohort.[8] Hence, it is possible that the temporal trend we detected in the digital space mirrors the same phenomenon that appears in population data.

Regarding the monthly pattern, classically analyzed trends of committed suicides show that suicides are more frequently happen during the spring and the summer (Massing and Angermeyer 1985; Zonda et al. 2005) and that these trends stand independently from the geographical location. Durkheim (1897, 2002), in his classic work, explained this trend with social factors. He observed an increase in those months (and within day periods) when social life is more active. Thus, we assumed that self-harm-related posts (topic 1) are also relatively more common during the spring and/or summer months. However, our results do not confirm this hypothesis. We could not capture the same temporal dynamic that we can observe in the case of committed suicides in the monthly trend of the posts related to self-harm. Although in January and February, the proportion of this topic (compared to all depression- and suicide-related posts) was lower than the average, in the rest of the year, there was no specific volatility between months. Thus, we could not capture the same temporal dynamics in social media as we can observe in the case of committed suicides. This confirms previous results (Kmetty et al. 2017), where the authors drew a similar conclusion analyzing Twitter data.

Weekly trends of committed suicides are also geographically and culturally independent as they show the same temporal dynamic in very different countries and continents. These trends show that suicide more frequently happens on Mondays

---

[7]Suicides in the UK: 2015 registrations: https://www.ons.gov.uk/peoplepopulationandcommunity/birthsdeathsandmarriages/deaths/bulletins/suicidesintheunitedkingdom/2015registrations (date of last access: 10th of April, 2019).

[8]American Foundation for Suicide Prevention: Suicide statistics: https://afsp.org/about-suicide/suicide-statistics/ (date of last access: 10th of April, 2019).

and Tuesdays and less often during the weekends. We did not find similar trends in our data; in fact, we observed an inverse tendency. The proportion of the topic (compared to all depression- and suicide-related posts) is the highest on the weekends, and the lowest in the middle of the week (Wednesday).

Our results reinforce the idea that posting about suicide is not equivalent to committing suicide. An intuitive explanation about the different temporal trends in the population and the social media can be, that there is a temporal shift between posting about- and committing suicide, however, a paper by Huang et al. (2017) does not support this explanation. They found that people who committed suicide shared the same amount of posts every day in their last month. Despite these results, we cannot say there is no relationship between suicide ideation in the social media and real-word commission of suicide. However, we can say that the link is not unconditionally direct. It is important to mention that social media companies continuously monitor these contents shared on their platform. This activity has two different effects on our data. If social media companies remove some posts or ban users (e.g., because they encourage suicide or self-harm), the composition of contents and the temporal dynamic of the posts can change. As the protocols of content moderation on the different platforms are not entirely transparent, we cannot control for this intervention. Instagram and Facebook use both algorithms and human operators to monitor the content of their platform and give signals when problematic contents appear (Chancellor et al. 2016a). Moreover, moderation has another consequence. Users, who would like to posts that will likely be moderated by the platform, start to use different expressions and words (Chancellor et al. 2016b; Moreno et al. 2016). It means that as they start using a coded language, it will be harder to capture these posts by researchers and even harder by platform mediators.

Although Instagram is primarily a photo and video sharing social media platform, we only concentrated on the texts and hashtags of the comments. A possible extension of this study may include the analysis of photos and videos, for example, with automated picture tagging techniques. Such research may have special importance, as earlier research, which dealt with the analysis of photos and videos of Instagram (Moreno et al. 2016; Reece and Danforth 2017; Brown et al. 2018), only analyzed a smaller number of cases and not large-scale data.

Wider application of our results would also be possible. Our study showed that in recent years, a mass amount of suicide- and depression-related content appeared on Instagram. Our research can provide added value in the detection of possible self-harm events, as we presented that with the utilization of complex techniques (such as topic modeling and word embedding methods), it is possible to identify the most problematic posts and most vulnerable users.

Our results also provide important input for further research on this topic. In the era of Big Data, based on the digital footprints of people, a massive amount of information is available for analysis. However, such data is very noisy and was not specifically made for scientific analyses. Thus, we have to understand the data generation mechanisms very well and also have knowledge on how to dig the valuable and useful information out from this data ocean. Our results can provide a starting point for researchers working with suicide-related social media data and can also help social scientists in the application of text analytics and machine-learning techniques.

Our paper is a good example of how the interpretative power of Social Science can be improved by incorporating the methods of Computational Social Science. It is overall really hard to research why people commit suicide. Only half of those who commit suicide have any mental illness. Every method/data which brings us closer to understand the ideation of suicide is really valuable in this research field. Moreover, social media is a perfect platform to understand people's thinking. It is also important that we did here a non-interventional study, which increases the internal validity of the results.

Appendix

Table 3 Number of likes and comments in the different topics of LDA analysis

	Likes	Comments
Self-harm	15,13	0,62
Memes	18,04	0,75
Art, photography	27,78	1,01
Fitness, diet	27,34	1,35
Mental health awareness	20,87	0,97
Illness and mental consequences	28,91	1,52
Mental illness, recovery	23,10	1,19
Grunge, Emo	19,06	0,71
LikeforLike	41,16	1,76
Poetry	18,50	0,85
Religion	23,69	0,91

Table 4 Words assigned to different clusters of the word embedding model

Cluster name	Words
Suicide, self-harm	#suicide, #suicidal, #death, #selfharm, #dead, #die, #help, #killme, #depressionquotes, #fuck, #kill, #blood, #rip, #kms, #selfhate, #helpme, #suicidegirls, #cut, #drugs, #selfharmmm, #worthless, #suicidalthoughts, #depressing, #ugly, #scars, #depressedquotes, #hopeless, #triggerwarning, #useless, #deadinside, #fat, #sadquotes, #dying, #cuts, #depressive, #numb, #cutter, #sadgirl, #cutting
Brokenheart	#depressed, #sad, #sadness, #alone, #hate, #cry, #hurt, #pain, #crying, #lost, #empty, #unhappy, #quotes, #tired, #lonely, #mood, #quote, #broken, #heartbroken, #tears, #feelings, #brokenheart, #heartbreak, #loneliness, #upset, #thoughts, #sadedits, #heart, #broke, #poetry
Stress, insomnia	#depression, #anxiety, #stress, #insomnia, #sleep, #anxietyrelief, #calm, #stressed, #relax
Memes, music, grunge	#emo, #memes, #meme, #funny, #music, #gay, #anime, #grunge, #tumblr, #aesthetic, #single
like4like	#love, #life, #girl, #happy, #me, #fear, #follow, #happiness, #smile, #friends, #like4like, #followforfollow, #l4l, #family, #instagood, #Repost, #mental, #emotional, #beautiful, #instagram, #photooftheday, #picoftheday, #people, #memories, #like, #crazy
Art, photography	#dark, #black, #darkness, #art, #blackandwhite, #night, #photography, #sky, #photo, #nature, #sunset, #cold, #travel, #winter, #blue, #sun, #evening, #rain, #clouds, #light, #shattered, #red, #old, #glass, #damaged, #hair, #design, #style, #haircut
Mental illness-awareness	#suicideprevention, #suicideawareness, #mentalhealthmatters, #endthestigma, #mentalillnessawareness, #mentalhealthrecovery, #stigma, #mentalhealthwarrior, #stopthestigma, #mentalhealthadvocate
Recovery, healing	#mentalillness, #mentalhealth, #mentalhealthawareness, #recovery, #hope, #selflove, #selfcare, #health, #healing, #therapy, #support, #loveyourself, #wellness, #positivevibes, #positivity, #peace, #mindfulness, #inspiration, #awareness, #motivation, #wellbeing, #healthy, #strength, #lifestyle, #healthylifestyle, #meditation, #psychology, #mind, #change, #yoga, #growth, #goals, #strong, #mindset, #mondaymotivation, #addiction, #progress, #work, #hardwork, #warrior, #freedom, #medication
Eating disorder	#anorexia, #anorexic, #bulimia, #ana, #eatingdisorder, #mia, #ed
Mental illness	#bipolar, #bpd, #anxious, #panicattack, #ocd, #socialanxiety, #adhd, #bipolardisorder, #borderlinepersonalitydisorder, #schizophrenia, #anxietydisorder, #manic, #mania, #borderline, #manicdepression, #psychosis, #bipolardepression
Ptsd recovery	#ptsd, #survivor, #ptsdawareness, #ptsdrecovery, #trauma, #cptsd, #veteran, #veterans, #military, #posttraumaticstressdisorder, #army, #ptsdsucks, #abuse, #PTSD, #tbi, #ptsdsurvivor
Fitness, diet	#fitness, #gym, #bodybuilding, #diet, #bulking, #workout, #gains, #abs, #fit. #shredded, #gymlife, #muscle, #training, #cardio, #fitfam, #food, #shredding, #protein, #fitspo, #fitnessmotivation, #gainz, #legday, #lifting, #gymrat, #dieting, #shred, #bodybuilder, #weightloss, #arms, #muscles, #physique
Chronic pain	#chronicpain, #chronicillness, #fibromyalgia, #invisibleillness, #spoonie

References

N. Andalibi, P. Ozturk, A. Forte, Sensitive self-disclosures, responses, and social support on Instagram: the case of #depression, in *Proceedings of the 2017 ACM Conference on Computer Supported Cooperative Work and Social Computing* (ACM, New York, 2017), pp. 1485–1500. https://doi.org/10.1145/2998181.2998243

R. Arun, V. Suresh, C.E.V. Madhavan, M.N.N. Murthy, M.J. Zaki, J. Xu Yu, et al., On finding the natural number of topics with latent Dirichlet allocation: some observations, in *Advances in Knowledge Discovery and Data Mining* (Springer, Berlin/Heidelberg, 2010), pp. 391–402. https://doi.org/10.1007/978-3-642-13657-3_43

G.A. Beauchamp, M.L. Ho, S. Yin, Variation in suicide occurrence by day and during major American holidays. J. Emerg. Med. **46**(6), 776–781 (2014). https://doi.org/10.1016/j.jemermed.2013.09.023

D.M. Blei, A.Y. Ng, M.I. Jordan, Latent Dirichlet allocation. J. Mach. Learn. Res. **3**(4–5), 993–1022 (2003)

S.R. Braithwaite, C. Giraud-Carrier, J. West, M.D. Barnes, C.L. Hanson, Validating machine learning algorithms for Twitter data against established measures of suicidality. JMIR Ment. Health **3**(2) (2016). https://doi.org/10.2196/mental.4822

R.C. Brown, T. Fischer, A.D. Goldwich, F. Keller, R. Young, P.L. Plener, #cutting: non-suicidal self-injury (NSSI) on Instagram. Psychol. Med. **48**(2), 337–346 (2018). https://doi.org/10.1017/S0033291717001751

J. Cao, T. Xia, J. Li, Y. Zhang, S. Tang, A density-based method for adaptive LDA model selection. Neurocomputing **72**(7), 1775–1781 (2009). https://doi.org/10.1016/j.neucom.2008.06.011

S. Chancellor, Z. Lin (Jerry), M. De Choudhury, "This post will just get taken down": characterizing removed pro-eating disorder social media content, in *Proceedings of the 2016 CHI Conference on Human Factors in Computing Systems* (ACM, New York, 2016a), pp. 1157–1162. https://doi.org/10.1145/2858036.2858248

S. Chancellor, J.A. Pater, T.A. Clear, E. Gilbert, M. De Choudhury, #thyghgapp: Instagram content moderation and lexical variation in pro-eating disorder communities, in *Proceedings of the 19th ACM Conference on Computer-Supported Cooperative Work & Social Computing – CSCW'16*. Presented at the 19th ACM Conference (ACM Press, San Francisco, 2016b), pp. 1199–1211. https://doi.org/10.1145/2818048.2819963

Q. Cheng, T.M. Li, C.-L. Kwok, T. Zhu, P.S. Yip, Assessing suicide risk and emotional distress in Chinese social media: a text mining and machine learning study. J. Med. Internet Res. **19**(7) (2017). https://doi.org/10.2196/jmir.7276

M. De Choudhury, E. Kiciman, M. Dredze, G. Coppersmith, M. Kumar, Discovering shifts to suicidal ideation from mental health content in social media, in *In Proceedings of the 2016 CHI Conference on Human Factors in Computing Systems*, (2016, May), pp. 2098–2110

R. Deveaud, E. Sanjuan, P. Bellot, Accurate and effective latent concept modeling for ad hoc information retrieval. Revue des Sciences et Technologies de l'Information – Série Document Numérique 61–84 (2014). https://doi.org/10.3166/DN.17.1.61-84

P. DiMaggio, M. Nag, D. Blei, Exploiting affinities between topic modeling and the sociological perspective on culture: Application to newspaper coverage of U.S. government arts funding. Poetics **41**(6), 570–606 (2013). https://doi.org/10.1016/j.poetic.2013.08.004

E. Durkheim, *Suicide: A Study in Sociology*, 2nd edn. (Routledge, London, 2002) [E. Durkheim (1897). Le Suicide. Etude De Sociologie, published by Ancienne Librairie Germer Bailliere et Cte, Felix Alcan, Editeur, Paris]

J.A. Evans, P. Aceves, Machine translation: mining text for social theory. Annu. Rev. Sociol. **42**(1), 21–50 (2016). https://doi.org/10.1146/annurev-soc-081715-074206

I. Feinerer, Introduction to the tm Package Text Mining in R (2018). https://cran.r-project.org/web/packages/tm/vignettes/tm.pdf

N. Garg, L. Schiebinger, D. Jurafsky, J. Zou, Word embeddings quantify 100 years of gender and ethnic stereotypes. Proc. Natl. Acad. Sci. **115**(16), E3635–E3644 (2018). https://doi.org/10.1073/pnas.1720347115

T.L. Griffiths, M. Steyvers, Finding scientific topics. Proc. Natl. Acad. Sci. **101**(suppl 1), 5228 (2004). https://doi.org/10.1073/pnas.0307752101

B. Grün, K. Hornik, Topicmodels: an R package for fitting topic models. J. Stat. Softw. **40**(1), 1–30 (2011). https://doi.org/10.18637/jss.v040.i13

X. Huang, L. Xing, J.R. Brubaker, M.J. Paul, Exploring timelines of confirmed suicide incidents through social media. Presented at the 2017 IEEE international conference on healthcare informatics (ICHI) (IEEE, 2017), pp. 470–477

G. Jessen, B.F. Jensen, E. Arensman, U. Bille-Brahe, P. Crepet, D. De Leo, et al., Attempted suicide and major public holidays in Europe: findings from the WHO/EURO multicentre study on Parasuicide. Acta Psychiatr. Scand. **99**(6), 412–418 (1999)

A. Joshi, V. Tripathi, K. Patel, P. Bhattacharyya, M. Carman, Are Word Embedding-Based Features Useful for Sarcasm Detection? (2016). arXiv:1610.00883 [cs]. http://arxiv.org/abs/1610.00883. Accessed 15 Apr 2019

A. Joulin, E. Grave, P. Bojanowski, T. Mikolov, Bag of Tricks for Efficient Text Classification (2016). arXiv:1607.01759 [cs]. http://arxiv.org/abs/1607.01759. Accessed 16 Apr 2019

Z. Kmetty, J. Koltai, E. Bokányi, K. Bozsonyi, Seasonality pattern of suicides in the US – a comparative analysis of a Twitter based bad-mood index and committed suicides. Intersect. East Eur. J. Soc. Polit. **3**(1), 56–75 (2017). https://doi.org/10.17356/ieejsp.v3i1.302

A.C. Kozlowski, M. Taddy, J.A. Evans, The Geometry of Culture: Analyzing Meaning Through Word Embeddings (2018). *arXiv:1803.09288 [cs]*. http://arxiv.org/abs/1803.09288. Accessed 30 Oct 2018

V. Kulkarni, R. Al-Rfou, B. Perozzi, S. Skiena, Statistically significant detection of linguistic change, in *Proceedings of the 24th International Conference on World Wide Web – WWW'15*. Presented at the the 24th International Conference (ACM Press, Florence, 2015), pp. 625–635. https://doi.org/10.1145/2736277.2741627

B. Liu, *Sentiment Analysis: Mining Opinions, Sentiments, and Emotions* (Cambridge University Press, Cambridge, 2015). https://doi.org/10.1017/CBO9781139084789

W. Massing, M.C. Angermeyer, The monthly and weekly distribution of suicide. Soc. Sci. Med. **21**(4), 433–441 (1985)

T. Mikolov, K. Chen, G. Corrado, J. Dean, Efficient Estimation of Word Representations in Vector Space (2013). arXiv:1301.3781 [cs]. http://arxiv.org/abs/1301.3781. Accessed 15 Apr 2019

D. Mimno, H. Wallach, E. Talley, M. Leenders, A. McCallum, Optimizing semantic coherence in topic models, in *Proceedings of the Conference on Empirical Methods in Natural Language Processing* (Association for Computational Linguistics, 2011), pp. 262–272

S.M. Mohammad, P.D. Turney, Crowdsourcing a word-emotion association lexicon. Comput. Intell. **29**(3), 436–465 (2013)

J.W. Mohr, P. Bogdanov, Introduction—topic models: what they are and why they matter. Poetics **41**(6), 545–569 (2013). https://doi.org/10.1016/j.poetic.2013.10.001

C.E. Moody, Mixing Dirichlet Topic Models and Word embeddings to Make lda2vec (2016). arXiv:1605.02019 [cs]. http://arxiv.org/abs/1605.02019. Accessed 16 Apr 2019

M.A. Moreno, A. Ton, E. Selkie, Y. Evans, Secret society 123: understanding the language of self-harm on Instagram. J. Adolesc. Health **58**(1), 78–84 (2016). https://doi.org/10.1016/j.jadohealth.2015.09.015

S. Muralidhara, M.J. Paul, #healthy selfies: exploration of health topics on Instagram. JMIR Public Health Surveill. **4**(2), e10150 (2018). https://doi.org/10.2196/10150

B. O'Dea, S. Wan, P.J. Batterham, A.L. Calear, C. Paris, H. Christensen, Detecting suicidality on Twitter. Internet Interv. **2**(2), 183–188 (2015). https://doi.org/10.1016/j.invent.2015.03.005

J. Pennington, R. Socher, C. Manning, Glove: global vectors for word representation, in *Proceedings of the 2014 Conference on Empirical Methods in Natural Language Processing (EMNLP)* (Association for Computational Linguistics, Doha, 2014), pp. 1532–1543. https://doi.org/10.3115/v1/D14-1162

A.G. Reece, C.M. Danforth, Instagram photos reveal predictive markers of depression. EPJ Data Sci. **6**(1), 15 (2017). https://doi.org/10.1140/epjds/s13688-017-0110-z

F. Schwendinger, *Package 'fastTextR.'* en (2017). https://cran.r-project.org/web/packages/fastTextR/fastTextR.pdf

D. Selivanov, Q. Wang, *Package 'text2vec.'* (2018). http://text2vec.org

J. Silge, D. Robinson, Tidytext: text mining and analysis using tidy data principles in R. J. Open Source Softw. **1**(3), 37 (2016). https://doi.org/10.21105/joss.00037

H. Sueki, The association of suicide-related Twitter use with suicidal behaviour: a cross-sectional study of young internet users in Japan. J. Affect. Disord. **170**, 155–160 (2015). https://doi.org/10.1016/j.jad.2014.08.047

H. Wickham, R. François, L. Henry, K. Müller, RStudio. *Package 'dplyr.'* en (2019). http://dplyr.tidyverse.org, https://github.com/tidyverse/dplyr

T. Zonda, K. Bozsonyi, E. Veres, D. Lester, M. Frank, The impact of holidays on suicide in Hungary. OMEGA J. Death Dying **58**(2), 153–162 (2009). https://doi.org/10.2190/OM.58.2.e

T. Zonda, K. Bozsonyi, E. Veres, Seasonal fluctuation of suicide in Hungary between 1970–2000. Arch. Suicide Res. **9**(1), 77–85 (2005)

Epilogue

Tamás Rudas and Gábor Péli

Changing Understanding in Algorithmic Societies: Exploring a New Social Reality with the Tools of Computational Social Science

1 Algorithmic Societies

The emergence of Big Data and the associated technologies reflect a fundamental transformation of modern societies. We consider the appearance of this huge stream of digitalized information a manifestation of a deeper change: the overwhelming algorithmization of our societies. While algorithms are syntactic objects, societal algorithmization has deep semantic consequences: it generates new ways of sense-making of our social world. The cumulative impacts of the algorithmization processes thus cohere into a new entity, the algorithmic society. The penetration of digital devices, with the ubiquitous presence of technical and societal applications enabled by them form necessary, but yet insufficient, conditions for the emergence of algorithmic societies. What makes them distinguished new social entities is an evolving novel perception of social reality: we learn to take for granted how algorithmization impacts change the ways we live our daily lives, forming our sense and concept of normality.

T. Rudas
Faculty of Social Sciences, Eötvös Loránd University, Budapest, Hungary
e-mail: trudas@elte.hu

G. Péli
Centre for Social Sciences, Hungarian Academy of Sciences Centre of Excellence

Department of Sociology, Gáspár Károli University of the Reformed Church in Hungary
e-mail: Peli.Gabor@tk.mta.hu

T. Rudas, G. Péli (eds.), *Pathways Between Social Science and Computational Social Science*, Computational Social Sciences,
https://doi.org/10.1007/978-3-030-54936-7

For many, the appearance of Big Data seems to be the main driving force of societal algorithmization. But importantly, it was the ongoing algorithmization process that produced the phenomenon labeled as Big Data, not the other way around. Consequently, we rather consider Big Data to be the well-visible surface of a deeper transformation process, the *ephemeral* level, the study of which enables understanding the societal changes induced by the algorithmization trends. Analyzing the Big Data *footprint* of these developments, in combination with finding new theoretical insights, can bring about a better understanding of how algorithmic societies emerge and can also hint to the directions in which they may develop.

In this chapter, we call a society algorithmic, if the processes framing and regulating the daily functioning of the society rely on algorithms. An algorithm is a sequence of commands, which takes some initial information (the input) and after performing the commands, each of which relies on the results of the previous steps, yields some modified information (the output). Anybody who has ever filled an income tax return has already worked with algorithms. Also, algorithms decide about the ads showing up on a webpage based on information like previous browsing history, and the availability of vendors who are ready to pay for their ads to be placed, etc. Algorithmic decision-making emerged as a response to the sometimes contradicting and certainly not exclusive requirements of accountability, objectivity, and automatization (for speed or low cost). In everyday language, the phrase "bureaucratic" often stands for being over-concerned with procedures at the expense of efficiency or common sense, something which clearly characterizes certain aspects of algorithmic decision-making. Nevertheless, the algorithmization of societies is neither good nor bad. Although algorithms are transparent by definition; yet, they may grow so complex that become de facto murky. They can also be fair or biased, rigid, or adaptive, depending on the context of their application.

This chapter visits some of the main aspects and processes that exemplify and facilitate the move toward algorithmic societies. Existing social science knowledge and theory can serve as a framework for this endeavor. But societal algorithmization produces new phenomena to study with an extreme velocity so that our societal knowledge on algorithmization becomes obsolete rapidly. This shift in relevant research subjects also impacts the frame: new concepts and new emergent theoretical constructs develop. The emergence of computational social science is facilitating a paradigm change in social research. We are still in the middle of this change, collecting insights in the hope of some future synthesis. This chapter addresses a number of aspects we deem relevant from the point of view of societal algorithmization; its genre is rather a diary of insights rather than a systematic inventory.

Section 2 discusses how penetrating algorithmization processes change our human environment and our perception of daily life, including our sense of normality. Section 3 is about transforming privacy. Section 4 surveys mechanisms, fueled by algorithmization, that can substantially limit or extend our actual access to information varieties. Section 5 is about the taming of algorithmization: the alignment of its impacts with desired societal outcomes. Section 6 assesses the new role and new task set of the social sciences in algorithmic societies.

2 The Changing Perception of Reality

In their path-breaking classic, Berger and Luckmann (1966) describe how our social reality is constructed in the course of ongoing interactions. The very perception of everyday life, of what we consider reality, is subject to radical change in algorithmic societies. An already broadly realized indicator of this process is how the mental borderlines between virtual life and reality have been shifting. A decade ago, public transport users were busy doing things we would now classify as belonging to reality in a traditional sense: watching the traffic, standing or sitting silently, musing about daily tasks or even deeper problems, or just reading newspapers. Now, looking around on a bus, tram, or metro, we see most passengers being merged in the "reality" conveyed through their smartphones. One may say that reading books was a form of virtual reality a century ago; or much earlier, developing our language and so the ability of more abstract thinking had been innovations opening "virtual worlds" beyond "tangible" reality. Still, the speed of change with which our use of algorithmic devices and social media tools alter our sense of reality is unprecedented, challenging the adaptive capabilities of individuals and societies.

Social scientists have pointed out long ago that our sense-making strongly relies on what we take for granted (Scott 2001). Taken-for-grantedness provides building blocks for the mental composition of everyday life: things deemed being "normal" parts of the picture so that we do not question their presence. The change in our perception of reality involves the appearance of new things becoming taken for granted. For example, the avatars in online virtual reality games may obtain personalities in the eyes of the user community, so that characters and assets that players develop are sold for real money at online auctions. Users do not question the "reality" of those with whom they communicate in chatrooms or at online dating sites, be they people or well-designed bots, just like how our grandparents did not question that the voices heard through telephone conversations stood for the person they "knew."

Another kind of taken-for-grantedness in algorithmic societies concerns the mechanisms how devices and computer programs, which we deem inanimate, function. In this respect, again, a kind of taken-for-grantedness develops: "I do not know how, but it works." This user attitude is not new at all. Housewives have never been too eager to learn the electro-mechanical functioning of their washing machines; they were rather focused on the comparison between the input (laundry) and the output (clean clothes). Similar input-output orientation ("what do I give, what do I get") applies to the functioning of sophisticated software systems. The data-misuse scandals of software giants like Facebook indicate, however, that the stake is much higher now than it had been in the time of non-smart household appliances.

The discrepancy between the everyday knowledge of individuals and the cutting-edge knowledge embodied in their devices has not been experienced before; it can only be compared to the discrepancy that had been long in existence with respect to everyday knowledge and scientific knowledge. The bridging of the gap between that two has been an inspiration for the best minds. According to a *bon mot* attributed to Albert Einstein "You do not really understand something unless you can explain it to your grandmother." However, he is also supposed to have said that things should be made/explained as simply as possible, but not simpler than that. These somewhat proverbial statements indicate that the essence of complex phenomena can be summarized in plain words for the non-expert audience by connecting it to the vested concepts, taken-for-granted things, in our mind sets. But before scientific knowledge would get idealized, it is worth remembering that John von Neumann has said: "In mathematics you don't understand things, you just get used to them." Similar holds, *a fortiori*, concerning the understanding chances of laymen users aiming to get insights into the working of complex algorithmic devices. Even knowing something at the system level does not imply having sufficient insights into the operational details: the functioning of modern digital tools and communication platforms is not fully transparent even for their creators. Still, having societal knowledge on the broadly meant functioning of such algorithmic entities can be vital in assessing the validity of information they channel for us. Identifying fake news in messages from online systems is gaining extra salience. Validity is also a focal issue at scientific information systems. In his essay, Scaruffi (2016) argues that a big deal of Wikipedia content has to be wrong; it either represents some kind of interest, for example, because of being motivated by the widespread access of the posted item, or it simply reflects majority views because of the easy access to content editing. Experts have no priority in editing Wikipedia articles, and readers do not have the knowledge to judge if the information conveyed is correct.

3 Algorithmic Awareness and Our Changing View on Privacy

In parallel with the increasing algorithmic awareness in society, social scientists have addressed an array of characteristic domains impacted upon by algorithmization, including free speech (Balkin 2018), social media and protests (Neumayer and Svensson 2016), data protection (Sappa 2019), algorithmic ideology (Mager 2012), accountability (Caplan et al. 2018), culture (Dourish 2016), and smart contracts between human stakeholders mediated by machines (Rahwan 2018).

Politics also expresses dedicated interest in these aspects. For example, algorithmic transparency is part of the recent Smarter London Together Roadmap (Mayor of London 2018). The main challenge, perhaps threat, comes from the fact that the depth of the societal changes involved remains hidden for most citizens. Thus far, the political system failed to realize and face the magnitude of the challenge societal algorithmization would pose on the system of representative democracy. While social media already possess the operational capacity to collect opinions of hundreds of millions of people in a matter of hours, democratic political entities, like the United States or the European Union, apparently make no effort to complement the system of periodic elections of representatives with more participatory forms of political decision-making. This lack of eagerness of the political system to utilize current technologies of information collection and processing may be partly justified by some bad experiences of the public with uses, and abuses, of the power of information technology; think of the case when the social media giant was forced to change its news feed algorithm (Zuckerberg 2018). Such highly publicized cases had been interpreted as warning signs by politicians and the public. Recent legislative responses, like the General Data Protection Regulations (GDPR) of the European Union may not provide, however, an effective response to such challenges. This is because these current measures fail to assess, and so appropriately address, the underlying fundamental societal changes of which the surfacing misuse cases are only consequences.

Privacy loss, a sensitive collateral damage of new digital technologies, is not a new phenomenon; it had been taking place recurrently in human history whenever some deeply going innovation extended the access to societal information substantially. Modern bureaucracies (Weber 1978 [1922]) have introduced organizational procedures with a degree of transparency and scalability that had been unprecedented before; they could radically improve the collection and processing of information about individuals, enabling running efficient registries in a broad variety of societal domains ranging from medical and housing registries to drafting conscripts. Recently, in several countries, even the decennial censuses have been replaced by a data collection exercise based on the existing registries (Nordholt 2018). Up until recently, however, societies could at least stand a chance to adapt to the privacy losses generated by the substantial societal changes. With time, "normal" life could go on as the increased scrutinizing capabilities of powerful tools and agents had earned a taken-for-granted status. People developed new ways how to hide vital pieces of information even under the new circumstances, so that the residual privacy could migrate to newly created safe havens. But considering the extreme scale and penetration power of digital technologies, the question arises: Can societies still duly adapt to the new privacy challenges they themselves will bring about?

Beyond its speed, another novelty of the present algorithmization-driven wave of privacy restrictions has been that people are willing to give up great chunks of their confidential life even voluntarily, at least at the outset. A possible explanation for this is that the emergent new generation of privacy-breaching innovations, especially those related to online social media, provide ample immediate benefits in the form of new communication possibilities and information access. Clearly, the bureaucratization of societies following the nineteenth-century industrial revolution had also led to fundamental daily life improvements; but then the process was much slower and so the coupling between cause and effect was also less manifest for the populace.

The emergent privacy change has implications for empirical social research as well. In traditional survey methodologies, researchers used to motivate respondents to participate. Beyond monetary incentives, this motivation might include explaining the importance of accurately exploring public views on an issue or highlighting the possibilities of influencing policy decisions. Computational social science typically works with data obtained without coupling with such societal motivations: the input is "found on the Internet," and coded electronically by some digital carriers. Data quality theory emphasizes the importance of understanding data-provider motivations as being indicative of possible intentional/unintentional survey biases (Kennett and Shmueli 2014). While respondent motivations are quite clear in case of data collected by traditional means, they are definitely much less understood in case of Big Data sources like online social media.

4 New Possibilities and New Barriers of Information Access

At the upsurge of online communication, the early and somewhat naïve understanding had been that Internet-use would support well-informed and democratic choice. Indeed, social media seemed to be a strong promoter of free speech by rendering a variety of opinions available, at least technically. However, a clear distinction needs to be made between the technical and actual access to information sources. While equalizing aspects of technical access, the Internet also magnifies actual information access differentials across the population. The fight intensifies for a scarce key resource: human information processing capacity. The beneficiaries of this fight are the large-scale media providers who possess the heavy weaponry to fight for our attention. The opinion-shaping efficiency of those in control of information channels is increasing. Below, we survey some – man-made or spontaneously developing – impacts facilitated by algorithmization that influence our actual access to information.

The emergent center-periphery structure in information markets can result in asymmetries in actual information access. It has been observed, and explained, since long that certain markets, including media markets, tend to develop a center-periphery structure (Carroll 1985). This tendency, which has been radically magnified by societal algorithmization, results in a serious asymmetry in news provider abilities in making their messages mainstream. Well-selected information contents, echoed by the large-scale media providers, are transmitted with high intensity in the market center constituted by mainstream audience tastes (Carroll et al. 2002). The surrounding market peripheries are characterized by thinner, atypical audience demands. Specialized, smaller news providers have all the technical abilities to offer an unprecedented variety of content for the diverse tastes at the peripheries. But the propagation of this information content from the periphery to the mainstream is very difficult, even in cases when the news is of general interest. Still, the fact that videos posted by unknown YouTube users may go viral demonstrates that such migration is not impossible either. But heavy selection mechanisms (Hannan et al. 2007) filter these partisan contents, allowing only an extremely small portion to reach the mass audience. As our scientific and user knowledge on the local properties of large-scale networks improve, new tricks are found to trespass this selection barrier, so that some content from peripheral positions may make its way, occasionally, to the mainstream. But the large media providers controlling the market center will soon spot the novel content and deploy their superb tools to handle it by assimilation: they transform and then transmit the new content to their mainstream audience according their own terms. Using economic language, the short-lived first mover advantage of the initiators gives way to massive efficient producer utilization (Péli and Masuch 1997).

In some cases, however, the news generated at peripheral positions is so characteristic or extreme that the mainstream is unable either to ignore or tame it by assimilation. This can be the case, for example, when the mainstream media systematically neglects certain important strays of information, thus increasing its novelty value, which some peripheral media providers can finally utilize. From then on, the easy technical access to information, provided by the Internet, begins to serve the peripheral news provider: a second, alternative opinion-center may develop, oppositional to the first one. Supporters of radical political views may now easily find comrades on the Internet (Behr et al. 2013; Jones 2018), while before the age of easy technical access to periphery-generated information, the central media could largely suppress peripheral views. This phenomenon may also contribute to the explanation of the growing visibility and weight of radical views in today's politics. Clearly, technical access to alternative information does not even require Internet; for example, it can also be granted by a capital injection, think of the 1996 foundation of American Fox News Channel by media tycoon Rupert Murdoch as a Republican-centered alternative to CNN. But for today, Internet has "democratized" this entry process by radically decreasing the entry costs. Online media platforms, blogs, newsletters can serve as crystallization seeds, from which selection processes will choose one or more alternative news centers, as a response to the increasing content convergence (Carroll 1985) of mainstream media providers.

Resistance to accept certain news on the receiver side constitutes another hurdle to free news propagation. Opinion segmentation processes can also be accentuated by societal algorithmization, so diminishing our ability to accept information not aligning with the prevailing opinions in our respective reference groups. It is known since long that social media can support the formation of opinion bubbles, little universes where voluntarily flocked-together people echo, and so further align, their similar views (Bozdag 2013; Spohr 2017). Bubbles foster the false perception for their encapsulated participants that their views constitute the mainstream. Bubble formation is facilitated by the news selection algorithms of social media tending to connect people according to the *homophily principle* (McPherson et al. 2001; Bakshy et al. 2015). This principle posits that we prefer interacting people similar to us along relevant social aspects. Though social media connects very different individuals at a superficial level, our frequently used ties have a lot to do with homophily along social characteristics that correlate with opinion formation. The social media algorithms that select content for us build upon our preferences identified by our clicking behavior. People with similar tastes are more likely to receive similar media content, thus accentuating homophily-based structure formation. Since our attention span is limited, grouping similar people into coherent and frequently visited bubbles also involves isolating them from the rest.

However, under specific conditions, the same homophily mechanism that leads to bubble formation may also explain landslide-like information propagation events. Opinion bubbles may be seen as compact echo-chambers, with sizes definitely smaller than that of the relevant societal segment or, in case of political mobilization, the size of the electorate. But this is not a necessity: decreasing the number of social characteristics that predict opinion tends to increase the pool of people that can potentially join the bubble. So, while being uniform along the few main bubble-defining social dimensions, this bigger pool of people is likely to represent a broader variety along other social characteristics and opinions. A decreasing coherence of views can certainly burst bubbles. But beyond the number of the bubble-defining dimensions, coherence also depends on the *salience* of these dimensions. In extreme cases, similarity along a single social dimension of very high societal importance can unite very heterogeneous people in one giant bubble. This can be the case, for example, when the deep resentment to an autocratic political rule unites very heterogeneous people in a giant bubble constituted by oppositional stance as salient dimension.

Importantly, the social mechanisms creating bubbles do not rely on the presence of online media; they had been in place well before the Internet age, and could have contributed to political changes which could not be foreseen and could not have happened without a widespread belief that they would be possible. Two such cases are mentioned next.

The 2012 Oscar nominee movie *No* exemplifies a giant bubble formation.[1] The movie is about the 1988 Chilean plebiscite aimed at deciding if Augusto Pinochet should stay in power for another 8 years. The opposition was very poorly endowed with media means; but due to external pressure on the incumbent government, the opposition still received a 15 min per day media spot during the campaign period. By filling these spots with well-crafted messages, the *No* side could turn the chances and finally win the plebiscite (see Rudas 2018, Chap. 11 for the technical analysis of relevant opinion poll data). One can interpret this outcome with the formation of a giant bubble encompassing the whole opposition side. In this context, the bubble was a dynamic phenomenon that could increase its size radically. Bubbles defined by several homophily dimensions usually cannot be as large as those defined by only a few. Here, homophily within the giant bubble was constituted along a single salient dimension: the deep oppositional sentiment toward the dictatorial political rule. When the mainstream channels of free information flow are restricted, the importance of scarce available channels soars. The smartly designed 15-min political ads, with a dedicated design and message each day, could finally change the political trend. More and more people were convinced that change was possible. Political opinions within the giant bubble have been aligned in a self-perpetuating manner, just like in case of the formation of smaller opinion bubbles.

[1] https://sonyclassics.com/no/

The authors of this chapter spent their younger years in Hungary's by-then senescent communist system, witnessing a similar extra-size bubble formation in 1988–1989. All legal media being monopolized by the state, alternative channels gained importance. The news broadcasted by the Munich-based Free Europe Radio spread by mouth-work way beyond its listeners. Politically devastating anti-communist jokes propagated through personal networks with extra speed and reach. Again, the oppositional stance to the repressive government had become the salient unifying dimension, leading to the formation of a giant bubble encompassing a substantial part of the Hungarian populace by 1989, the fall-down year of the communist rule. Clearly, the once Soviet-imposed regime did not only fall because of the sentiments against it; but still, those in the bubble tended to believe that this had been the case.

5 The Social Installation of Algorithmic Entities

There is an increasing information asymmetry between those who produce and deploy algorithmic tools and the users of physical and symbolic products generated by means of these tools. While many of the benefits associated with algorithmic tools are known for the users from the beginning, their disadvantages might be identified only later, in the course of application. Algorithmic service providers obtain an unprecedented amount of information on their users, which can be turned against them, can be used to manipulate their purchases or political choices. Minimizing this liberty-threatening information asymmetry between provider and user is essential for the sustainability of democratic societies. Therefore, the technical installation of digital information processing systems should be complemented with their *social installation*, a carefully designed procedure to diminish the potentially detrimental societal impacts of algorithmic entities.

It had been observed even before the Internet-age that the introduction of computer-based information systems to organizational units can raise surprisingly harsh resistance, a kind of immune response, on behalf of the people populating these units. Since the 1980s, local area networks (LANs) have been effective tools of mapping and systematizing the information flows of organizational units. In the language of organization science, LANs had been transparent electronic manifestations of the formal structure of their host organization. But the increased emphasis on formal structural elements brought about by the LANs have changed the delicate balance between the formal and *informal* aspects of organizational functioning, thus making the acceptance of LANs difficult for the people running the organizational unit (Péli 2004). The so-called formal organization comprises the explicit, codified ways how the organization should function. It always exists in convivialism with the *informal organization* constituted by the norms, practical rules, and tacit understandings that develop around the formal elements in the course of daily functioning. As formal specifications are normally far from being complete, organizations cannot function exactly as their formal rules prescribe; they need informal practices that fill gaps in regulations or even resolve potential conflicts between misaligned regulations.

Similarly, the social installation of nowadays algorithmic systems should somewhat restrain the excessive power of formal procedures, so aligning the impacts of their functioning with the needs for their societal use. While the technical installation specifies what algorithmic systems should do, their social installation also includes a not-to-do list as well. Internet-based algorithmic services are typically voluntary. So their social installation has to take into account that people take these services if their perceived benefits surpass their perceived disadvantages: for example, the extra information and social access the service provides compensate for the inconveniences of giving out personal information. Political philosophy makes a distinction between *positive* and *negative liberties* (Berlin 1969). Positive liberties are things we can do, while negative liberties are things others cannot do to us without our consent. Algorithmic devices, especially the social media, expand positive liberties substantially: users gain extra access to people and information. In parallel, their exposure to external scrutiny also expands at the cost of their negative liberties. This is aggravated by a temporal phase difference: the boost in positive liberties comes almost immediately after joining the service, making the user addicted to the system, while the impacts breaching negative liberties are only detected in the longer run. As a consequence, an asymmetry develops between positive and negative liberties at the detriment of the latter: we can do new things, but at the cost of also being much more vulnerable in respect what can be done to us. Little seems to have been done to develop effective and overarching social installation rules for pervasive algorithmic devices to lessen these asymmetries. The social installation tasks cannot be left solely to the providers. Well-aligned state-level and international surveillance seems necessary. The 2018 introduction of the General Data Protection Regulation (GDPR) by the European Union was just a preliminary step in this direction. The societal norms in an algorithmic society need to be operationalized in a way that makes them appropriate for application. Just like it happened in the very distant past of mankind's history when instead of just prescribing that "Thou shalt not steal," detailed specifications of types of economic crimes and elaborate preventive measures had been developed, also providing an early example of (yet not electronical device-based) societal algorithmization.

6 The Changing Role of Social Science

The scientific endeavor of understanding the societal impacts of algorithmization and the functioning of the emergent algorithmic societies can make use of the transdisciplinary methods of computational social science. The chapters in this book present computational social science not simply as an exercise in applying efficient methodologies to answer standard questions of Sociology, but report progress in transforming a big deal of traditional social science into computational social science. In theory, this transformation could happen in a way that the present generation of (traditional) social scientists is replaced by a new generation of computational social scientists. This book illustrates, on the contrary, how social science itself transforms. Social scientists have multiple roles in the research process of charting and understanding the societal algorithmization processes. Beyond participating in the non-trivial alignment of the variety of applied transdisciplinary methods, another core task for them is contributing to the interpretation of the research results. The large-scale availability of organic or process produced data, in combination with new methodologies, opens up the possibility of answering several questions of Sociology that could not have been answered even 10 years ago. Many of the chapters in this volume contribute to such developments. New questions also arise, oftentimes emerging jointly with new sociological concepts like algorithmization-driven perception. Traditional key concepts of social sciences may earn radically new connotations, and new meanings accordingly. The concept of identity is a prime example for such developments. Identity is inherently related to our social roles in a variety of daily life contexts and to our memberships in different groups in these social roles. Algorithmization, perhaps most importantly through online social media platforms, allows for a larger and more diffuse set of group memberships than before. Our participation in these oftentimes virtual groups, however, need not be intense. Resorting to the terminology of fuzzy set representations, our *grades of membership* in some new online formations might be low, due to our limited span of social attention.

Because of the powerful new method combinations, and because of the abundance of easy-to-access Big Data inputs, computational social science research produces quantitative results with an increasing velocity. An unwelcome consequence of this, however, is that the interpretation of the results oftentimes lags behind the production of outputs. This opening gap between producing and processing research outputs indicates that the interpretative capacity of the results is becoming the number one bottle neck of computational social science research. Social scientists will have a prime role in the interpretation. Beyond applying/adjusting existing social theories, their interpretative tasks include experimenting with the setting up of new powerful mental frames that help organizing information, and so understanding the evolving contexts of algorithmic societies. Clearly, the interpretation of the results of research in computational social science is not the exclusive task of social scientists. New interpretative frameworks have to be developed that draw on the powerful model-building experiences of several contributing disciplines, including physics, biology, linguistics, and data science. A major task for computational social science is building together these disciplinary contributions into coherent and empirically validated theoretical insights.

References

E. Bakshy, S. Messing, L.A. Adamic, Exposure to ideologically diverse news and opinion on Facebook. Science **348**(6239), 1130–1132 (2015). https://doi.org/10.1126/science.aaa1160

J.M. Balkin, Free Speech in the Algorithmic Society: Big Data, Private Governance, and New School Speech Regulation (2018). https://digitalcommons.law.yale.edu/fss_papers/5160/

I.V. Behr, A. Reding, C. Edwards, L. Gribbon, Radicalisation in the digital era. The use of the internet in 15 cases of terrorism and extremism. RAND Europe (2013). https://www.rand.org/content/dam/rand/pubs/research_reports/RR400/RR453/RAND_RR453.pdf

P.L. Berger, T. Luckmann, *The Social Construction of Reality: A Treatise in the Sociology of Knowledge* (Anchor Books, Garden City, 1966)

I. Berlin, Two concepts of liberty, in *Four Essays on Liberty* (Oxford University Press, Oxford, 1969), pp. 118–172

E. Bozdag, Bias in algorithmic filtering and personalization. Ethics Inf. Technol. **15**(3), 209–227 (2013)

R. Caplan, J. Donovan, L. Hanson, J. Matthews, Algorithmic Accountablility: A Primer (2018). https://datasociety.net/output/algorithmic-accountability-a-primer/

G.R. Carroll, Concentration and specialization: dynamics of niche width in populations of organizations. Am. J. Sociol. **90**(6), 1262–1283 (1985)

G.R. Carroll, S.D. Dobrev, A. Swaminathan, Organizational processes of resource partitioning. Res. Organ. Behav. **24**, 1–40 (2002)

P. Dourish, Algorithms and their others: algorithmic culture in context. Big Data Soc. **3** (2016). https://doi.org/10.1177/2053951716665128

M.T. Hannan, L. Pólos, G.R. Carroll, *Logics of Organization Theory Audiences, Codes, and Ecologies* (Princeton University Press, Princeton, 2007)

S.G. Jones, The rise of far-right extremism in the United States. CSIS Briefs, Nov 2018. https://csis-prod.s3.amazonaws.com/s3fs-public/publication/181119_RightWingTerrorism_layout_FINAL.pdf

C. Neumayer, J. Svensson J, Activism and radical politics in the digital age: towards a typology. *Convergence: The Int. J. of Research into New Media Technologies* **22**(2), 131–146 (2016)

R.S. Kennett, G. Shmueli, On information quality. J. R. Stat. Soc. Ser. A **177**(1), 3–38 (2014)

A. Mager, Algorithmic ideology: how capitalist society shapes search engines. Inf. Commun. Soc. **15** (2012). https://doi.org/10.1080/1369118X.2012.676056

Mayor of London, Smarter London Together. The Mayor's roadmap to transform London into the smartest city in the world. *Published by the Greater London Authority* (2018). https://www.london.gov.uk/sites/default/files/smarter_london_together_v1.66_-_published.pdf

M. McPherson, L. Smith-Lovin, J.M. Cook, Birds of a feather: homophily in social networks. Annu. Rev. Sociol. **27**(1), 415–444 (2001)

E.S. Nordholt, The usability of administrative data for register-based censuses. Stat. J. IAOS **34**, 487–498 (2018)

G. Péli, Whose transparency? The impact of computer systems on informal organization, in *Hide or Confide? The dilemma of transparency*, ed by G.-J. Hofstede, H. Schepers, L. Spaans, J. Trienekens, A. Beulens (Reed Business Information, Den Haag, 2004) pp. 111–115

G. Péli, M. Masuch, The logic of propagation strategies: axiomatizing a fragment of organizational ecology in first-order logic. Organ. Sci. **8**, 310–331 (1997)

I. Rahwan, Society-in-the-loop: programming the algorithmic social contract. Ethics Inf. Technol. **20**(1), 5–14 (2018) https://www.media.mit.edu/publications/society-in-the-loop-programming-the-algorithmic-social-contract/

T. Rudas, *Lectures on Categorical Data Analysis* (Springer, New York, 2018)

C. Sappa, How data protection fits with the algorithmic society via two intellectual property rights – a comparative analysis. J. Intellect. Prop. Law Pract. **14**(5), 407–418 (2019). https://doi.org/10.1093/jiplp/jpz022/5369198?redirectedFrom=fulltext

P. Scaruffi, Wikipedia as a Force for Evil (2016). https://www.scaruffi.com/politics/wikipedi.html

R.W. Scott, *Institutions and Organizations. Ideas, Interests and Identities* (Sage Publications, Thousand Oaks, 2001)

D. Spohr, Fake news and ideological polarization: filter bubbles and selective exposure on social media. Bus. Inf. Rev. **34**(3), 150–160 (2017)

M. Weber, *Economy and Society: An Outline of Interpretive Sociology* (University of California Press, Berkeley, 1978 [1922])

M. Zuckerberg, Post on Facebook (2018). https://www.facebook.com/zuck/posts/10104413015393571

Index

A
Agent-based computational modeling (ABM), 179, 180, 188–190
Akaike information criterion (AIC), 165, 168
Algorithmic awareness, 262–264
Algorithmic entities, 267–268
Algorithmic societies, 259–261
Analysis approaches, 199–200
Assessing model efficiency, 80–81
Attitude dynamics, 133–140
Attitudes, 24, 26, 59, 65, 66, 124, 125, 129, 130, 132–134, 140, 145, 146
Automated text analytics
  challenges, 50–51
  methodological
    new data sources, 52–53
    NLP methods, 53–55
    supervised methods, 56–58
    unsupervised methods, 58–62
  natural language processing, 50
  new possibilities, sociological research, 63–66

B
Ballot design, 223–224
Bayesian information criterion (BIC), 165, 168

C
Classification methods, 203–204
Clustering, 6, 9, 14, 208
Comparative agendas project (CAP), 203–204
Computation, 180–182
Computational social choice, 217, 218, 221, 232
Computational social sciences (CSS), 164–166, 173, 198
Computer simulation model, 21, 22
Convolutional neural networks (CNNS), 111–114

D
Deffuant-Weisbuch model, 130–133
Degree distribution, 5
Depression, 102–104
Discrete participatory budgeting, 226–229
Divisible participatory budgeting, 230–232

E
Emergence, 11, 12, 125

F
Face detection, 110–111
Fading, 7
Formal design methods
  agent-based simulation, 23–31
  similar approaches, 32–33
Fully connected layer, 112

G
Gender data revolution, 160–162
Generative processes, 190–191
German Internet Panel (GIP), 77–79

T. Rudas, G. Péli (eds.), *Pathways Between Social Science and Computational Social Science*, Computational Social Sciences,
https://doi.org/10.1007/978-3-030-54936-7

German Socio-Economic Panel (GSOEP), 133–136
GLES version model, 141–142
Global gender gap index (GGGI), 167
Granovetterian community structure, 6–7

H
Hegselmann-Krause model, 130–133
Histogram of oriented gradients (HOG), 106, 108
Homophily, 13–16
Human development index (HDI), 166
Hungarian Workers' Party (HWP), 199, 200
Hybrid models, 233

I
Image processing, 104–106
Industrial organization (IO), 180–182
Industry dynamics, 185–188
Initial endowments, 233

L
Large-scale social networks
  characterization
    assortative mixing, 5–6
    clustering, 6
    degree distribution, 5
    fading, 7
    granovetterian community structure, 6–7
    multiplex structure, 7–8
    overlapping communities, 7–8
    tie formation, 7
Latent Dirichlet Allocation, 241, 243–245
Link expiration, 11
Local area networks (LANs), 267

M
Market-based approaches, 233
Markov chain Monte Carlo (MCMC), 80, 165
Mathematical formulation, 218–225
Max pooling, 112
Mean-squared errors (MSEs), 83, 85–90
Mental health, 102–104
Millennium Development Goals (MDGs), 159
Model-based methods, 108
Model evaluation, 79–80
Modeling granovetterian community structure
  homophily, 13–16
  link expiration, 11
  multiplex structure, 11–13
  structural changes, 13–16
  weighted social network, 9–10
Multi-dimensional constraints, 232
Multimodal distributions, 134
Multiple attitude dynamics, 128–130
Multiplex structure, 7–8, 11–13
Multivariate latent Markov models (MLMMs), 160, 163–165, 172

N
Natural language processing (NLP) methods, 53–55
New simulation model
  translation
    designing agent types, 36–39
    functions, 36–39
    relations, 36–39
    results, 39
NK models, 185–188
Non scientific surveys
  Bayesian, 73–75
  German internet panel
    assessing model efficiency, 80–81
    model evaluation, 79–80
    nonprobability data, 77–79
    probability data, 77–79
  methodology, 75–77
  modeling approach, 75–77
  results, 81–90
Non-statement view, 33–36

O
Odd protocol, 23–31
One-dimensional models, 127–128
Optical character recognition (OCR), 200
Organizational ecology
  agent-based computational modeling, 179, 180
  computation, 180–182
  explanation, 190–191
  generative processes, 188–190–191
  industrial organization, 180–182
  industry dynamics, 185–188
  NK models, 185–188
  simulation modeling, 182–185
Overlapping communities, 7–8

P
Participatory budgeting (PB)
  ballot design, 223–224
  decision space, 219–221

discrete participatory budgeting, 226–229
divisible model, 230–232
extensions, 232–234
mathematical formulation, 218–219
popular PB models, 219–221
preference modeling, 221–222
vote aggregation, 224–226
Pixel-level features, 107–109
Pledges, 233
Political elite
data analysis
AI-based classification, 206–207
analysis approaches, 199–200
classification methods, 203–204
discursive elite, 211
ontology-based classification, 204–206
relational network, 207–210
text network analysis, 201–203
textual sources, 199–200
Preference elicitation, 217, 223

R
Reality, changing perception, 261–262
Referendum campaign, 24
ReLU, 112

S
Scale-invariant feature transform (SIFT), 104
Scientific surveys
Bayesian, 73–75
German internet panel
assessing model efficiency, 80–81
model evaluation, 79–80
nonprobability data, 77–79
probability data, 77–79
methodology, 75–77
modeling approach, 75–77
results, 81–90
Sigmoid, 112
Simulation modeling, 182–185
Social media images
computational social science, 96–104
traditional social science, 96–104
Socio-economic panel (SOEP), 133
Softmax, 112
Stable unit treatment value assumption (SUTVA), 126
Statistical methods, 108
Strategic agent models, 233
Structural methods, 108
Structural validity, 125
Suicide
data and methods, 240–243
recent studies, 239–240
topic model, 243–243
word-embedding, 249–250, 255
Supervised methods, 56–58

T
Text mining, 251
Text network analysis (TNA), 201–203
Textual sources, 199–200
Tie formation, 7
Topic model, 54, 58–60, 62
Topic modelling, 238, 241–249
Transform-based methods, 108
Translation
new simulation model
designing agent types, 36–39
functions, 36–39
relations, 36–39
results, 39

U
United Nations Human Development (UNDP), 167
Unsupervised methods, 58–62

V
Validating simulation models
attitude dynamics, 133–140
data transformations results, 148–152
GLES version model, 141
GSOEP version model, 141–142
multiple attitude dynamics, 128–130
one-dimensional models, 127–128
opinions, 133–140
original versions model, 142–145
two-dimensional model, 130–133
types, 125–127
Visual features, 103
Visual themes, 103
Vote aggregation, 224–225
Voter group entitlements, 233

W
Weighted Social Network (WSN), 9–10
Word Embedding Model, 61